Mathematics and Visualization

The series *Mathematics and Visualization* is intended to further the fruitful relationship between mathematics and visualization. It covers applications of visualization techniques in mathematics, as well as mathematical theory and methods that are used for visualization. In particular, it emphasizes visualization in geometry, topology, and dynamical systems; geometric algorithms; visualization algorithms; visualization environments; computer aided geometric design; computational geometry; image processing; information visualization; and scientific visualization. Three types of books will appear in the series: research monographs, graduate textbooks, and conference proceedings.

More information about this series at http://www.springer.com/series/4562

Noemi Gyori · Jana Hutter · Vishwesh Nath ·
Marco Palombo · Marco Pizzolato · Fan Zhang
Editors

Computational Diffusion MRI

International MICCAI Workshop, Lima, Peru, October 2020

MICCAI

Springer

Editors
Noemi Gyori
University College London
London, UK

Jana Hutter
Centre for Medical Engineering
King's College London
London, UK

Vishwesh Nath
Nvidia
Nashville, TN, USA

Marco Palombo
University College London
London, UK

Marco Pizzolato
Department of Applied Mathematics
and Computer Science
Technical University of Denmark
Kgs. Lyngby, Denmark

Fan Zhang
Laboratory of Mathematics in Imaging
Harvard Medical School
Boston, MA, USA

ISSN 1612-3786 ISSN 2197-666X (electronic)
Mathematics and Visualization
ISBN 978-3-030-73017-8 ISBN 978-3-030-73018-5 (eBook)
https://doi.org/10.1007/978-3-030-73018-5

Mathematics Subject Classification: 00B25, 00A66, 00A72, 42B35, 60J60, 60J65, 62P10, 65CXX, 65DXX, 65Z05, 68R10, 68T99, 92BXX

This Springer imprint is published by the registered company Springer Nature Switzerland AG
The registered company address is: Gewerbestrasse 11, 6330 Cham, Switzerland

Programme Committee

Nagesh Adluru, University of Wisconsin-Madison, USA
Maryam Afzali, Cardiff University, UK
Manisha Aggarwal, Johns Hopkins University, USA
Suyash P. Awate, Indian Institute of Technology, Bombay, India
Dogu Baran Aydogan, Aalto University School of Science, Finland
Yael Balbastre, University College London, UK
Ryan Cabeen, University of Southern California, USA
Erick Jorge Canales-Rodríguez, EPFL, Switzerland and FIDMAG Research Foundation, Spain
Emmanuel Caruyer, IRISA, France
Maxime Chamberland, Cardiff University Brain Research Imaging Centre, UK
Alessandro Daducci, University of Verona, Italy
Samuel Deslauriers-Gauthier, Irnia Sophia Antipolis—Mediterranee, France
Rutger Fick, Therapanacea, France
Gabriel Girard, University Hospital Center (CHUV), Switzerland
Francesco Grussu, University College London, UK
Tom Dela Haije, University of Copenhagen, Denmark
Colin Hansen, Vanderbilt University, US
Muge Karaman, University of Illinois at Chicago, USA
Suheyla Karayumak, Harvard Medical School, USA
Jan Klein, Fraunhofer MEVIS, Germany
Christophe Lenglet, University of Minnesota, USA
Emma Muñoz Moreno, IDIBAPS, Spain
Sudhir Pathak, University of Pittsburgh, US
Tomasz Pieciak, AGH University of Science and Technology, Poland
Alonso Ramirez-Manzanares, CIMAT A.C., Mexico
Gabriel Ramos-Llordén, Harvard Medical School, USA
Torben Schneider, DeepSpin UG, Germany
Thomas Schultz, University of Bonn, Germany
Farshid Sepehrband, USC, USA

Preface

It is our great pleasure to present the proceedings of the 2020 International Workshop on Computational Diffusion MRI (CDMRI'20). CDMRI'20 was held under the auspices of the International Conference on Medical Image Computing and Computer Assisted Intervention (MICCAI), which took place virtually on October 8th 2020, having originally been planned to take place in Lima, Peru.

This volume presents the latest developments in the highly active and rapidly growing field of diffusion MRI. The reader will find numerous contributions covering a broad range of topics, from the mathematical foundations of the diffusion process and signal generation, to new computational and machine learning methods and estimation techniques. The contributions are focused toward the in vivo recovery of microstructural and connectivity features, as well as combined diffusion-relaxometry acquisitions techniques. This edition includes chapters from high-profile researchers with the specific focus on four topics that are gaining momentum or have received increasing interest within the diffusion MRI community: (i) diffusion MRI signal acquisition, (ii) orientation processing: tractography and visualization, (iii) microstructure modeling and representation, (iv) signal augmentation and super resolution, and (v) diffusion MRI applications.

This volume offers the opportunity to share new perspectives on the most recent research challenges for those currently working in the field, and a valuable starting point for anyone interested in learning computational techniques in diffusion MRI. The book includes rigorous mathematical derivations, a large number of rich, full-color visualizations, and clinically relevant results. As such, it will be of interest to researchers and practitioners in the fields of computer science, MRI physics, and applied mathematics.

Each contribution in this volume has been peer-reviewed by multiple members of the international Program Committee. We would like to express our gratitude to all CDMRI'20 authors and reviewers for ensuring the quality of the presented work. We are grateful to the MICCAI 2020 chairs for providing a platform to present and discuss the work collected in this volume. We also would like to thank the editors of the Springer book series Mathematics and Visualization as well as Leonie Kunz and Martin Peters (Springer, Heidelberg) for their support to publish this collection as part of their series.

Finally, we express our sincere congratulations to the winners of the prizes that were awarded during CDMRI'20, including:

- Prize for the best CDMRI paper: "Learning Anatomical Segmentations for Tractography from Diffusion MRI", Christian Ewert et al. German Center for Neurodegenerative Diseases (DZNE), Bonn, Germany
- Prize for the best CDMRI oral presentation: "Longitudinal Parcellation of the Infant Cortex Using Multi-Modal Connectome Harmonics", H. Partick Taylor et al. University of North Carolina at Chapel Hill, Chapel Hill, North Carolina, USA
- Prize for the best CDMRI power-pitch presentation: "Image Reconstruction from Accelerated Slice-Interleaved Diffusion Encoding Data", Tiantian Xu et al. University of North Carolina at Chapel Hill, Chapel Hill, North Carolina, USA
- Joint First Prize for the best Super-MUDI method: Kurt Schilling et al. Vanderbilt University, Nashville, Tennessee, USA.
- Joint First Prize for the best Super-MUDI method: Haoyu Lan et al. University of Southern California, Los Angeles California, USA.

Lima, Peru
October 2020

Noemi Gyori, MSci.
UCL, UK

Jana Hutter, Ph.D.
KCL, UK

Vishwesh Nath
Nvidia, USA

Marco Palombo, Ph.D.
UCL, UK

Marco Pizzolato, Ph.D.
DTU, Denmark

EPFL, Switzerland

Fan Zhang, Ph.D.
Harvard, USA

Contents

Diffusion MRI Signal Acquisition

Image Reconstruction from Accelerated Slice-Interleaved Diffusion Encoding Data

Tiantian Xu, Ye Wu, Yoonmi Hong, Khoi Minh Huynh, Weili Lin, Wei-Tang Chang, and Pew-Thian Yap

Abstract We present a reconstruction scheme for diffusion MRI data acquired using slice-interleaved diffusion encoding (SIDE). We show that, when combined with multi-band imaging, the method is capable of reducing the amount of data that needs to be acquired by as much as 25 times, therefore remarkably speeding up acquisition and making high angular resolution diffusion imaging much more feasible, particularly for pediatric, elderly, and claustrophobic patients. In contrast to the conventional approach of acquiring a full diffusion-weighted (DW) volume for each diffusion wavevector, SIDE acquires for each repetition time (TR) a volume consisting of interleaved slice groups, each corresponding to a different diffusion wavevector. This allows SIDE to rapidly acquire information associated with a larger number of wavevectors within a short period of time. We formulate the inverse problem involved in recovering the full DW images as a constrained variational problem regularized by multidimensional total variation. The problem can be solved efficiently using the alternating direction method of multipliers (ADMM). Experiment results based on SIDE data of adults indicate that DW images can be recovered with high fidelity despite high undersampling for multifold acceleration.

1 Introduction

Diffusion magnetic resonance imaging (dMRI) offers a unique probe into white matter pathways in the living human brain in association with development and disorders in a non-invasive manner [2, 9]. However, dMRI is characterized by long acquisition time due to the need to acquire a large number of diffusion-weighted (DW) images with different diffusion encodings for sufficient coverage of the

T. Xu · P.-T. Yap (✉)
Department of Computer Science, University of North Carolina, Chapel Hill, NC, USA
e-mail: ptyap@med.unc.edu

Y. Wu · Y. Hong · K. Minh Huynh · W. Lin · W.-T. Chang · P.-T. Yap
Department of Radiology and BRIC, University of North Carolina, Chapel Hill, NC, USA

N. Gyori et al. (eds.), *Computational Diffusion MRI*, Mathematics and Visualization,
https://doi.org/10.1007/978-3-030-73018-5_1

q-space over a range of diffusion gradient directions. Speeding up dMRI acquisition is needed to reduce motion artifacts [6], patient discomfort, and imaging costs.

Acceleration of dMRI acquisition typically involves a sampling scheme and a reconstruction strategy [5, 7]. Ning et al. [7] proposed a q-space undersampling scheme using three sets of intersecting thick slices, each corresponding to a different set of gradient directions. The high-resolution slices are then recovered for the full set of gradient directions. The authors claimed to reduce the acquisition time by at least 3 times. Gabriel et al. [8] proposed a novel method, gSlider-SR, to accelerate high-resolution dMRI acquisition by downsampling in both radio frequency (RF) encoding space and q-space. Reconstruction was then performed with the help of sparse spherical-ridgelets. Sun et al. [10] proposed a method for improved reconstruction of ensemble average propagators from highly undersampled k - space and q - space. Hong et al. [5] proposed to accelerate dMRI acquisition via slice-interleaved diffusion encoding (SIDE) and deep learning reconstruction. In contrast to the conventional approach of acquiring a full DW volume for each diffusion wavevector, SIDE acquires for each repetition time (TR) a volume consisting of interleaved slice groups, each corresponding to a different diffusion wavevector. This allows SIDE to rapidly acquire information associated with a large number of wavevectors within a short period of time.

In this paper, we propose a reconstruction method for SIDE, which unlike [5], does not rely on training data. We show that, when combined with multi-band imaging, our method is capable of reducing the amount of data that needs to be acquired by as much as 25 times, therefore significantly speeding up acquisition and making high angular resolution diffusion imaging much more feasible, particularly for pediatric, elderly, and claustrophobic patients. We formulate the inverse problem involved in recovering the full DW images as a constrained variational problem regularized by multidimensional total variation. The problem can be solved efficiently using the alternating direction method of multipliers (ADMM). Experiment results based on SIDE data of adults indicate that DW images can be recovered with high fidelity despite high undersampling for multifold acceleration.

2 Methods

2.1 SIDE Acquisition

Slice-interleaved diffusion encoding (SIDE) accelerates acquisition of DW images. SIDE involves simultaneous multi-slice (SMS) excitation of a slice group (SG) [1]. Let N_g denote the number of SGs in a volume and N_d the total number of wavevectors. In conventional acquisition schemes, all SGs in a volume share the same diffusion encoding. In SIDE, each SG is associated with a different diffusion wavevector. Each SIDE cycle covers every wavevector with a SG, amounting to acquiring N_d/N_g volumes in N_d/N_g TRs. We assume N_d is a multiple of N_g for convenience. With each transition to the next cycle, the wavevector table is offset by 1. N_g cycles cover all the slices of all diffusion wavevectors. A subset of the cycles can be selectively acquired to achieve a certain acceleration factor R. Sorting the slice groups in the SIDE volumes according to their associated wavevectors gives N_d volumes with subsampled slices. We provide an illustration of SIDE acquisition in Fig. 1 for $N_d = 16$ and $N_d/N_g = 4$. When the acceleration factor is $R = 2$, only half of the cycles need to be acquired. More details on SIDE are provided in [5].

2.2 Reconstruction

For convenience, we reshape the DW images into a matrix $\mathbf{X} \in \mathbb{R}^{N_v \times N_d}$, where N_v is the number of voxels in each DW image and $N_v = N_x N_y N_z$, where N_x, N_y, and N_z are the image dimensions in the x, y, and z directions, respectively. N_d is the number of wavevectors. Then, a set of SIDE image volumes $\{\mathbf{X}_k^{\mathrm{S}}\}_{k=1}^{N_d}$ can be represented as

$$\mathbf{X}_k^{\mathrm{S}}(:, l) = \mathbf{H}_l \mathbf{X} \mathbf{q}_{kl} + \mathbf{n}_{kl}, \tag{1}$$

where $\mathbf{X}_k^{\mathrm{S}}(:, l)$ is the l-th column vector of $\mathbf{X}_k^{\mathrm{S}}$ and represents the l-th observed slice along the z-direction, k is the volume index, and $\mathbf{n}_{kl}$ is the measurement noise. Matrix $\mathbf{H}_l$ is a slice selector along the z direction and vector $\mathbf{q}_{kl}$ is a gradient selector. $\mathbf{X}_k^{\mathrm{S}}$ is a partial volume that does not cover all the slices due to subsampling by a factor of R.

Our goal is to reconstruct the full volumes $\mathbf{X}$ from the subsampled DW images. The simplest way to obtain $\mathbf{X}$ is to solve

$$\min_{\mathbf{X}} \frac{1}{2} \sum_{k,l} \left\| \mathbf{H}_l \mathbf{X} \mathbf{q}_{kl} - \mathbf{X}_k^{\mathrm{S}}(:, l) \right\|_{\mathrm{F}}^2 . \tag{2}$$

However, the inverse problem is ill-posed and can be regularized spatially by using the total variation (TV) semi-norm of $\mathbf{X}$, giving

$$\min_{\mathbf{X}} \frac{1}{2} \sum_{k,l} \left\| \mathbf{H}_l \mathbf{X} \mathbf{q}_{kl} - \mathbf{X}_k^{\mathrm{S}}(:, l) \right\|_{\mathrm{F}}^2 + \lambda \left\| \mathbf{X} \right\|_{\mathrm{TV}} . \tag{3}$$

Parameter λ is nonnegative and controls the tradeoff between data fidelity and the degree of regularization. For angular regularization, $\mathbf{X}$ is represented as real and symmetric spherical harmonics (SH) of maximum order $\ell_{\max} = 8$ [11] via $\mathbf{X}^\top = \mathbf{A}\mathbf{V}$, where $\mathbf{A}$ is a matrix of the SH basis functions and $\mathbf{V}$ is the corresponding SH coefficients. Therefore,

$$\min_{\mathbf{X}} \frac{1}{2} \sum_{k,l} \left\| \mathbf{H}_l \mathbf{X} \mathbf{q}_{kl} - \mathbf{X}_k^{\mathrm{S}}(:, l) \right\|_{\mathrm{F}}^2 + \lambda \left\| \mathbf{X} \right\|_{\mathrm{TV}} \quad \text{s.t.} \quad \mathbf{X}^\top = \mathbf{A}\mathbf{V}. \tag{4}$$

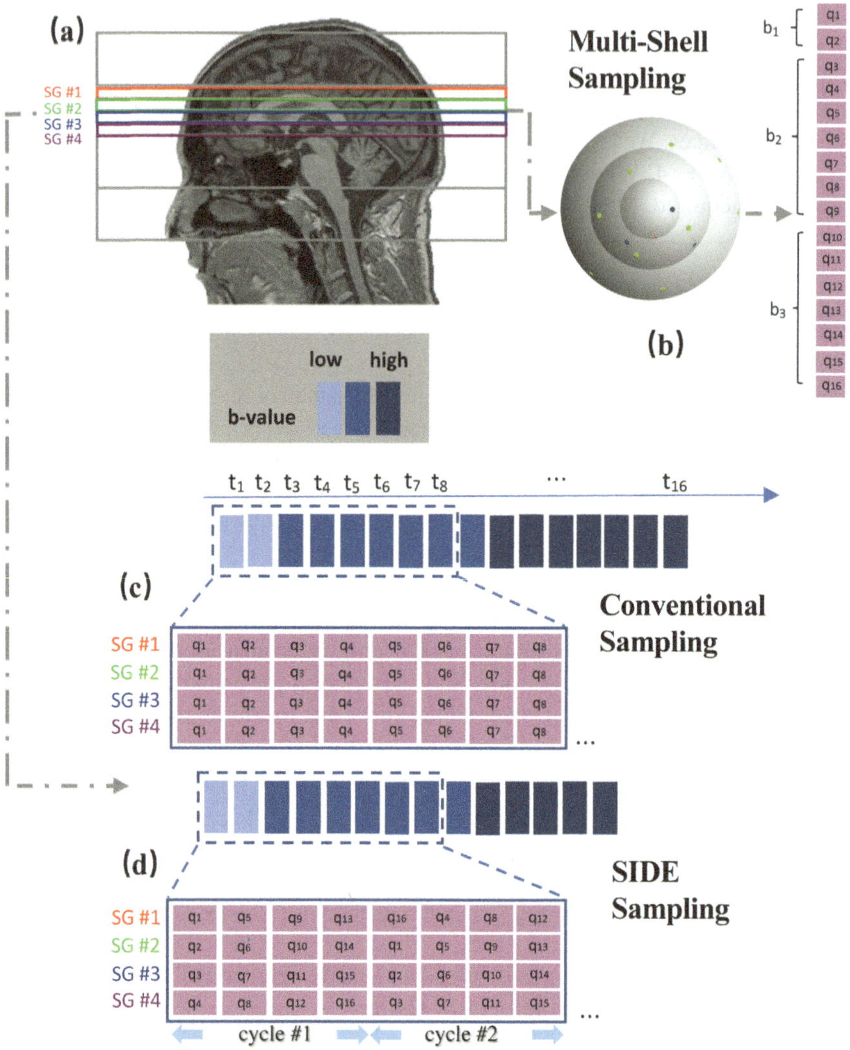

Fig. 1 SIDE acquisition scheme. **(a)** SMS acquistion. **(b)** Multi-shell sampling with 16 diffusion wavevectors covering 3 shells. **(c)** Conventional one-volume-one-encoding sampling with SMS slice groups in each volume associated with the same diffusion encoding. **(d)** SIDE acquisition with SMS slice groups in each volume associated with different wavevectors

2.3 Optimization

We utilize the alternating direction method of multipliers (ADMM) [3, 4, 12] for solving (4), which involves high-dimensional variables and non-differentiable convex optimization. To regularize the ill-posed problem, we will utilize the TV norm:

$$\sum_{k=1}^{N_d} \|\mathbf{X}_{(k)}\|_{\text{TV}}, \tag{5}$$

where $\mathbf{X}_{(k)}$ is the k-th row of $\mathbf{X}$ reshaped into a tensor of size $N_x \times N_y \times N_z$. The TV regularization term encourages $\mathbf{X}$ to be spatially smooth. $\|\cdot\|_{\text{TV}}$ can be the anisotropic or isotropic TV norm.

ADMM decomposes a large global problem into a series of smaller local subproblems, and attempts to combine the benefits of augmented Lagrangian methods and dual decomposition for constrained optimization problems. The constrained optimization problem (4) can be reformulated as the following unconstrained optimization problem:

$$L(\mathbf{X}, \mathbf{V}, \mathbf{Y}_{\{1,\dots,N_d\}}, \boldsymbol{\Psi}_{\{1,\dots,N_d\}}, \boldsymbol{\Phi}) = \min_{\mathbf{X}} \frac{1}{2} \sum_{k,l} \left\| \mathbf{H}_l \mathbf{X} \mathbf{q}_{kl} - \mathbf{X}_k^{\text{S}}(:, l) \right\|_{\text{F}}^2 + \lambda \sum_{k=1}^{N_d} \|\mathbf{Y}_k\|_{\text{TV}}$$
$$+ \sum_{k=1}^{N_d} \left\langle \boldsymbol{\Psi}_k, \mathbf{Y}_k - \mathbf{X}_{(k)} \right\rangle + \frac{\rho_1}{2} \sum_{k=1}^{N_d} \left\| \mathbf{Y}_k - \mathbf{X}_{(k)} \right\|_{\text{F}}^2 + \langle \boldsymbol{\Phi}, \mathbf{X} - \mathbf{A}\mathbf{V} \rangle + \frac{\rho_2}{2} \left\| \mathbf{X}^{\top} - \mathbf{A}\mathbf{V} \right\|_{\text{F}}^2, \tag{6}$$

where $\langle \cdot, \cdot \rangle$ denotes the Frobenius inner product. Applying ADMM, we carry out the following steps to minimize $L(\mathbf{X}, \mathbf{V}, \mathbf{Y}_{\{1,\dots,N_d\}}, \boldsymbol{\Psi}_{\{1,\dots,N_d\}}, \boldsymbol{\Phi})$ over $\mathbf{X}$, $\mathbf{V}$, and auxiliary variables $\mathbf{Y}_{\{1,\dots,N_d\}}$, and update Lagrangian multipliers $\boldsymbol{\Psi}_{\{1,\dots,N_d\}}$ and $\boldsymbol{\Phi}$ in each iteration t:

Subproblem 1

Update $\mathbf{V}$ by minimizing

$$\mathbf{V}^{(t+1)} = \arg\min_{\mathbf{V}} \frac{\rho_2}{2} \left\| \mathbf{X}^{\top^{(t)}} - \mathbf{A}\mathbf{V} + \boldsymbol{\Phi}^{(t)} \right\|_2^2. \tag{7}$$

Subproblem 2

Update $\mathbf{Y}_{1,\dots,N_d}$ by minimizing

$$\mathbf{Y}_k^{(t+1)} = \arg\min_{\mathbf{Y}_k} \rho_1 \sum_{k=1}^{N_d} \left\| \mathbf{Y}_k - \mathbf{X}_{(k)}^{(t)} + \boldsymbol{\Psi}_k^{(t)} \right\|_2^2 + \lambda \sum_{k=1}^{N_d} \|\mathbf{Y}_k\|_{\mathrm{TV}}. \tag{8}$$

Subproblem 3

Update $\mathbf{X}$ by minimizing

$$\begin{aligned} \mathbf{X}^{(t+1)} =& \arg\min_{\mathbf{X}} \frac{1}{2} \sum_{k,l} \left\| \mathbf{H}_l \mathbf{X} \mathbf{q}_{kl} - \mathbf{X}_k^{\mathrm{S}}(:, l) \right\|_{\mathrm{F}}^2 + \\ & \frac{\rho_1}{2} \sum_{k=1}^{N_d} \left\| \mathbf{Y}_k^{(t+1)} - \mathbf{X}_{(k)} + \boldsymbol{\Psi}_k^{(t)} \right\|_{\mathrm{F}}^2 + \frac{\rho_2}{2} \left\| \mathbf{X}^\top - \mathbf{A}\mathbf{V}^{(t+1)} + \boldsymbol{\Phi}^{(t)} \right\|_{\mathrm{F}}^2. \end{aligned} \tag{9}$$

Subproblem 4

Update $\boldsymbol{\Psi}$ and $\boldsymbol{\Phi}$ by

$$\boldsymbol{\Psi}_k^{(t+1)} = \boldsymbol{\Psi}_k^{(t)} + (\mathbf{Y}_k^{(t+1)} - \mathbf{X}_{(k)}^{(t+1)}), \tag{10}$$

$$\boldsymbol{\Phi}^{(t+1)} = \boldsymbol{\Phi}^{(t)} + (\mathbf{X}^{\top(t+1)} - \mathbf{A}\mathbf{V}^{(t+1)}). \tag{11}$$

3 Experiments

3.1 Materials

We scanned seven healthy subjects using a 3T Siemens whole-body Prisma scanner (Siemens Health-care, Erlangen, Germany). A monopolar diffusion-weighted spin-echo EPI sequence was utilized with imaging parameters as follows: Resolution = 1.5 mm isotropic; FOV = $192 \times 192 \times 150\,\mathrm{mm}^3$; image dimensions = $128 \times 128 \times 100$; partial Fourier = 6/8; bandwidth = 1776 Hz/Px; 160 gradient directions distributed over 4 b-value shells of $b = 500, 1000, 2000$ and $3000\,\mathrm{s/mm}^2$, plus one $b = 0\,\mathrm{s/mm}^2$; TR/TE = 3120/90 ms; 32-channel head array coil; SMS factor is 5. The total acquisition time is 8.19 minutes for each phase-encoding direction. No in-plane acceleration was used. The ground truth datasets were collected with 20 SIDE cycles. For $R = 2$, the 1st to 10th cycles from the 20 cycles were selected. For $R = 5$, the 1st, 4th, 6th and 8th cycles were selected.

3.2 Results

We compared our method with bicubic interpolation and spherical interpolation, for three different undersampling factors and three b-shells. Spherical interpolation was performed using spherical harmonics basis with order up to 8. Figure 2 shows the accuracy of reconstructed signal quantified by using normalized mean square error (NMSE), peak signal-to-noise (PSNR), and structural similarity (SSIM). Figure 3 shows the quantitative comparison results for GFA when $b = 2000\,\mathrm{s/mm^2}$. We compared reconstruction accuracy of different methods with respect to the ground truth. The reconstructed signal and computed GFA under two acceleration factors $R = 2, 5$ and $b = 2000\,\mathrm{s/mm^2}$ are shown in Fig. 4. The proposed method yields results that are closest to the ground truth.

Evaluation was also performed based on microstructure scalars computed by using neurite orientation dispersion and density imaging (NODDI) [13]. The quantitative results for $R = 2, 5$ and $b = 2000\,\mathrm{s/mm^2}$ for intra-cellular volume fraction (ICVF) and orientation dispersion (OD) are shown in Fig. 5. The proposed method shows better results that are closer to the ground truth. Figure 6 shows the fiber orientation distribution functions (ODFs) in three representative regions with crossing, bifurcat-

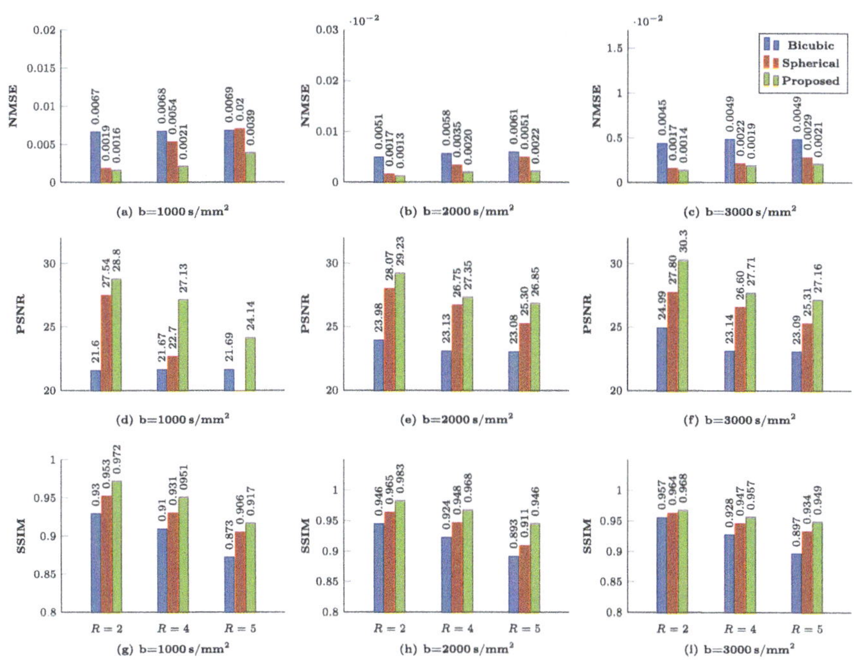

Fig. 2 Quantitative comparison of reconstructed signal difference via NMSE, PSNR, and SSIM for different undersampling factors and differenct b-values

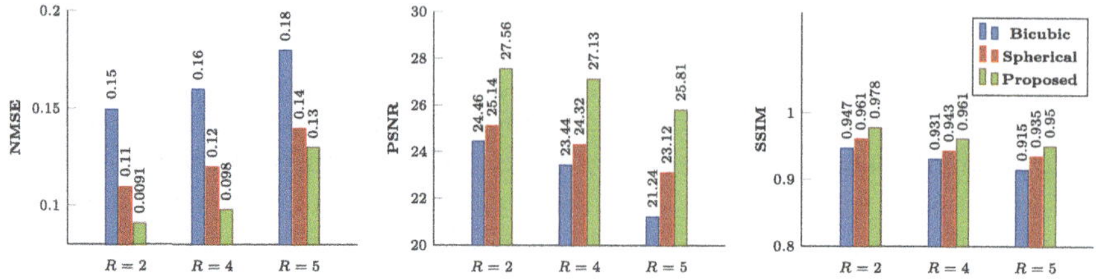

Fig. 3 Quantitative comparison of GFA maps via NMSE, PSNR, and SSIM ($b = 2000\,\text{s/mm}^2$)

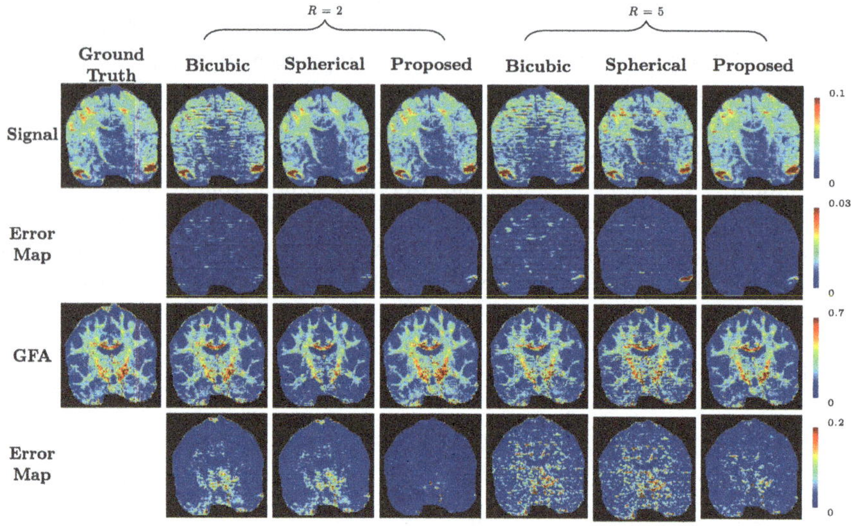

Fig. 4 Predicted DW images, GFA images, and error maps ($b = 2000\,\text{s/mm}^2$)

ing and unidirectional fibers, indicating that our method yields more accurate results with greater coherence.

4 Conclusion

We have demonstrated that dMRI acquisition can be significantly accelerated by leveraging SIDE acquisition in combination with the reconstruction scheme proposed in this paper. Experimental results show that our method can yield acceleration factor as high as 25 folds when combined with SMS imaging, allowing more than a hundred DW images (1.5 mm isotropic resolution) to be acquired within a couple of minutes.

Acknowledgements This work was supported in part by NIH grants NS093842 and EB006733.

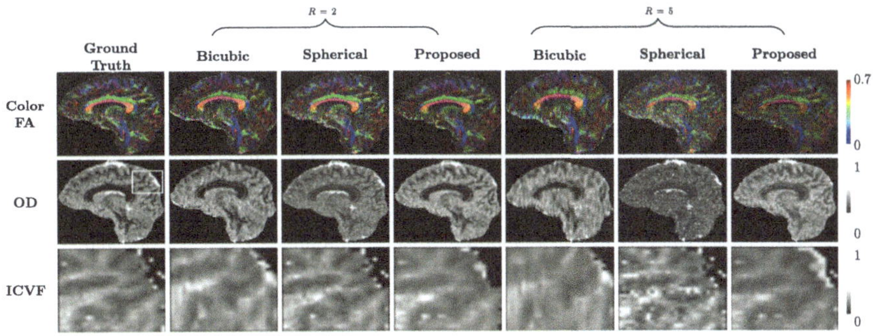

Fig. 5 Representative color FA, OD, and ICVF maps ($b = 2000\,\mathrm{s/mm^2}$)

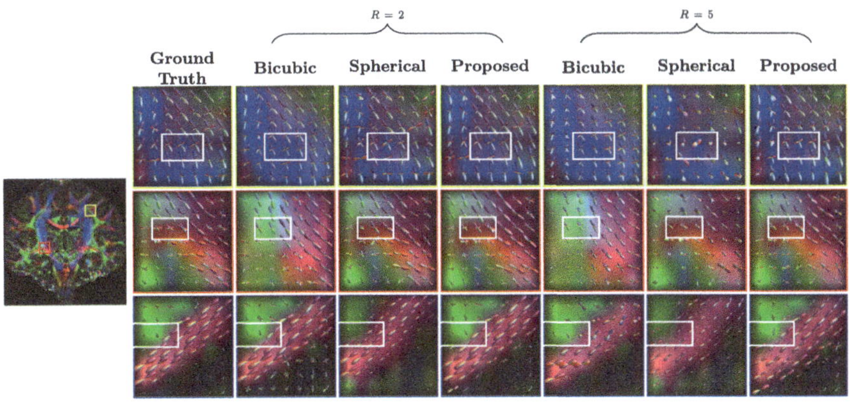

Fig. 6 Representative fiber ODFs in different regions with crossing, bifurcating, and unidirectional fibers

References

1. Barth, M., Breuer, F., Koopmans, P.J., Norris, D.G., Poser, B.A.: Simultaneous multislice (SMS) imaging techniques. Magnetic resonance in medicine 75(1), 63–81 (2016)
2. Basser, P.J., Mattiello, J., LeBihan, D.: Estimation of the effective self-diffusion tensor from the NMR spin echo. J. Magn. Reson., Ser. B **103**(3), 247–254 (1994)
3. Chan, S.H., Khoshabeh, R., Gibson, K.B., Gill, P.E., Nguyen, T.Q.: An augmented lagrangian method for total variation video restoration. IEEE Trans. Image Process. **20**(11), 3097–3111 (2011)
4. Chengbo, L.: An efficient algorithm for total variation regularization with applications to the single pixel camera and compressive sensing. Master's thesis, Rice University, Houstan, Texas (2009)
5. Hong, Y., Chang, W.T., Chen, G., Wu, Y., Lin, W., Shen, D., Yap, P.T.: Multifold acceleration of diffusion MRI via slice-interleaved diffusion encoding (SIDE) arXiv preprint arXiv:2002.10908 (2020)
6. Hutter, J., Christiaens, D.J., Schneider, T., Cordero-Grande, L., Slator, P.J., Deprez, M., Price, A.N., Tournier, J.D., Rutherford, M., Hajnal, J.V.: Slice-level diffusion encoding for motion and distortion correction. Medical Image Analysis 48, 214–229 (2018)

7. Ning, L., Setsompop, K., Michailovich, O., Makris, N., Shenton, M.E., Westin, C.F., Rathi, Y.: A joint compressed-sensing and super-resolution approach for very high-resolution diffusion imaging. NeuroImage 125, 386–400 (2016)
8. Ramos-Llordén, G., Ning, L., Liao, C., Mukhometzianov, R., Michailovich, O., Setsompop, K., Rathi, Y.: High-fidelity, accelerated whole-brain submillimeter in-vivo diffusion MRI using gSlider-Spherical Ridgelets (gSlider-SR). Magn. Reson. Med. **84**(4), 1781–1795 (2020)
9. Shenton, M.E., Hamoda, H., Schneiderman, J., Bouix, S., Pasternak, O., Rathi, Y., Vu, M.A., Purohit, M.P., Helmer, K., Koerte, I., et al.: A review of magnetic resonance imaging and diffusion tensor imaging findings in mild traumatic brain injury. Brain imaging and behavior 6(2), 137–192 (2012)
10. Sun, J., Entezari, A., Vemuri, B.C.: Exploiting structural redundancy in q-space for improved EAP reconstruction from highly undersampled (k, q)-space in DMRI. Medical image analysis 54, 122–137 (2019)
11. Tournier, J.D., Yeh, C.H., Calamante, F., Cho, K.H., Connelly, A., Lin, C.P.: Resolving crossing fibres using constrained spherical deconvolution: validation using diffusion-weighted imaging phantom data. NeuroImage 42(2), 617–625 (2008)
12. Yang, S., Wang, J., Fan, W., Zhang, X., Wonka, P., Ye, J.: An efficient ADMM algorithm for multidimensional anisotropic total variation regularization problems. In: Proceedings of the 19th ACM SIGKDD International Conference on Knowledge Discovery and Data Mining, pp. 641–649 (2013)
13. Zhang, H., Schneider, T., Wheeler-Kingshott, C.A., Alexander, D.C.: NODDI: practical in vivo neurite orientation dispersion and density imaging of the human brain. NeuroImage 61(4), 1000–1016 (2012)

Towards Learned Optimal q-Space Sampling in Diffusion MRI

Tomer Weiss, Sanketh Vedula, Ortal Senouf, Oleg Michailovich, and Alex Bronstein

Abstract Fiber tractography is an important tool of computational neuroscience that enables reconstructing the spatial connectivity and organization of white matter of the brain. Fiber tractography takes advantage of diffusion Magnetic Resonance Imaging (dMRI) which allows measuring the apparent diffusivity of cerebral water along different spatial directions. Unfortunately, collecting such data comes at the price of reduced spatial resolution and substantially elevated acquisition times, which limits the clinical applicability of dMRI. This problem has been thus far addressed using two principal strategies. Most of the efforts have been extended towards improving the quality of signal estimation for any, yet *fixed* sampling scheme (defined through the choice of diffusion-encoding gradients). On the other hand, optimization over the sampling scheme has also proven to be effective. Inspired by the previous results, the present work consolidates the above strategies into a unified estimation framework, in which the optimization is carried out with respect to both estimation model and sampling design *concurrently*. The proposed solution offers substantial improvements in the quality of signal estimation as well as the accuracy of ensuing analysis by means of fiber tractography. While proving the optimality of the learned estimation models would probably need more extensive evaluation, we nevertheless claim that the learned sampling schemes can be of immediate use, offering a way to improve the dMRI analysis without the necessity of deploying the neural network used for their estimation. We present a comprehensive comparative analysis based on the Human Connectome Project data. Code and learned sampling designs available at https://github.com/tomer196/Learned_dMRI.

T. Weiss (✉) · S. Vedula · O. Senouf · A. Bronstein
Technion, Israel
e-mail: weiss@cs.technion.ac.il

O. Michailovich
University of Waterloo, Waterloo, Canada

N. Gyori et al. (eds.), *Computational Diffusion MRI*, Mathematics and Visualization,
https://doi.org/10.1007/978-3-030-73018-5_2

1 Introduction

Fiber tractography has long become a standard tool of computational neuroscience which makes it possible to delineate the structure of neural fiber tracts within white matter, thus facilitating quantitative assessment of its integrity and connectivity in application to clinical diagnosis [2, 3].

The accuracy of fiber tractography, however, depends on the quality of diffusion MRI (dMRI) data used for estimation of the local directions of fiber tracts at each spatial voxel. Such data is usually available as a collection of 3D MRI volumes, known as diffusion-encoded images, which represent signal attenuation due to water diffusion along various directions and different levels of diffusion sensitization. In particular, high angular resolution diffusion imaging (HARDI) [19], were each diffusion-encoding image representing a single sampling point on the spherical shell. In this case, collection of a relatively large number of samples (as it is often required by more advanced methods of diffusion data analysis) entails longer imaging sessions, which tend to be avoided in clinical settings due to a number of practical constraints. Consequently, HARDI data typically suffer from relatively poor spatial resolution and other effects of undersampling, thus undermining the adequacy of ensuing data analysis by means of fiber tractography.

In addition to the use of parallel imaging [4], the problem of long acquisition times in dMRI has been addressed using a range of post-processing solutions. In particular, most works in this direction has focused on the development of post-processing methods capable of reconstructing the dMRI signals from their partial measurements in either the k-space [15] or q-space. In the latter case, accelerated imaging is achieved through the use of a deliberately smaller number of diffusion encodings then necessary, while compensating for the effects of undersampling by means of properly regularized inverse solvers, including the recent use of deep neural networks: [9] reconstructed the DWI from undersampled k-space using graph CNNs, [6] suggested to learn diffusion metrics directly from undersampled q-space. In virtually all such cases, however, the directions of diffusion encoding gradients is assumed to be given and fixed, being typically defined by means of the electrostatic repulsion (Thomas) algorithm [11] or tessellation of platonic solids [18].

Recent studies in computational imaging for inverse problems have uncovered significant benefits of simultaneously learning the forward and inverse operators, implying a concurrent estimation of the inverse (reconstruction) model along with the parameters of the acquisition system in use [8, 13, 22]. Lately, these ideas have been used to accelerate structural MRI imaging through the use of convolutional neural networks (CNNs) that are capable of learning the optimal reconstruction model *together with* optimal k-space sampling, either point [7, 24, 26] or trajectory-based [23]. Inspired by these results, the present work introduces a general estimation framework which allows one to learn the optimal directions of diffusion-encoding gradients for HARDI along with an optimal reconstruction model required for signal reconstruction from under-sampled data.

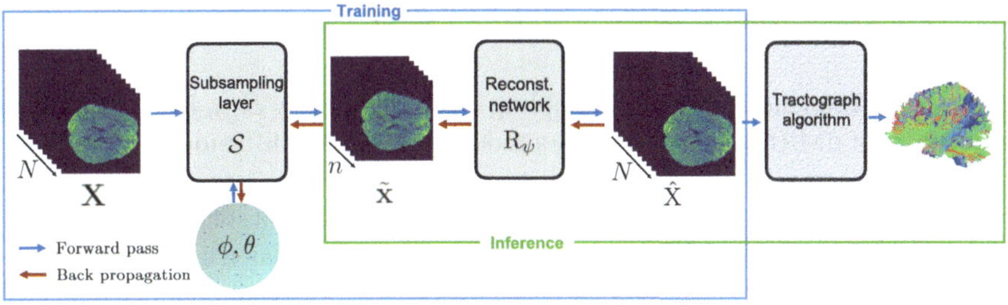

Fig. 1 The data flow pipeline of our method. Notation is explained in the text

1.1 Main Contributions

In this paper, we demonstrate that:

- using learned directions of diffusion encoding along with a learned reconstruction model leads to substantial improvements in the quality of estimated dMRI signals;
- the proposed solution leads to improved performance of fiber tractography as an important example of dMRI-related end-tasks;
- the learned directions of diffusion-encoding gradients generalize to dMRI datasets other than the data used for learning the directions.

2 Method

The proposed method can be viewed as an end-to-end pipeline combining the forward (acquisition) and the inverse (reconstruction) models which undergo simultaneous optimization (see Fig. 1 for a schematic depiction). The input to the forward model, denoted as $\mathbf{X}$, is formed by a total of N 3D diffusion-encoded volumes arranged into a 4D numerical array. The input layer is followed by a sub-sampling layer, which only uses a subset of $n \ll N$ dMRI volumes (or, equivalently, n diffusion-encoding directions). The inverse model is represented by a CNN which learns the inverse mapping that goes from the measurement space (of n directions) to the target space (of N directions). All components of our end-to-end model are differentiable with respect to the directions of diffusion encoding (represented by their spherical coordinates θ and ϕ), which makes the latter trainable with respect to the performance of desired end-task (e.g. fully-sampled signal reconstruction).

2.1 Forward Model: Sub-Sampling Layer

The principal function of the sub-sampling layer, denoted by $S_{\phi,\theta}$, is to emulate the acquisition of dMRI measurements at a reduced number of diffusion encoding (i.e., n), which is referred below to as $\tilde{\mathbf{X}}$. To this end, the dMRI signals used for training (and acquired at N spherical points) were fit with a truncated basis of spherical harmonics. In this case, it is convenient to describe the directions of diffusion encoding in terms of their azimuth $0 \leq \phi \leq 2\pi$ and elevation $0 \leq \theta \leq \pi$, which can both be collected into vectors of length n. In what follows, we refer to the ratio N/n as the *acceleration factor* (AF).

2.2 Reconstruction Model

The goal of the reconstruction model is to extract the latent image $\mathbf{X}$ from the limited set of diffusion directions $\tilde{\mathbf{X}}$. The resulting approximation is henceforth denoted as $\hat{X} = R_{\psi}(\tilde{\mathbf{X}})$, where R is the reconstruction model and ψ represent its learnable parameters. Although in our implementation we chose the off-the-shelf U-NET as the reconstruction model architecture, architectural search of the optimal reconstruction model is not within the scope of this work. Our proposed algorithm can be used with any differentiable reconstruction model.

2.3 Optimization

Training of the proposed pipeline is performed by simultaneously learning the diffusion gradient directions (ϕ, θ) and the parameters of the reconstruction model ψ. The training is carried out by minimizing the discrepancy between the model output image $\hat{\mathbf{X}}$ and the ground-truth image $\mathbf{X}$, meaning, by solving the optimization problem

$$\min_{\psi,\phi,\theta} \sum_{X} \|R_{\psi}(S_{\phi,\theta}(X)) - X\|_2 \qquad (1)$$

where the loss is summed over a training set comprising of 4D diffusion volumes of the full set of diffusion directions $\mathbf{X}$.

3 Experimental Evaluation

3.1 Dataset

The experiments reported in this paper have been obtained using the dMRI data provided by the Human Connectome Project (HCP) [21]. The HCP database contains 1065 brain MRI volumes acquired with 288 diffusion-encoding directions including 90 directions for each of the b values 1000, 2000, 3000 s/mm^2 and the rest 18 volumes with b value of 0. For the sake of simplicity, we choose to use only the diffusion directions pertaining to the spherical shell defined by $b = 1000$ s/mm^2. We used 868 volumes (102,000 slices) for training and 100 volumes (12,000 slices) for validation. To maintain consistency across the dataset that was originally acquired with different gradient directions schemes, we first resample the DWI data into $N = 90$ predefined directions evenly distributed on the unit hemisphere using spherical harmonics.

3.2 Training Settings

The sub-sampling layer and the reconstruction network were trained with the Adam optimizer [14]. The learning rate was set to 0.001 for the reconstruction model, while the sub-sampling layer was trained with a learning rate of 0.0001. For the reconstruction model, we used a multi-resolution encoder-decoder network with symmetric skip connections, also known as the U-Net architecture [20]. U-Net is widely-used in medical imaging tasks with many application in structural MRI reconstruction [25] and segmentation [10]. The reconstruction network has been trained in a slice-wise manner, with the image slices being concatenated back to 3D volume during the stage of inference. The input to the model is given by a slice where each of the diffusion directions n is a channel and the output is the slice with $N = 90$ channels. We emphasize that the scope of this work is not directed toward building the best reconstruction method, but rather demonstrating the benefit of simultaneous optimization of the acquisition-reconstruction pipeline. We perform experiments for the following acceleration factors (AF): 3, 5, 10, 15, *and* 30 that correspond to acquisitions performed with $n = 30, 18, 9, 6, 3$ diffusion directions, respectively.

3.3 Results and Discussion

Baselines. We consider the following baselines: *learned directions* (joint optimization of diffusion directions and the reconstruction network, ϕ and θ were initialized randomly), and *fixed directions* (optimizing the reconstruction network alone). For the fixed regime, we use the electrostatic repulsion algorithm [11] to design the diffusion gradient and only train the reconstruction network. In order to demonstrate

Table 1 **Quantitative results in the signal space**. The presented baselines are networks trained with *fixed* and *learned* directions across different acceleration factors (AF). Presented are the PSNR between the reconstructed and groundtruth DWI volumes. n = number of diffusion directions

AF/n	3/30	5/18	10/9	15/6	30/3
Fixed dirs.	48.99±1.62	48.73±1.51	45.13±1.67	42.04±1.52	39.67±1.45
Learned dirs.	**49.18±1.52**	**48.94±1.52**	**45.23±1.49**	**42.20±1.42**	**39.70±1.44**

the robustness of the learned diffusion directions and their practical applicability in the absence of reconstruction network, we present two additional baselines, namely, *fixed w/o reconstruction* and *learned w/o reconstruction*. Effectively, this implies rendering the operator R_{ψ} to be identity.

Metrics in signal space. First, we evaluate the algorithm performance by comparing directly $\hat{\mathbf{X}}$ the ground-truth $\mathbf{X}$ using the PSNR (peak signal-to-noise ratio). In Table 1, we present the results of the fixed and learned directions baselines across several acceleration factors. As expected, we notice that the PSNR of the reconstructed images degrade as the acceleration factor increases. Interestingly, we can notice improvement in the case of learned directions over the fixed one demonstrating the merit of joint optimization of diffusion directions and reconstruction network.

Metrics in tractography space. The performance of the proposed method has been also tested in application to fiber tractography. To this end, we applied the same tractography reconstruction to both $\hat{\mathbf{X}}$ and $\mathbf{X}$ (concatenate with 6 originals volumes of $b = 0$) and compare their tractograms. To perform tractography, we first fit the DWI to constant solid angle ODF model [12] and then invoke fiber tracking using EuDX [5] using the implementation available in the `dipy` package. To compare the tractograms, we used the Bhattacharyya distance [1] over *bundles* (a collection of individual fibres), we compute it over 15 different bundles and compute the average distance. Table 2 presents a comparison of the results obtained from fixed and learned directions baselines. We evaluate the quality of the learned directions both with and without using the reconstruction network across few acceleration factors. Visual results of the entire tractogram can be seen in Fig. 2 and for specific bundles in Fig. 3 (visualization for all acceleration factor are presented in the supplementary material, see Figs. 5, 6, 7 and 8). Based on the obtained quantitative and qualitative results, the following are the observations (Fig. 4).

- The use of supervised learning for reconstruction gave significant improvement of 0.186 – 1.346 points in Bhattacharyya distance and a clear noticeable improvement in the qualitative results presented in Figs. 5, 6, 7 and 8.
- Joint optimization of the diffusion directions and the reconstruction gave a further sizeable improvement of 0.015–0.478 points in Bhattacharyya distance.
- A particularly appealing result is that, we notice even in the absence of reconstruction network, the learned directions yield an improvement of 0.076–0.735 points in Bhattacharyya distance when compared to the fixed directions. A possible reason for this improvement is because joint optimization of the reconstruction

Table 2 Quantitative results in the tractography space. The presented baselines are networks trained with *fixed* and *learned* directions across different acceleration factors (AF). Presented below are the Bhattacharyya distance (BD) between the tractographies obtained from the groundtruth and reconstructed DWI volumes. The distance is averaged over 15 bundles. n = number of diffusion directions

	Bhattacharyya distance				
AF/n	3/30	5/18	10/9	15/6	30/3
Fixed w/o reconst.	0.329±0.032	0.539±0.057	0.685±0.058	1.802±0.190	–
Learned w/o reconst.	0.253±0.024	0.327±0.032	0.514±0.055	1.067±0.092	–
Fixed w/ reconst.	0.143±0.029	0.172±0.032	0.264±0.031	0.456±0.048	0.860±0.097
Learned w/ reconst.	**0.128±0.022**	**0.151±0.027**	**0.205±0.029**	**0.335±0.035**	**0.382±0.036**

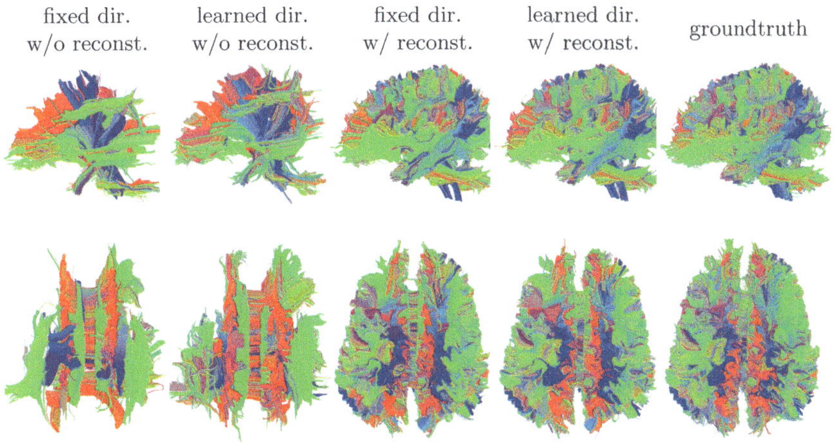

Fig. 2 Qualitative evaluation of the tractogram of a test sample from the human connectome project. Presented below are the tractograms achieved using 6 gradient directions (AF = 15) for fixed/learned directions with/without the reconstruction network, and the corresponding groundtruth tractogram. Digital zoom-in is recommended. Top: side-view, bottom: top-view

network together with the diffusion directions enforces metric that is learned from the data in the direction space. This metric is more meaningful than minimizing the electrostatic energy. Our learned diffusion directions can be simply translated into diffusion gradients and readily deployed onto real MRI scanners to achieve sizeable acceleration.

Learned directions A comparison of the fixed (red) and learned directions (blue) plotted on a hemi-sphere is visualized in the adjacent figure. The left and right figures correspond to AF = 3 (n = 30) and AF = 10 (n = 9) respectively. We can notice

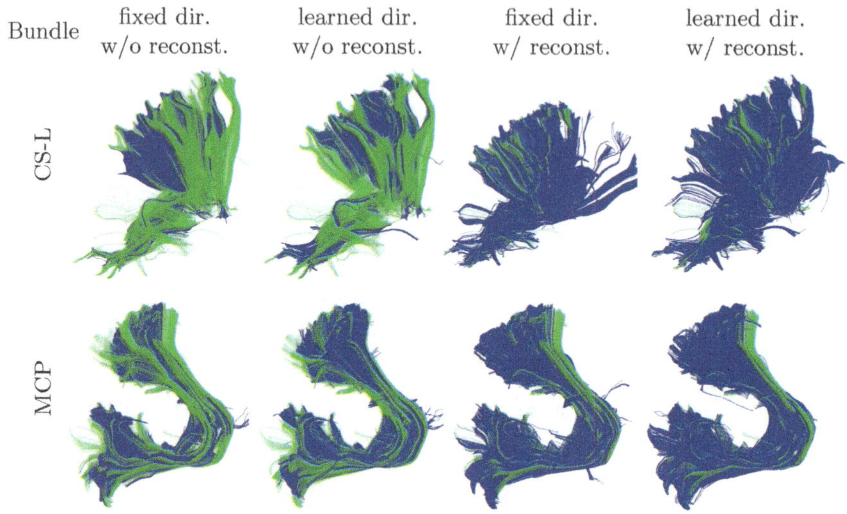

Fig. 3 Visualization of CS-L and middle cerebellar peduncle bundles of a test volume in the HCP dataset after tractography. Results depicted are obtained using fixed/learned diffusion directions (AF = 3, $n = 30$), both with/without reconstruction. Colored in blue and green are the reconstructed and groundtruth bundles respectively

that although both fixed and learned directions look similar uniformly covering the hemisphere, our results presented above results suggest that the learned directions hit the right spots that yield better reconstruction and tractography performance. This implies that the joint optimization leads to an importance sampling in the diffusion space.

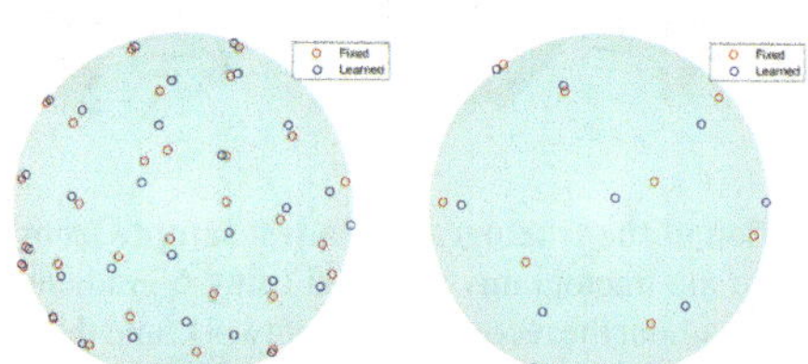

Tractometer, ISMRM challenge. To further demonstrate the robustness, deployability and generalization capability of our learned diffusion directions, we test our learned diffusion directions on the ISMRM tractography challenge DWI phantom dataset [16]. We evaluated the results using the *tractometer tool* [17]. For the evaluation, we first sub-sampled the original DWI volume using the fixed and learned diffusion directions that we obtain from the human connectome dataset. Then, we applied the tractography algorithms to the resulting sub-sampled DWI volumes and quantitative evaluate using the tractometer tool. The results depicted in Table 3 demonstrate that the learned directions outperform the fixed directions in most of the metrics

Table 3 Quantitative results on the ISMRM challenge dataset using the tractometer tool. Depicted below are the results obtained using the fixed and learned directions *without* the reconstruction network across different number of gradient directions. VC: valid connections (+), IC: invalid connections (–), NC: non-connections (–), VB: valid bundles (+), IB: invalid bundles (–), OL: overlap (+), OR: overreach (–), F1 score (+). +/– indicate higher or lower score is better

n		VC (+)	IC (–)	NC (–)	VB (+)	IB (–)	OL (+)	OR (–)	F1(+)
30	Fixed	45.32%	54.67%	**0.000%**	20	**71**	15.74%	**23.59%**	22.88%
	Learned	**52.48%**	**47.51%**	0.004%	**23**	73	**23.51%**	29.66%	**32.46%**
18	Fixed	56.01%	43.98%	**0.000%**	21	60	17.32%	26.83%	25.82%
	Learned	**60.26%**	**39.73%**	**0.000%**	**22**	**60**	**18.32%**	**24.67%**	**26.89%**
9	Fixed.	58.86%	41.13%	**0.000%**	19	43	11.55%	18.39%	18.81%
	Learned	**61.42%**	**38.57%**	**0.000%**	**21**	**42**	**12.13%**	**17.51%**	**19.54%**
6	Fixed	26.15%	73.84%	**0.000%**	**8**	20	**1.26%**	6.22%	**2.35%**
	Learned	**28.06%**	**71.93%**	**0.000%**	5	**17**	0.76%	**3.21%**	1.45%

across multiple acceleration factors. Visual results of this experiments are presented in Fig. 4 in the supplementary material.

4 Conclusion

We demonstrated, as a proof-of-concept, that the learning-based design of diffusion gradient design in diffusion MR imaging leads to better tractography when compared to the off-the-shelf designs. To the best of our knowledge, this is the first attempt of data-driven design of diffusion gradient directions in diffusion MRI. We trained and evaluated the performance of the learned designs and reconstruction networks on the human connectome project dataset, both in signal and tractography domains. We evaluated the generalization of the learned designs on the ISMRM challenge phantom, and observed that the learned designs consistently outperform the handcrafted ones at different acceleration factors. Therefore, we believe that the learned directions can be deployed standalone onto real machines without the reconstruction network to already improve the end-task performance (full signal reconstruction, tractography). We defer the following important aspects to future work:

- In this work, we limited our attention to only diffusion directions lying on one shell ($b = 1000$). Allowing the diffusion gradient directions on multiple shell might lead to further improvement in the end-task performance.
- As can be inferred from the PSNR results presented in Table 1 and the tractogram metrics in Tables 2 and 3, the discrepancy in the DWI domain is not directly related to the end-task performance (in our case, tractography). This leads us to assume that optimizing directly for the end-task (or at least, closer to it) will allow for better sampling designs that are needed for the end-task and can further push the acceleration factor vs. performance trade-offs.

5 Supplementary Materials

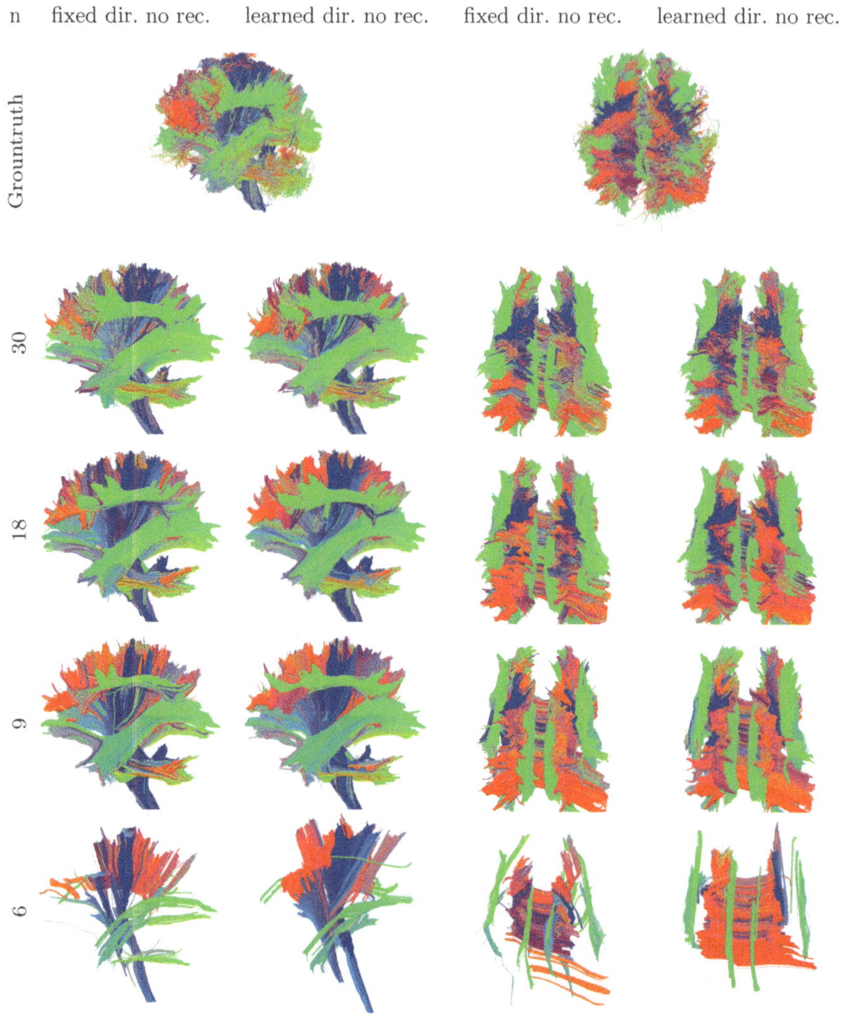

Fig. 4 ISMRM fiber cup tractogram. Visual comparison of the learned and fixed diffusion parameters applied on diffusion data from the ISMRM fiber cup challenge. Note that we only use the learned directions and not the reconstruction network. Presented in figures are the side and top views, without reconstruction. Left column depicts the number of diffusion directions used. The first two columns and last two columns depict the side-view and top-view of the same tractogram. Note that the above depicted figures are obtained using just the learned diffusion sampling without the reconstruction network

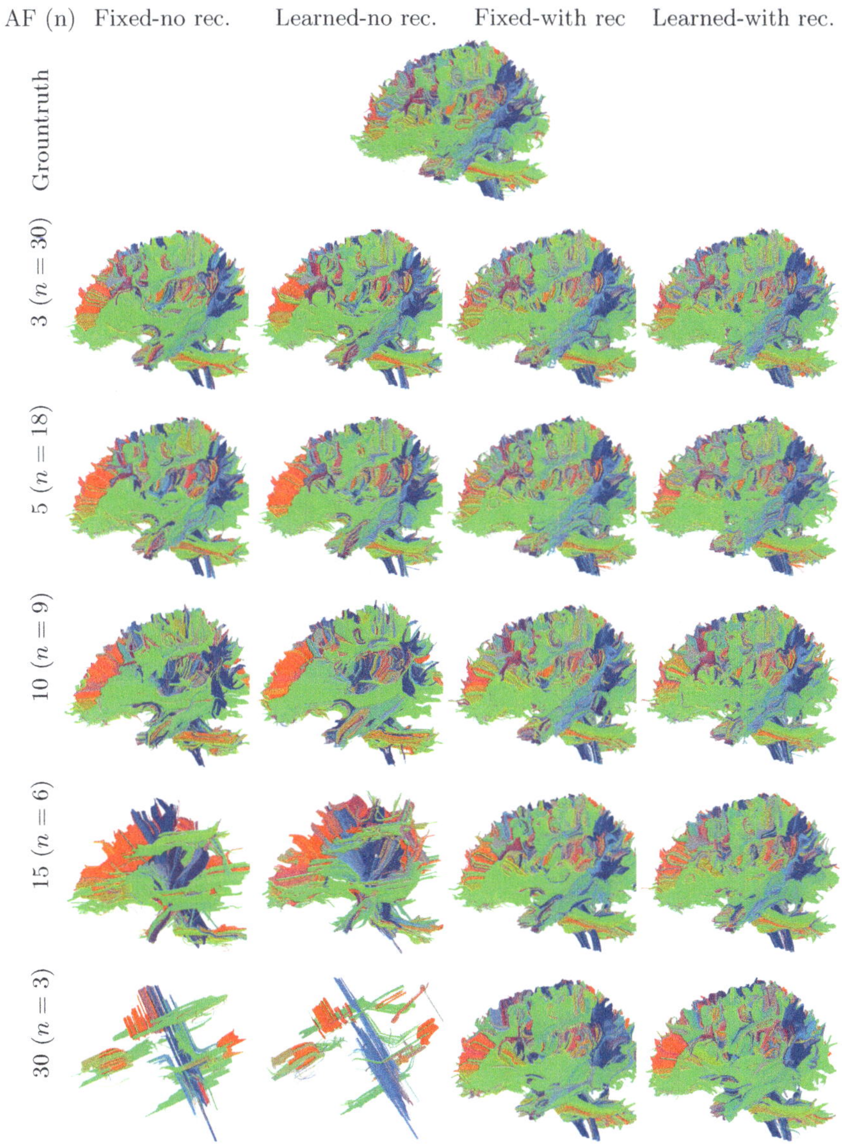

Fig. 5 Tractogram: side-view. Visual comparison of the full tractogram achieved from the DWI volume using fixed and learned directions, with and without reconstruction. Left-most column depicts the corresponding acceleration factor and the corresponding n (number of diffusion directions)

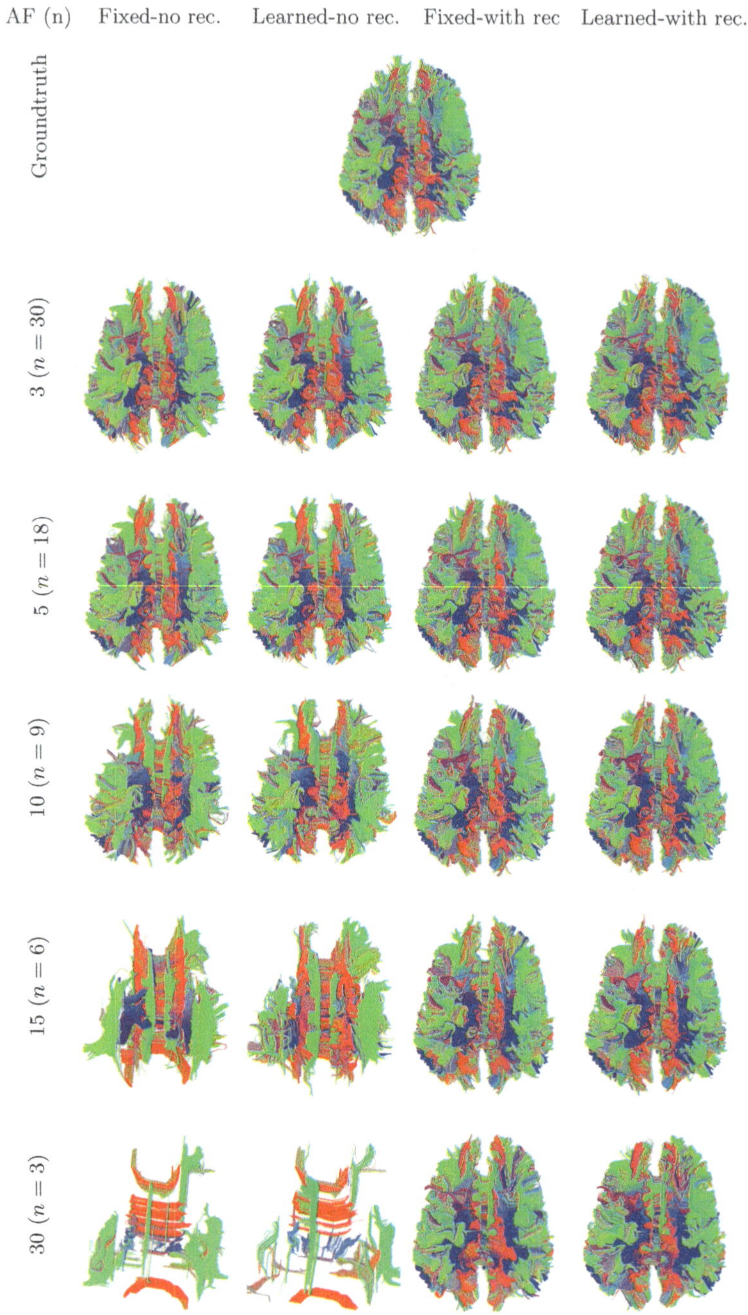

Fig. 6 Tractogram top-view. Visual comparison of the full tractogram achieved from the DWI volume using fixed and learned directions, with and without reconstruction. Left-most column depicts the corresponding acceleration factor and the corresponding n (number of diffusion directions)

Fig. 7 Middle cerebellar peduncle bundle. Visual comparison of the bundle achieved from applying tractography on the DWI volume using fixed and learned directions with and without reconstruction. Colored in green is the groundtruth bundle. Left-most column depicts the corresponding acceleration factor and the corresponding n (number of diffusion directions)

Fig. 8 CS Left bundle. Visual comparison of the bundle achieved from applying tractography on the DWI volume using fixed and learned directions with and without reconstruction. Colored in green is the groundtruth bundle. Left-most column depicts the corresponding acceleration factor and the corresponding n (number of diffusion directions)

References

1. Bhattacharyya, A.: On a measure of divergence between two multinomial populations. Sankhyā: Indian J. Stat. (1933–1960) (1946)
2. Bullmore, E., Sporns, O.: Complex brain networks: graph theoretical analysis of structural and functional systems. Nat. Rev. Neurosci. (2009)
3. Duffau, H.: Brain Mapping: From Neural Basis of Cognition to Surgical Applications. Springer, Vienna (2011)
4. Feinberg, D.A., Setsompop, K.: Ultra-fast mri of the human brain with simultaneous multi-slice imaging. J. Mag. Reson. (2013)
5. Garyfallidis, E.: Towards an Accurate Brain Tractography. University of Cambridge (2013)
6. Golkov, V., Dosovitskiy, A., Sperl, J.I., Menzel, M.I., Czisch, M., Sämann, P., Brox, T., Cremers, D.: Q-space deep learning: twelve-fold shorter and model-free diffusion mri scans. IEEE Trans. Med. Imaging **35**(5), 1344–1351 (2016)
7. Gözcü, B., Mahabadi, R.K., Li, Y.H., Ilıcak, E., Çukur, T., Scarlett, J., Cevher, V.: Learning-based compressive MRI. IEEE Trans. Med. Imaging (2018)
8. Haim, H., Elmalem, S., Giryes, R., Bronstein, A.M., Marom, E.: Depth estimation from a single image using deep learned phase coded mask. IEEE Trans. Comput. Imaging (2018)
9. Hong, Y., Chen, G., Yap, P.T., Shen, D.: Reconstructing high-quality diffusion mri data from orthogonal slice-undersampled data using graph convolutional neural networks. In: International Conference on Medical Image Computing and Computer-Assisted Intervention, pp. 529–537. Springer, Berlin (2019)
10. Isensee, F., Jaeger, P.F., Full, P.M., Wolf, I., Engelhardt, S., Maier-Hein, K.H.: Automatic cardiac disease assessment on cine-MRI via time-series segmentation and domain specific features. In: International Workshop on Statistical Atlases and Computational Models of the Heart. Springer, Berlin (2017)
11. Jones, D.K., Horsfield, M.A., Simmons, A.: Optimal strategies for measuring diffusion in anisotropic systems by magnetic resonance imaging. Mag. Reson. Med. Official J. Int. Soc. Mag. Reson. Med. (1999)
12. Kamath, A., Aganj, I., Xu, J., Yacoub, E., Ugurbil, K., Sapiro, G., Lenglet, C.: Generalized constant solid angle odf and optimal acquisition protocol for fiber orientation mapping. In: MICCAI Workshop on Computational Diffusion MRI (2012)
13. Kellman, M., Bostan, E., Chen, M., Waller, L.: Data-Driven Design for Fourier Ptychographic Microscopy (2019)
14. Kingma, D.P., Ba, J.: Adam: A Method for Stochastic Optimization (2014). arXiv:1412.6980
15. Lustig, M., Donoho, D., Pauly, J.M.: Sparse MRI: The application of compressed sensing for rapid MR imaging. Mag. Reson. Med. Official J. Int. Soc. Mag. Reson. Med. (2007)
16. Maier-Hein, K.H.e.a.: Tractography-based connectomes are dominated by false-positive connections. BioRxiv (2016)
17. Marc-Alexandre Côté, G.e.a.: Tractometer: Towards validation of tractography pipelines. Med. Image Anal. (2013)
18. Meschkowski, H.: Unsolved and unsolvable problems in geometry. F. Ungar (1966)
19. Poupon, C., Rieul, e.a.: New diffusion phantoms dedicated to the study and validation of high-angular-resolution diffusion imaging (hardi) models. Mag. Reson. Med. Official J. Int. Soc. Mag. Reson. Med. (2008)
20. Ronneberger, O., Fischer, P., Brox, T.: U-net: Convolutional networks for biomedical image segmentation. In: MICCAI (2015)
21. Van Essen, D., Ugurbil et al., K.: The Human Connectome Project: A data acquisition perspective. NeuroImage (2012). https://doi.org/10.1016/j.neuroimage.2012.02.018,connectivity
22. Vedula, S., Senouf, O., Zurakhov, G., Bronstein, A., Michailovich, O., Zibulevsky, M.: Learning beamforming in ultrasound imaging. In: Proceedings of The 2nd International Conference on Medical Imaging with Deep Learning. Proceedings of Machine Learning Research, PMLR (2019)

23. Weiss, T., Senouf, O., Vedula, S., Michailovich, O., Zibulevsky, M., Bronstein, A.: PILOT: Physics-Informed Learned Optimal Trajectories for Accelerated MRI (2019). arXiv:1909.05773
24. Weiss, T., Vedula, S., Senouf, O., Bronstein, A., Michailovich, O., Zibulevsky, M.: Learning fast magnetic resonance imaging (2019). arXiv:1905.09324
25. Zbontar, J., Knoll, F., Sriram, A., Muckley, M.J., Bruno, M., Defazio, A., Parente, M., Geras, K.J., Katsnelson, J., Chandarana, H., et al.: fastMRI: An open dataset and benchmarks for accelerated MRI (2018). arXiv:1811.08839
26. Zhang, Z., Romero, A., Muckley, M.J., Vincent, P., Yang, L., Drozdzal, M.: Reducing uncertainty in undersampled MRI reconstruction with active acquisition (2019). arXiv:1902.03051

A Signal Peak Separation Index for Axisymmetric B-Tensor Encoding

Gaëtan Rensonnet, Jonathan Rafael-Patiño, Benoît Macq, Jean-Philippe Thiran, Gabriel Girard, and Marco Pizzolato

Abstract Diffusion-weighted MRI (DW-MRI) has recently seen a rising interest in planar, spherical and general B-tensor encodings. Some of these sequences have aided traditional linear encoding in the estimation of white matter microstructural features, generally by making DW-MRI *less* sensitive to the orientation of axon fascicles in a voxel. However, less is known about their potential to make the signal *more* sensitive to fascicle orientation, especially in crossing-fascicle voxels. Although planar encoding has been commended for the resemblance of its signal with the voxel's orientation distribution function (ODF), linear encoding remains the near undisputed

---

Gabriel Girard and Marco Pizzolato—These senior authors contributed equally.

G. Rensonnet (✉) · J. Rafael-Patiño · J.-P. Thiran · G. Girard · M. Pizzolato
Signal Processing Lab (LTS5), École polytechnique fédérale de Lausanne, Lausanne, Switzerland
e-mail: gaetan.rensonnet@epfl.ch

J. Rafael-Patiño
e-mail: jonathan.patinolopez@epfl.ch

J.-P. Thiran
e-mail: jean-philippe.thiran@epfl.ch

G. Girard
e-mail: gabriel.girard@epfl.ch

M. Pizzolato
e-mail: marco.pizzolato@epfl.ch

G. Rensonnet · B. Macq
ICTEAM Institute, Université catholique de Louvain, Louvain-la-Neuve, Belgium
e-mail: benoit.macq@uclouvain.be

J.-P. Thiran · G. Girard
Center for BioMedical Imaging (CIBM), Lausanne, Switzerland

Radiology Department, Centre hospitalier universitaire vaudois and University of Lausanne, Lausanne, Switzerland

M. Pizzolato
Department of Applied Mathematics and Computer Science, Technical University of Denmark, Kgs. Lyngby, Denmark

N. Gyori et al. (eds.), *Computational Diffusion MRI*, Mathematics and Visualization,
https://doi.org/10.1007/978-3-030-73018-5_3

method of choice for orientation estimation. This paper presents a theoretical framework to gauge the sensitivity of axisymmetric B-tensors to fascicle orientations. A signal peak separation index (SPSI) is proposed, motivated by theoretical considerations on a simple multi-tensor model of fascicle crossing. Theory and simulations confirm the intuition that linear encoding, because it maximizes B-tensor anisotropy, possesses an intrinsic advantage over all other axisymmetric B-tensors. At identical SPSI however, oblate B-tensors yield higher signal and may be more robust to acquisition noise than their prolate counterparts. The proposed index relates the properties of the B-tensor to those of the tissue microstructure in a straightforward way and can thus guide the design of diffusion sequences for improved orientation estimation and tractography.

1 Introduction

Diffusion-weighted magnetic resonance imaging (DW-MRI) is based on the application of time-varying external magnetic-field gradients $\mathbf{g}(t)$ to probe water diffusion [21]. In the brain white matter, it is mainly used for two purposes. One is a necessary first step for tractography consisting in estimating the orientation of the main fascicles of axons in a voxel, often via the orientation distribution function (ODF), and is referred to as *orientation estimation*. Another is to estimate finer microstructural properties of those fascicles such as the morphology of their axons, generally referred to as *microstructure estimation* [3].

In both tasks, the focus has traditionally been on *linear encoding* in which $\mathbf{g}(t) \in \mathbb{R}^3$ is parallel to a fixed direction $\hat{\mathbf{u}} \in \mathbb{S}^2$ for all t, the best-known example of which being the pulsed-gradient spin-echo (PGSE) [21]. More recently, there has been growing interest in more general gradient waveforms living in a 2D plane, known as *planar encoding* [19, 26], or the whole 3D space, referred to as general multidimensionnal or *B-tensor encoding* [8, 17, 27, 28]. As the name indicates, such sequences are often studied through their associated symmetric, positive-definite B-tensor defined as $\mathbf{B} := \gamma^2 \int_0^T \int_0^t \int_0^t \mathbf{g}(t_1) \cdot \mathbf{g}(t_2)^\top \,\mathrm{d}t_1 \,\mathrm{d}t_2 \,\mathrm{d}t \in \mathbb{R}^{3\times3}$, where γ is the gyromagnetic ratio of protons and T the duration of the sequence. B-tensors with $1, 2$ and 3 identical, strictly positive eigenvalues refer to linear, planar and spherical encoding respectively [27, 28].

Those general waveforms have been mostly used for microstructure estimation [12, 14, 17], especially to resolve degeneracies wherein different microstructural properties are difficult to estimate simultaneously using conventional linear encoding only; e.g., extracting axonal microstructural properties irrespective of their orientation with 3D B-tensor encoding [2, 24]; disentangling volume or signal fractions and diffusivities with (planar) double diffusion encoding (DDE) [5] or (spherical) triple diffusion encoding [11]; separating microscopic anisotropy from orientation dispersion using DDE [12, 15] or spherical encoding [6, 14, 22].

How well these general waveforms may perform at orientation estimation is still an open question. Linear encoding maximizes B-tensor anisotropy [9] and is thus expected to provide high signal sensitivity to the orientation of anisotropic structures. Spherical encoding on the other hand minimizes B-tensor anisotropy [2, 6, 9, 11, 14, 22, 24] and probably offers little benefit. Planar encoding DW-MRI data directly reflects the ODF of the voxel without needing the post-processing or modeling typically required in linear encoding [19, 26]. It has been shown to compare to [26] and potentially outperform [19] linear encoding.

This paper proposes a framework to assess the potential of waveforms characterized by an axisymmetric B-tensor for providing DW-MRI data suited to orientation estimation. A signal peak separation index (SPSI) is proposed, motivated by theoretical considerations on a toy model of fascicle crossing assuming a superposition of diffusion tensors. The index relates the properties of the B-tensors to microstructural properties such as the crossing angle, the NMR-apparent volume fraction and the microscopic anisotropy of the fascicles, to quantify the directional information content of the signal. Theoretical predictions are made about the respective merits of linear, planar and intermediate encodings which are then verified in simulation experiments.

2 Theory

In this work, a diffusion sequence is entirely characterized by an axisymmetric tensor $\mathbf{B}$ with eigenvalues $\left\{\frac{b_\perp}{2}, \frac{b_\perp}{2}, b_\parallel\right\}$. The orientation $\hat{\mathbf{u}}_B$ of $\mathbf{B}$ is defined as the eigenvector associated with $b_\parallel$, which is not necessarily the largest eigenvalue, and the b-value, a measure of diffusion weighting, as $b := \mathrm{tr}(\mathbf{B}) = b_\perp + b_\parallel$. The linearity coefficient c_L is defined as $c_L := \frac{b_\parallel}{b}$ and characterizes B-tensor encodings as planar ($c_L = 0$), planar-like or oblate ($0 < c_L < \frac{1}{3}$), spherical ($c_L = \frac{1}{3}$), linear-like or prolate ($\frac{1}{3} < c_L < 1$) and linear ($c_L = 1$). It is related to the tensor anisotropy b_Δ [9] via $b_\Delta = \frac{3}{2}\left(c_L - \frac{1}{3}\right) \in \left[-\frac{1}{2}, 1\right]$. In this setting, the exact temporal profiles of the physically-applied magnetic-field gradients are thus ignored.

The diffusion of water within a fascicle of bundled axons is represented by an axisymmetric diffusion tensor $\mathbf{D}$, referred to as "zeppelin", with eigenvalues $\left\{\lambda_\perp, \lambda_\perp, \lambda_\parallel\right\}$, where $\lambda_\parallel > \lambda_\perp$ is enforced, and with orientation $\hat{\mathbf{u}}_D$ defined as its principal eigenvector. The normalized DW-MRI signal S_{sing} arising from a fascicle characterized by a zeppelin $\mathbf{D}$ subject to $\mathbf{B}$ is $\exp(-\mathbf{B} : \mathbf{D})$ [18], where : denotes the Frobenius inner product. This yields

$$S_{\mathrm{sing}}(\mathbf{B}; \mathbf{D}) = \exp\left(-\frac{b_\perp}{2}\lambda_\perp - \cos^2(\varphi_D - \varphi_B)\left(\frac{b_\perp}{2}\lambda_\perp + b_\parallel\lambda_\parallel\right) - \sin^2(\varphi_D - \varphi_B)\left(b_\parallel\lambda_\perp + \frac{b_\perp}{2}\lambda_\parallel\right)\right), \quad (1)$$

showing that the signal is effectively a function of the angle $|\varphi_D - \varphi_B|$ between $\hat{\mathbf{u}}_D$ and $\hat{\mathbf{u}}_B$, with φ_D and φ_B their azimuthal coordinates in their common plane, defined from an arbitrary common reference. In the spherical case $c_L = \frac{1}{3}$, all values of φ_B identically lead to $S_{\text{sing}} = \exp\left(-\frac{b}{3}\left(2\lambda_\perp + \lambda_\parallel\right)\right)$.

2.1 A Toy Model of Fascicle Crossing Under B-Tensor Encoding

The voxel-level signal S_{cros} resulting from the crossing of two populations characterized by $\mathbf{D}_1$ and $\mathbf{D}_2$ with NMR-apparent volume fractions, referred to as signal fractions, ν_1 and $\nu_2 = 1 - \nu_1$ under the diffusion-encoding tensor $\mathbf{B}$ is approximated by the following superposition [20]

$$S_{\text{cros}}\left(\mathbf{B}; \mathbf{D}_1, \mathbf{D}_2\right) = \nu_1 S_{\text{sing}}\left(\mathbf{B}; \mathbf{D}_1\right) + \nu_2 S_{\text{sing}}\left(\mathbf{B}; \mathbf{D}_2\right). \tag{2}$$

Our toy model assumes identical microstructural properties ($\mathbf{D}_1$ and $\mathbf{D}_2$ have identical eigenvalues) and Fascicle 1 as the dominant fascicle ($\nu_1 \geq \nu_2$).

The in-plane signal S_{ip} of a crossing is defined as the signal for $\hat{\mathbf{u}}_B$ lying in the plane spanned by $\hat{\mathbf{u}}_1$ and $\hat{\mathbf{u}}_2$, assumed non collinear. It is the only signal contribution relevant to peak detection because the out-of-plane signal, defined when $\hat{\mathbf{u}}_B \cdot \hat{\mathbf{u}}_1 = \hat{\mathbf{u}}_B \cdot \hat{\mathbf{u}}_2 = 0$, is equal to $\exp\left(-\frac{b_\perp}{2}\left(\lambda_\parallel + \lambda_\perp\right) - b_\parallel \lambda_\perp\right)$ and therefore holds no information about the fascicles' orientations $\hat{\mathbf{u}}_1$ and $\hat{\mathbf{u}}_2$. Setting without loss of generality $\varphi_1 = 0$, $\varphi_2 = \alpha$ with α the crossing angle between $\hat{\mathbf{u}}_1$ and $\hat{\mathbf{u}}_2$, and noting φ_B the in-plane azimuthal coordinate of $\hat{\mathbf{u}}_B$, S_{ip} is computed as

$$\begin{aligned} S_{\text{ip}}\left(\varphi_B\right) &= \nu_1 S_{\text{sing}}\left(\varphi_B\right) + \nu_2 S_{\text{sing}}\left(\alpha - \varphi_B\right) \\ &= \nu_1 \exp\left(-\frac{b_\perp}{2}\lambda_\perp - \cos^2\left(\varphi_B\right)\left(\frac{b_\perp}{2}\lambda_\perp + b_\parallel \lambda_\parallel\right) - \sin^2\left(\varphi_B\right)\left(b_\parallel \lambda_\perp + \frac{b_\perp}{2}\lambda_\parallel\right)\right) \\ &+ \nu_2 \exp\left(-\frac{b_\perp}{2}\lambda_\perp - \cos^2\left(\alpha - \varphi_B\right)\left(\frac{b_\perp}{2}\lambda_\perp + b_\parallel \lambda_\parallel\right) - \sin^2\left(\alpha - \varphi_B\right)\left(b_\parallel \lambda_\perp + \frac{b_\perp}{2}\lambda_\parallel\right)\right), \end{aligned} \tag{3}$$

where the single argument to S_{sing} refers to the angle between $\hat{\mathbf{u}}_D$ and $\hat{\mathbf{u}}_B$, with the dependence on all other properties of $\mathbf{B}$ and $\mathbf{D}$ implied.

Radial plots of S_{ip} are shown in Fig. 1 for various values of b, α and c_L and display the characteristic butterfly shape of DW-MRI signals. Signals from planar-like ($c_L < \frac{1}{3}$) and linear-like ($c_L > \frac{1}{3}$) encodings are 90° out of phase, with high signal obtained respectively in the range $\varphi_B \in [0, \alpha]$ and $\varphi_B \in \left[-\frac{\pi}{2}, \alpha - \frac{\pi}{2}\right]$. The signal increases when magnetic-field gradients are applied perpendicular to $\hat{\mathbf{u}}_1$ and $\hat{\mathbf{u}}_2$, "against" the fascicles (see Fig. 2a).

Mathematically, voxel-level fascicles are distinguishable when the high-signal range of S_{ip} exhibits two distinct maxima separated by one minimum. Intuitively, the maxima of $S_{\text{ip}}\left(\varphi_B\right)$ are expected to be around 0 and α (corresponding to $\hat{\mathbf{u}}_1$ and $\hat{\mathbf{u}}_2$) in planar-like encoding, and around $-\frac{\pi}{2}$ and $\alpha - \frac{\pi}{2}$ (corresponding to the directions $\hat{\mathbf{n}}_1$ and $\hat{\mathbf{n}}_2$ normal to $\hat{\mathbf{u}}_1$ and $\hat{\mathbf{u}}_2$) in linear-like encoding. Likewise, the

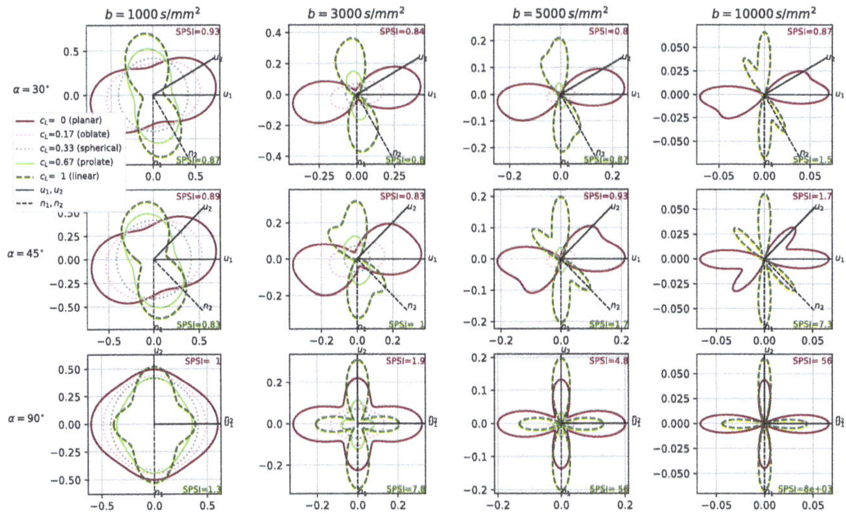

Fig. 1 Linear encoding is more sensitive to fascicle orientation than planar encoding while intermediate cases offer little benefit. The peaks and troughs in the in-plane signal S_{ip} of our toy model, especially those associated with the smaller peak (here $\nu_2 = 0.4$), only become visible at sufficiently high b, large α and extremal value of c_L, which is captured by values of our proposed SPSI staying below or exceeding 1

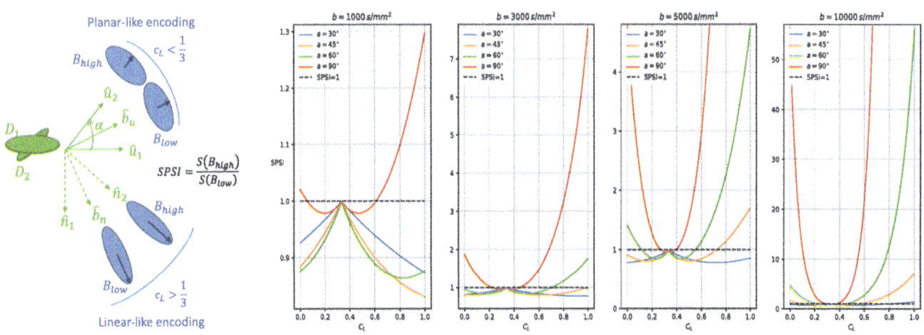

(a) Schematic definition of SPSI.

(b) SPSI vs shape of the B-tensor c_L with $\nu_1 = 0.6$. The y-axis was trimmed for clarity at $b = 5000$ and $b = 10\,000$.

Fig. 2 The proposed signal peak separation index (SPSI) is a ratio of signal "against" the smaller fascicle to signal between the fascicle. It is maximized by linear, rather than planar, encoding

minima of $S_{\text{ip}}(\varphi_B)$ are expected to lie about halfway between the maxima, along the bisector $\hat{\mathbf{b}}_u$ for planar- and $\hat{\mathbf{b}}_n$ for linear-like encoding. In planar-like encoding for instance, if S_{ip} has a lower value along $\hat{\mathbf{u}}_2$ than along $\hat{\mathbf{b}}_u$, it suggests that the second peak in the high-signal range of S_{ip} is too small or hasn't appeared yet (see Fig. 1 at $b = 3000\,\text{s/mm}^2$ and $\alpha = 45°$ and that the signal does not contain the orientational information required for accurate ODF or fascicle orientation estimation. The small fascicle ($\nu_2 < \nu_1$) is only detectable when its associated signal maximum (if present at all) exceeds the signal dip (if present at all) between the fascicles, as apparent in Fig. 1. These considerations motivate the definition of a signal index based on (approximate) peaks and troughs of the signal, as presented in the following section.

2.2 The Signal Peak Separation Index

The signal peak separation index (SPSI) of an axisymmetric B-tensor encoding is defined as the ratio of the signal "against" the smaller fascicle to the signal acquired along the bisector of the fascicles in a toy model of identical intersecting zeppelins (see Fig. 2a) and reads

$$\text{SPSI} := \begin{cases} \frac{S_{\text{ip}}(\alpha)}{S_{\text{ip}}(\alpha/2)} & \text{for } c_L \leq \frac{1}{3} \\ \frac{S_{\text{ip}}(\alpha-\pi/2)}{S_{\text{ip}}(\alpha/2-\pi/2)} & \text{for } c_L > \frac{1}{3}. \end{cases} \tag{4}$$

Using Eq. (3), Eq. (4) becomes

$$\begin{aligned} \text{SPSI}\,(c_L; b, \alpha, \nu_1, \varepsilon_D) = \nu_1 \exp\left(-\frac{3}{2}\sin\left(\frac{\alpha}{2}\right)\sin\left(\frac{3\alpha}{2}\right)\left|c_L - \frac{1}{3}\right| b\varepsilon_D\right) \\ + \nu_2 \exp\left(\frac{3}{2}\sin^2\left(\frac{\alpha}{2}\right)\left|c_L - \frac{1}{3}\right| b\varepsilon_D\right), \end{aligned} \tag{5}$$

where $\varepsilon_D := \left(\lambda_{\parallel} - \lambda_{\perp}\right)$ represents the microscopic anisotropy of each fascicle. The numerator in Eq. (4) is in general not equal, mathematically, to the true maximum of the in-plane signal S_{ip} associated with $\hat{\mathbf{u}}_2$ and therefore underestimates the value of the signal peak. Similarly, the denominator is in general an overestimation of the true trough of S_{ip} between the fascicles. The proposed SPSI is therefore a conservative underestimate of the true peak-to-trough ratio, which makes SPSI > 1 *a sufficient but not always necessary condition for signal peaks to be separated*, in noiseless settings. As discussed in more details below, the true peaks and troughs of S_{ip}, when they exist, are actually found along directions fairly close to those selected in Eq. (4). When SPSI > 1, higher SPSI should indicate better orientation estimation performance; when SPSI < 1, the fascicles are often indistinguishable in the signal and changes in SPSI values become less interpretable.

Linear encoding maximizes signal contrast. As shown in Fig. 2b, SPSI (c_L) is symmetric around $c_L = \frac{1}{3}$ and strictly convex on either side of $c_L = \frac{1}{3}$. It can thus only be maximized at the boundaries of the subintervals, i.e., at $c_L = 0$, $\frac{1}{3}$ or 1. Assuming SPSI(0) > 1 leads to SPSI($\frac{2}{3}$) $=$ SPSI(0) > 1, by symmetry. Based on the strict convexity for $c_L \in \left[\frac{1}{3}, 1\right]$ and because SPSI($\frac{1}{3}$) $= 1$ (spherical encoding), SPSI(c_L) must then be increasing, not decreasing, for $c_L \geq \frac{2}{3}$. Therefore, SPSI(1) $>$ SPSI($\frac{2}{3}$) $=$ SPSI(0) holds, i.e., *in all cases where* $\exists c_L$ s.t. SPSI $(c_L) > 1$, *linear encoding always* *achieves a higher* SPSI *than planar encoding*. Since maximizing $\left|c_L - \frac{1}{3}\right|$ maximizes SPSI, intermediate oblate and prolate encodings are sub-optimal.

Planar encoding provides higher signal. At equal SPSI, Eq. (3) reveals that planar-like encoding yields higher signal than linear-like encoding with

$$\frac{S_{\text{ip}}\left(\varphi_B; c_L = \frac{1}{3} - \Delta_L\right)}{S_{\text{ip}}\left(\varphi_B - \frac{\pi}{2}; c_L = \frac{1}{3} + \Delta_L\right)} = \exp\left(\frac{b}{2}\Delta_L \varepsilon_D\right) > 1 \quad \forall \varphi_B, \tag{6}$$

valid for any α, ν_1, for $0 < \Delta_L \leq \frac{1}{3}$ (the "deviation" from spherical), which is reminiscent of the link between ε_D and a ratio of linear and spherical signal derived in [6]. The signal from fully planar ($c_L = 0$) encoding turns out to also exceed that of fully linear ($c_L = 1$) encoding $\forall \varphi_B$, for $b\varepsilon_D > 0$ and $\nu_2 > 0$

$$\begin{aligned} &\frac{S_{\text{ip}}\left(\varphi_B; c_L = 0\right)}{S_{\text{ip}}\left(\varphi_B - \frac{\pi}{2}; c_L = 1\right)} \\ &= \exp\left(\sin^2\left(\varphi_B\right) \frac{b}{2} \varepsilon_D\right) \cdot \left[\frac{\nu_1 + \nu_2 \exp\left(\sin\left(\alpha\right) \sin\left(2\varphi_B - \alpha\right) \frac{b}{2} \varepsilon_D\right)}{\nu_1 + \nu_2 \exp\left(\sin\left(\alpha\right) \sin\left(2\varphi_B - \alpha\right) b\varepsilon_D\right)}\right] > 1. \end{aligned} \tag{7}$$

Theoretical support for SPSI. The true locations of the extrema of $S_{\text{ip}}\left(\varphi_B\right)$ are roots of its first derivative $\frac{\partial S_{\text{ip}}}{\partial \varphi_B}$, i.e. solutions of the non-linear equation

$$\begin{aligned} &\nu_2 \sin\left(2\varphi_B\right) = \\ &\nu_1 \sin\left(2\left(\alpha - \varphi_B\right)\right) \exp\left(\frac{3}{2} \sin\left(\alpha\right) \sin\left(\alpha - 2\varphi_B\right) \left(c_L - \frac{1}{3}\right) b\varepsilon_D\right). \end{aligned} \tag{8}$$

As hinted at in the previous section and shown by the circles in Fig. 3, Eq. (8) is in general *not* satisfied by the values α, $\frac{\alpha}{2}$, $\alpha - \frac{\pi}{2}$ and $\frac{\alpha-\pi}{2}$ used in our definition of SPSI (Eq. (4)). In this section, two particular cases are studied in which Eq. (8) admits closed-form solutions. Whether these solutions are minima or maxima of S_{ip} is then determined by the sign of the second derivative $\frac{\partial^2 S_{\text{ip}}}{\partial \varphi_B^2}$.

Equal fascicle contributions. The first special case is $\nu_1 = \nu_2 = 0.5$. Equation (8) is solved by $\varphi_B = \frac{\alpha}{2}$ and $\varphi_B = \frac{\alpha-\pi}{2}$, yielding candidate extrema along the bisectors $\hat{\mathbf{b}}_u$ and $\hat{\mathbf{b}}_n$. The second derivative is computed as

$$\left.\frac{\partial^2 S_{\text{ip}}}{\partial \varphi_B^2}\right|_{\varphi_B = \frac{\alpha}{2}} = \frac{3}{2} S_{\text{sing}}\left(\frac{\alpha}{2}\right)\left(c_L - \frac{1}{3}\right) b\varepsilon_D \left[\frac{3}{2}\left(c_L - \frac{1}{3}\right) b\varepsilon_D \sin^2\left(\alpha\right) + 2\cos\left(\alpha\right)\right]$$

$$\left.\frac{\partial^2 S_{\text{ip}}}{\partial \varphi_B^2}\right|_{\varphi_B = \frac{\alpha-\pi}{2}} = \frac{3}{2} S_{\text{sing}}\left(\frac{\alpha-\pi}{2}\right)\left(c_L - \frac{1}{3}\right) b\varepsilon_D \left[\frac{3}{2}\left(c_L - \frac{1}{3}\right) b\varepsilon_D \sin^2\left(\alpha\right) - 2\cos\left(\alpha\right)\right],$$

which actually holds for any value of ν_1. This shows that $\hat{\mathbf{b}}_u$ in planar-like ($c_L < \frac{1}{3}$) and $\hat{\mathbf{b}}_n$ in linear-like ($c_L > \frac{1}{3}$) encoding are minima, not maxima, when

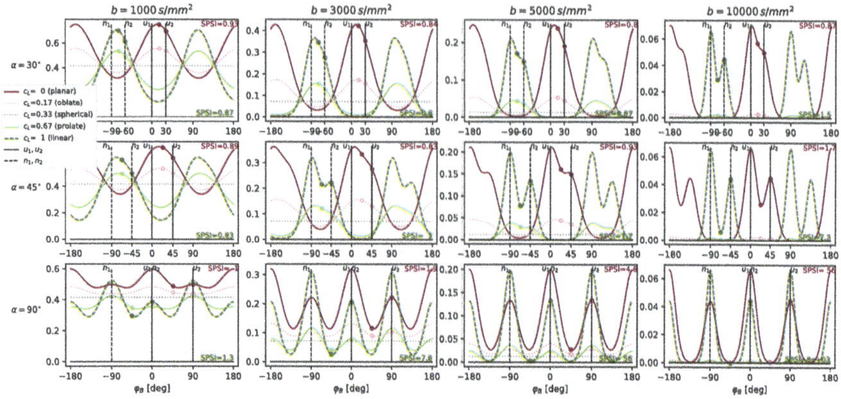

Fig. 3 Our proposed (SPSI) accurately approximates the ratio of the true signal peaks and troughs. The in-plane signal S_{ip} acquired "against" the smaller fascicle (see filled circles) and along the main or normal bisector (empty circles), used to define SPSI in Eq. (4), seem very close to the true extrema, when they exist, under the hypotheses of our toy model ($\nu_1 = 0.6$ here).

$$\frac{3}{2}\left|c_L - \frac{1}{3}\right| b\varepsilon_D > \frac{2\cos(\alpha)}{\sin^2(\alpha)}. \tag{10}$$

Equation (10) is satisfied for large enough b, microscopic anisotropy ε_D, crossing angle α and anisotropy $\left|c_L - \frac{1}{3}\right|$, giving linear encoding (with $c_L = 1$) an advantage over planar encoding (with $c_L = 0$) as it can reach a value double that of planar encoding ($\frac{2}{3}$ vs. $\frac{1}{3}$). It can reasonably be assumed that these conclusions extend to the general case $\nu_1 \neq \nu_2$ unless $\nu_1 \gg 0.5$ and that, in general, the signal troughs occur near the locations of the bisectors when Eq. (10) holds, thus justifying the denominator of Eq. (4) defining SPSI.

Right-angle crossing. The second special case is $\alpha = \frac{\pi}{2}$. Using Eq. (8), the roots of $\frac{\partial S_{\text{ip}}}{\partial \varphi_B}$ are found to be

$$\sin(2\varphi_B) = 0 \Leftrightarrow \varphi_B = k\pi/2 \;\; (k \in \mathbb{Z}), \tag{11a}$$

$$\text{or} \qquad \cos(2\varphi_B) = \frac{2\log(\nu_1/\nu_2)}{3\left(c_L - \frac{1}{3}\right)b\varepsilon_D}. \tag{11b}$$

Equation (11a) indicates candidate extrema along ($\varphi_B = -\frac{\pi}{2}$, 0) and perpendicular ($\varphi_B = 0$, $\frac{\pi}{2}$) to the fascicles. The second derivative of S_{ip}

$$\left.\frac{\partial^2 S_{\text{ip}}}{\partial \varphi_B^2}\right|_{\varphi_B = 0} = 3\left(c_L - \tfrac{1}{3}\right) b\varepsilon_D \left(\nu_1 S_{\text{sing}}(0) - \nu_2 S_{\text{sing}}\left(\tfrac{\pi}{2}\right)\right)$$

$$\left.\frac{\partial^2 S_{\text{ip}}}{\partial \varphi_B^2}\right|_{\varphi_B = \pm\frac{\pi}{2}} = 3\left(c_L - \tfrac{1}{3}\right) b\varepsilon_D \left(\nu_2 S_{\text{sing}}(0) - \nu_1 S_{\text{sing}}\left(\tfrac{\pi}{2}\right)\right)$$

reveals, after using Eq. (1), that acquisitions with physical gradients against the larger fascicle ($\varphi_B = \alpha$ in planar-like and $\varphi_B = \alpha - \frac{\pi}{2}$ in linear-like encoding, $\alpha = \frac{\pi}{2}$)

always correspond to signal maxima ($\frac{\partial^2 S_{\text{ip}}}{\partial \varphi_B^2} < 0$) and that acquisitions against the smaller fascicle are also signal maxima whenever

$$\log\left(\frac{\nu_1}{\nu_2}\right) < \frac{3}{2}\left|c_L - \frac{1}{3}\right| b\varepsilon_D, \tag{13}$$

which again occurs with sufficiently large b and ε_D, c_L far enough from $\frac{1}{3}$ and ν_1 close enough to ν_2. Equation (11a) and (13) likely describe general trends extending to the case $\alpha \leq \frac{\pi}{2}$, justifying the numerator in Eq. (4) defining SPSI.

Condition (13) also guarantees that the cosine in Eq. (11b) takes a value in the feasible range $[-1, 1]$. Solving Eq. (11b) then leads to candidate extrema at

$$\varphi_B = \pm\frac{1}{2}\arccos\left(\frac{2\log(\nu_1/\nu_2)}{3\left(c_L - \frac{1}{3}\right) b\varepsilon_D}\right), \tag{14}$$

corresponding to locations between the fascicles. Those extrema are minima of S_{ip} when $\frac{\partial^2 S_{\text{ip}}}{\partial \varphi_B^2} > 0$, which is ensured by the sufficient (possibly too restrictive) condition

$$\frac{9}{4}\left(c_L - \frac{1}{3}\right)^2 b^2 {\varepsilon_D}^2 > 1, \tag{15}$$

easily attained in practice. Equation (14) further reveals that those signal troughs are obtained with $\hat{\mathbf{u}}_B$ close to the bisectors $\hat{\mathbf{b}}_u$ and $\hat{\mathbf{b}}_n$, in line with the conclusions of the previous particular case $\nu_1 = \nu_2 = 0.5$ and justifying the denominator of Eq. (4). For a relatively extreme case $\nu_1 = 0.8$, with $\frac{3}{2}\left(c_L - \frac{1}{3}\right)$ b=3000 s/mm^2 and $\varepsilon_D = 2.0\,\mu\text{m}^2/\text{ms}$, the signal troughs are located at $|\varphi_B| = 38.3°$ instead of 45°, i.e. a difference of only 14.8%.

In general, a small error on the location of an extremum of a continuously differentiable function leads to a limited error on the signal value since the derivative is almost zero, and the function thus more or less flat, in the vicinity of an extremum.

3 Methods

Verification of theoretical predictions. The proposed SPSI was computed as a function of c_L, for various values of b and α with ν_1 =0.6 to verify that SPSI was systematically maximized, in so far as it ever reached the threshold value of 1, by linear rather than planar encoding. The in-plane signal S_{ip} was then computed for various values of b, α, c_L with ν_1 =0.6 to visually verify whether $S_{\text{ip}}(\alpha)$ and $S_{\text{ip}}(\alpha - \pi/2)$ were close to true signal peaks and $S_{\text{ip}}(\alpha/2)$ and $S_{\text{ip}}(\alpha/2 - \pi/2)$ close to true signal troughs. The corresponding SPSI was computed for each scenario to assess that (i) SPSI < 1 was associated to in-plane signals that did not exhibit two clear, separate peaks; (ii) in regimes where SPSI > 1, higher SPSI was linked to sharper separation of signal peaks; (iii) at fixed SPSI and $\left|c_L - \frac{1}{3}\right| < \frac{1}{3}$, signal for $c_L < \frac{1}{3}$ (oblate, planar-like) was higher than for $c_L > \frac{1}{3}$ (prolate, linear-like).

Robustness of SPSI beyond our toy model. The goal of this experiment was to (i) verify that SPSI correlated with accuracy of fascicle orientation estimation even when the assumptions of the toy model, which motivated its definition, were not met; (ii) evaluate different types of B-tensor encoding at orientation estimation in the presence of acquisition noise.

In order to account for an intra- and extra-axonal compartment, each fascicle was modeled by a "stick" ($\lambda_\perp = 0$, $\lambda_\parallel = 2.2\,\mu\text{m}^2/\text{ms}$) aligned with a zeppelin ($\lambda_\perp = 0.4\,\mu\text{m}^2/\text{ms}$, $\lambda_\parallel = 1.5\,\mu\text{m}^2/\text{ms}$) with lower parallel $\lambda_\parallel$ following evidence on small animals [13], with intra-fascicle signal fractions 0.65 and 0.35 respectively. Values of $\nu_1 = 0.6$ and $\alpha = 60°$ were used. The signal was simulated for B-tensors with $b = 3000\,\text{s/mm}^2$, varying $c_L \in [0, 1]$ and 200 orientations $\hat{\mathbf{u}}_B$ uniformly distributed on the sphere. The signal was corrupted by Rician noise with signal-to-noise ratio (SNR) defined for all c_L as SNR $= \frac{1}{\sigma_g}$, with σ_g the standard deviation of the Gaussian noise process in the receiver coils. For each value of SNR and c_L, 90 independent crossing-fascicle voxels were simulated.

The estimation of fascicle orientation was performed with DIPY [10] routines by first reconstructing the ODF from the noisy signal and then extracting the maxima from the ODF. Two strategies were considered to reconstruct the ODF. In the first one, for planar-like encoding ($c_L < 1/3$), following [19], the ODF was directly computed as a spherical harmonics (SH) fit of the DW-MRI signal with maximum degree $l = 10$ and a Laplace–Beltrami regularization factor $\lambda = 0.001$ [7] while for linear-like encoding ($c_L > 1/3$) the constant solid angle (CSA) model [1] was used. In the second strategy, constrained spherical deconvolution (CSD) [25] was identically applied to all encodings. The single-fascicle response function required by CSD was recursively calibrated in a data-driven way from an initial rotational harmonics (RH) fit of the signal from a "fat" zeppelin [23] (FA = 0.20, trace = $2.2\,\mu\text{m}^2/\text{ms}$) for each specific B-tensor type. To perform that calibration automatically, 10 single-fascicle voxels using the *same* stick-and-zeppelin fascicle model were included. In theory, this calibration algorithm thus had access to the ideal, although corrupted by noise, single-fascicle response. In both cases the ODF was computed on 724 points uniformly spread on the unit sphere. An ODF value was considered a peak if it exceeded the ODF minimum by at least 15% of the total ODF range. A minimum angular separation of 15° between detected maxima was enforced and only the 3 maxima with largest ODF values were kept. The angular error θ was computed by iterating over the true orientations $\hat{\mathbf{u}}_k$ ($k = 1, \ldots, M_{\text{true}}$) as [4]

$$\theta = \frac{1}{M_{\text{true}}} \sum_{k=1}^{M_{\text{true}}} \min_m \left\{ \arccos\left(\left| \hat{\mathbf{e}}_m \cdot \hat{\mathbf{u}}_k \right| \right) \right\}, \tag{16}$$

where $\hat{\mathbf{e}}_m$ is the unit orientation of the m-th detected peak and $M_{\text{true}} = 2$ the true number of fascicles. The mean angular error was then computed as the mean $\mathbb{E}[\theta]$

over all noise repetitions for each type of encoding at each SNR level. Finally, SPSI was computed for each tested value of c_L using the groundtruth values for b, ν_1 and α but setting ε_D, which is not defined in the case of a multi-compartment fascicle model, to a generic value of $1.8\,\mu^2\text{ms}^{-1}$.

4 Results

Verification of theoretical predictions. In Fig. 2b, all SPSI curves crossing the SPSI $= 1$ threshold were systematically maximized by linear encoding at $c_L = 1$, as predicted by theory. SPSI was particularly sensitive to α and b. Figure 3 confirms that the signal values used in Eq. (4) for the SPSI were close to the true peaks and troughs of S_{ip} when those existed at all and that two distinct signal peaks became apparent when SPSI ≈ 1. At $\alpha = 45°$ for instance, the second peak appeared at $b \approx 3000$ with SPSI ≈ 1 for linear encoding ($c_L = 1$, green dashes) while for planar encoding ($c_L = 0$, pink lines) the transition occurred between $b = 5000$ and $b = 10\,000$ with SPSI increasing from 0.93 to 1.7. In regimes where SPSI > 1, higher SPSI was associated with sharper signal peaks. Planar encoding signal ($c_L = 0$) was higher than signal from prolate encoding with $c_L = \frac{2}{3}$ (i.e., $\left|c_L - \frac{1}{3}\right| = \frac{1}{3}$, continuous pink and green lines), at identical SPSI.

Robustness of SPSI beyond our toy model. Even in the context of a more complex multi-fascicle, multi-compartment signal model, lower mean angular error (MAE) in orientation estimation was linked to higher SPSI, i.e. higher $|c_L - 1/3|$, as evidenced in Fig. 4. At equal SPSI or $|c_L - 1/3|$ ($c_L \in \left\{0, \frac{2}{3}\right\}$ and $c_L \in \left\{\frac{1}{6}, \frac{1}{2}\right\}$), oblate performed slightly better than prolate encoding. Linear encoding, which maximizes SPSI, consistently outperformed all other encodings across all SNR values.

5 Discussion and Conclusion

Limitations. Because different gradient waveforms $\mathbf{g}(t)$ (e.g., pulsed and oscillating gradients) may lead to the same $\mathbf{B}$, the B-tensor is a convenient but incomplete representation of a gradient waveform. The zeppelin is also an incomplete picture of diffusion in a fascicle of aligned axons, ignoring among others undulation, orientation dispersion and diffusion restriction. In practice, the actual physics of the gradient waveforms and biological features of tissues likely affect the ability of a sequence to estimate fascicle orientations. Although Fig. 4 suggested robustness to deviations from model assumptions, our SPSI could be improved with a more realistic interaction between tissue and sequence via $\varepsilon_D = \left(\lambda_\parallel - \lambda_\perp\right)$, by making $\lambda_\parallel$ and $\lambda_\perp$ functions of, for instance, the sequence's oscillating frequency, diffusion times, gradient separation angle and of the tissue's dispersion, undulation or density. The proposed index is a signal ratio unaffected by the actual signal intensities, which Eqs. (6) and (7) have shown can be different at identical SPSI. This facilitated rigor-

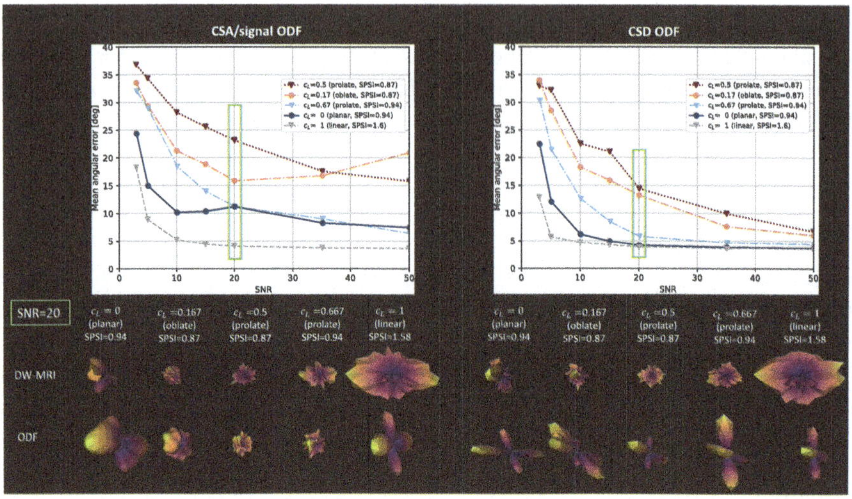

Fig. 4 High SPSI and high signal correlate with better orientation estimation in multi-fascicle, multi-compartment tissue. Top row: mean angular error (MAE) versus SNR for various types of axisymmetric B-tensor encodings at $b = 3000\ \mathrm{s/mm^2}$. Bottom row: example of ODF reconstruction from noisy DW-MRI. Left: peaks extracted from the signal ODF [19] (planar-like encoding) or the CSA ODF [1] (linear-like encoding). Right: peaks extracted from the CSD ODF [25] (all encodings)

ous mathematical analysis but may become problematic when the signal approaches the noise floor (e.g., high b and low SNR). Finally, future analyses should consider non axisymmetric B-tensors (i.e. with a non-zero asymmetry factor as in [9]), possibly via their decomposition into simpler planar, spherical and linear B-tensors, as well as an extension to three-way fascicle crossings.

SPSI as a tool for sequence design. The proposed SPSI (Eq. (5)) allows encoding parameters (c_L, b) to be tuned to target tissue properties such as the expected crossing angle α, fascicles' signal fractions ν_1 and microscopic anisotropy ε_D. Pathological WM pathways may for instance have low volume and thus low signal contribution ($\nu_2 \approx 0$) or low bundle coherence (low ε_D), which may decrease the SPSI at given c_L, b and negatively impact fascicle orientation or ODF estimation. SPSI is a conservative metric as values less than 1 may sometimes be related to in-plane signal already exhibiting two distinct peaks. Ensuring SPSI > 1 should thus generally provide a robust safety margin.

Linear versus planar encoding. Linear encoding uniquely maximizes signal contrast via its tensor anisotropy $\left|c_L - \frac{1}{3}\right|$ (equivalently, $|b_\Delta|$ [9]), as shown by Eqs. (5), (10), (13) and (15). However, Eq. (7) shows that planar always provides higher signal intensities than linear encoding, which might make planar encoding more robust to acquisition noise, ignoring any dependence of the SNR on the type of encoding. Our experiments (Fig. 4) showed that linear systematically outperformed planar encoding, suggesting that the benefits of signal contrast outweigh those of higher signal values. Consistent with Eq. (6), oblate performed better than prolate at fixed SPSI (i.e., identical $\left|c_L - \frac{1}{3}\right|$), possibly owing to their increased robustness to noise. Moderately prolate B-tensors ($c_L \in \left[\frac{1}{3}, \frac{2}{3}\right]$) therefore seem to offer little benefit for fascicle orientation estimation.

Moreover, a practical drawback of general gradient waveforms compared to linear encoding such as the PGSE is the long durations that they still require to achieve the high b-values [2, 5, 6, 11, 24] needed to resolve difficult crossings with small angle or a very dominant fascicle (Eqs. (10), (13), (15)). Longer sequences are subject to important T2-decay, which adversely affects SNR. Hybrid protocols combining different types of B-tensor encodings [16], possibly spread over multiple b-values, may advantageously combine robustness to noise and high signal contrast in fascicle crossings.

Acknowledgements This work was supported by the Swiss National Science Foundation Spark grant number 190297 and has received funding from the European Union's Horizon 2020 research and innovation programme under the Marie Skodowska-Curie grant agreement No. 754462.

References

1. Aganj, I., Lenglet, C., Sapiro, G., Yacoub, E., Ugurbil, K., Harel, N.: Reconstruction of the orientation distribution function in single-and multiple-shell q-ball imaging within constant solid angle. Magn. Reson. Med. **64**(2), 554–566 (2010)
2. Avram, A.V., Sarlls, J.E., Basser, P.J.: Measuring non-parametric distributions of intravoxel mean diffusivities using a clinical MRI scanner. Neuroimage **185**, 255–262 (2019)
3. Basser, P.J., Jones, D.K.: Diffusion-tensor MRI: theory, experimental design and data analysis-a technical review. NMR in Biomed.: Int. J. Devoted Dev. Appl. Magn. Reson. In Vivo **15**(7–8), 456–467 (2002)
4. Canales-Rodríguez, E.J., Legarreta, J.H., Pizzolato, M., Rensonnet, G., Girard, G., Rafael-Patino, J., Barakovic, M., Romascano, D., Alemán-Gómez, Y., Radua, J., et al.: Sparse wars: a survey and comparative study of spherical deconvolution algorithms for diffusion MRI. Neuroimage **184**, 140–160 (2019)
5. Coelho, S., Pozo, J.M., Jespersen, S.N., Jones, D.K., Frangi, A.F.: Resolving degeneracy in diffusion MRI biophysical model parameter estimation using double diffusion encoding. Magn. Reson. Med. **82**(1), 395–410 (2019)
6. Cottaar, M., Szczepankiewicz, F., Bastiani, M., Hernandez-Fernandez, M., Sotiropoulos, S.N., Nilsson, M., Jbabdi, S.: Improved fibre dispersion estimation using b-tensor encoding. NeuroImage 116832 (2020)
7. Descoteaux, M., Angelino, E., Fitzgibbons, S., Deriche, R.: Regularized, fast, and robust analytical Q-ball imaging. Magn. Reson. Med. **58**(3), 497–510 (2007)
8. Drobnjak, I., Alexander, D.C.: Optimising time-varying gradient orientation for microstructure sensitivity in diffusion-weighted MR. J. Magn. Reson. **212**(2), 344–354 (2011)
9. Eriksson, S., Lasic, S., Nilsson, M., Westin, C.F., Topgaard, D.: NMR diffusion-encoding with axial symmetry and variable anisotropy: distinguishing between prolate and oblate microscopic diffusion tensors with unknown orientation distribution. J. Chem. Phys. **142**(10), 104201 (2015)
10. Garyfallidis, E., Brett, M., Amirbekian, B., Rokem, A., Van Der Walt, S., Descoteaux, M., Nimmo-Smith, I.: Dipy, a library for the analysis of diffusion MRI data. Front. Neuroinformatics **8**, 8 (2014)
11. Jensen, J.H., Helpern, J.A.: Characterizing intra-axonal water diffusion with direction-averaged triple diffusion encoding MRI. NMR Biomed. **31**(7), e3930 (2018)

12. Jespersen, S.N., Lundell, H., Sønderby, C.K., Dyrby, T.B.: Orientationally invariant metrics of apparent compartment eccentricity from double pulsed field gradient diffusion experiments. NMR Biomed. **26**(12), 1647–1662 (2013)
13. Kunz, N., da Silva, A.R., Jelescu, I.O.: Intra-and extra-axonal axial diffusivities in the white matter: which one is faster? Neuroimage **181**, 314–322 (2018)
14. Lasic, S., Szczepankiewicz, F., Eriksson, S., Nilsson, M., Topgaard, D.: Microanisotropy imaging: quantification of microscopic diffusion anisotropy and orientational order parameter by diffusion MRI with magic-angle spinning of the q-vector. Front. Phys. **2**, 11 (2014)
15. Lawrenz, M., Finsterbusch, J.: Double-wave-vector diffusion-weighted imaging reveals microscopic diffusion anisotropy in the living human brain. Magn. Reson. Med. **69**(4), 1072–1082 (2013)
16. Lundell, H., Sønderby, C.K., Dyrby, T.B.: Diffusion weighted imaging with circularly polarized oscillating gradients. Magn. Reson. Med. **73**(3), 1171–1176 (2015)
17. Mori, S., Van Zijl, P.C.: Diffusion weighting by the trace of the diffusion tensor within a single scan. Magn. Reson. Med. **33**(1), 41–52 (1995)
18. Neeman, M., Freyer, J.P., Sillerud, L.O.: Pulsed-gradient spin-echo diffusion studies in NMR imaging. Effects of the imaging gradients on the determination of diffusion coefficients. J. Magn. Reson. **90**(2), 303–312 (1969, 1990)
19. Özarslan, E., Memiç, M., Avram, A.V., Afzali, M., Basser, P.J., Westin, C.F.: Rotating field gradient (RFG) MR offers improved orientational sensitivity. In: 2015 IEEE 12th International Symposium on Biomedical Imaging (ISBI), pp. 955–958 (2015)
20. Rensonnet, G., Scherrer, B., Warfield, S.K., Macq, B., Taquet, M.: Assessing the validity of the approximation of diffusion-weighted-MRI signals from crossing fascicles by sums of signals from single fascicles. Magn. Reson. Med. **79**(4), 2332–2345 (2018)
21. Stejskal, E.O., Tanner, J.E.: Spin diffusion measurements: spin echoes in the presence of a time-dependent field gradient. J. Chem. Phys. **42**(1), 288–292 (1965)
22. Szczepankiewicz, F., Lasic, S., van Westen, D., Sundgren, P.C., Englund, E., Westin, C.F., Ståhlberg, F., Lätt, J., Topgaard, D., Nilsson, M.: Quantification of microscopic diffusion anisotropy disentangles effects of orientation dispersion from microstructure: applications in healthy volunteers and in brain tumors. Neuroimage **104**, 241–252 (2015)
23. Tax, C.M., Jeurissen, B., Vos, S.B., Viergever, M.A., Leemans, A.: Recursive calibration of the fiber response function for spherical deconvolution of diffusion MRI data. Neuroimage **86**, 67–80 (2014)
24. Topgaard, D.: Diffusion tensor distribution imaging. NMR Biomed. **32**(5), e4066 (2019)
25. Tournier, J.D., Calamante, F., Connelly, A.: Robust determination of the fibre orientation distribution in diffusion MRI: non-negativity constrained super-resolved spherical deconvolution. Neuroimage **35**(4), 1459–1472 (2007)
26. Wedeen, V.J., Rosene, D.L., Wang, R., Dai, G., Mortazavi, F., Hagmann, P., Kaas, J.H., Tseng, W.Y.I.: The geometric structure of the brain fiber pathways. Science **335**(6076), 1628–1634 (2012)
27. Westin, C.F., Szczepankiewicz, F., Pasternak, O., Özarslan, E., Topgaard, D., Knutsson, H., Nilsson, M.: Measurement tensors in diffusion MRI: generalizing the concept of diffusion encoding. In: International Conference on Medical Image Computing and Computer-Assisted Intervention, pp. 209–216. Springer (2014)
28. Westin, C.F., Knutsson, H., Pasternak, O., Szczepankiewicz, F., Özarslan, E., van Westen, D., Mattisson, C., Bogren, M., O'Donnell, L.J., Kubicki, M., et al.: Q-space trajectory imaging for multidimensional diffusion MRI of the human brain. Neuroimage **135**, 345–362 (2016)

Orientation Processing: Tractography and Visualization

Improving Tractography Accuracy Using Dynamic Filtering

Matteo Battocchio, Simona Schiavi, Maxime Descoteaux, and Alessandro Daducci

Abstract Based on diffusion-weighted magnetic resonance imaging data, tractography allows studying the complex structure of the brain white matter. During the last decade, different approaches showed the benefits of using microstructural information to help tractography reconstruction. Filtering methods, in particular, exploit this information to find the contribution of an input set of streamlines and keep only the tracts needed to explain the diffusion signal. However, this implies that the accuracy of the reconstructions is bounded to the quality of the input tractogram. Here, we introduce a novel approach able to adapt the position and the shape of the input streamlines. Our novel approach promotes higher spatial coverage and lower number of IBs while not affecting VBs, allowing microstructure-informed filtering and tractography techniques to overcome the limitation imposed by poorly reconstructed input tractograms.

1 Introduction

The complex organization of the white matter (WM) in the brain can be studied in vivo using tractography algorithms, which estimate the macroscopic trajectories of the fibers by exploiting the information in diffusion-weighted magnetic resonance imaging (DW-MRI) acquisitions. Early *tractography algorithms* were based on integration procedures on the fiber orientations estimated in each voxel to infer fiber

M. Battocchio (✉) · S. Schiavi · A. Daducci
Computer Science department, University of Verona, Verona, Italy
e-mail: matteo.battocchio@univr.it

S. Schiavi
e-mail: simona.schiavi@univr.it

A. Daducci
e-mail: alessandro.daducci@univr.it

M. Battocchio · M. Descoteaux
Sherbrooke Connectivity Imaging Lab, Université de Sherbrooke, Sherbrooke, Canada
e-mail: Maxime.Descoteaux@USherbrooke.ca

N. Gyori et al. (eds.), *Computational Diffusion MRI*, Mathematics and Visualization,
https://doi.org/10.1007/978-3-030-73018-5_4

trajectories; these approaches are very fast but also rather sensitive to estimation errors of the local orientations. Probabilistic algorithms cope with such local inaccuracies by using probability distribution functions and propagating the streamlines by sampling from these distributions. More recent approaches estimate the optimal configuration of streamlines, i.e., tractogram, by fitting parametric models of their trajectories to the measured DW-MRI data [11, 12] using Monte Carlo Markov Chain (MCMC) stochastic procedures. These methods are computationally heavier but show improved tractogram reconstruction quality compared to previous approaches Tractography represents an invaluable tool to study in vivo a wide spectrum of neurological conditions, but suffers from some critical limitations that affect the accuracy of the reconstructions [5, 9].

Microstructure Informed Tractography [2] has been recently proposed as a novel paradigm to improve the quality of the reconstruction, leveraging the assumption that the microstructural properties of the fibers remain constant along their path. These methods follow a top-down approach and attempt to identify the optimal subset of streamlines from an initial set of candidates reconstructed with tractography that are more compatible with the measured DW-MRI data. For this reason, they are also referred to as *filtering techniques*. The Spherical-deconvolution Informed Filtering of Tractograms (SIFT) and its evolution (SIFT2) [15, 16] use the fiber orientation distributions in each voxel to filter out implausible streamlines. MicroTrack [14] proposed to use more advanced biophysical models of the neuronal tissue to characterize the streamline contributions, but the complexity of the algorithm was too high for its practical application. A big step forward to reduce the dimensionality of the problem was represented by the Convex Optimization Modeling for Microstructure Informed Tractography (COMMIT) [1] and the Linear Fascicle Evaluation (LiFE) [10]. In fact, both approaches use similar formulations and formulate the problem as a linear system. Notably, COMMIT is a very flexible framework and was recently improved to consider also anatomical priors in the filtering, e.g., COMMIT2 [13].

These filtering procedures have shown very promising results [13, 17], but the *accuracy of the reconstructions is indissolubly bound to the quality of the input tractogram*. In fact, unlike MCMC-based methods, they assume a static configuration of streamlines throughout the filtering process, whose position and shape is assumed to be correctly estimated by the tractography algorithm.

We propose an *hybrid procedure* with the aim of exploiting the advantages of both filtering and MCMC-based approaches. Starting from an initial set of streamlines estimated with classical tractography , our method can adapt the configuration of streamlines by alternating the filtering to short runs of classical MCMC-based techniques. In this work, in particular, we present results obtained using the COMMIT framework. Our experiments clearly indicate that the possibility to dynamically adapt the spatial configuration of the streamlines during the filtering allows overcoming the limitations due to a static input and further improves the reconstructions. The presented method promotes higher spatial coverage and lower number of IBs while not affecting VBs, which is shows great potential for microstructure-informed filtering and tractography techniques.

2 Materials and Methods

Our method is inspired by the work of Lemkaddem et al. [7] and can be divided in *three main steps* (see flowchart in Fig. 1):

1. Construction of the initial set of streamlines using tractography . This tractogram is. first, pre-filtered to remove streamlines prematurely stopping inside the WM and, then, randomly divided into two groups: the first group, called M, is the initial set of streamlines we want to optimize and the other one, called A, is an auxiliary set used as support in the following phase.
2. Parameterization of the streamlines using splines, in order to drastically reduce the complexity for their representation as well as to guarantee smoothness of their trajectories.
3. Alteration and evaluation of the streamline configuration following an MCMC-based optimization approach.

2.1 Initial Set of Streamlines

The input tractogram for our proposed approach can be reconstructed with any arbitrary tracking technique. In this work, we used local deterministic SD_STREAM tracking, implemented in the MRtrix3 toolbox [18]. As mentioned above, the first operation is the removal of streamlines that do not reach any cortical and/or subcortical structures. Practically, this is done using labeled regions of interest (ROIs).

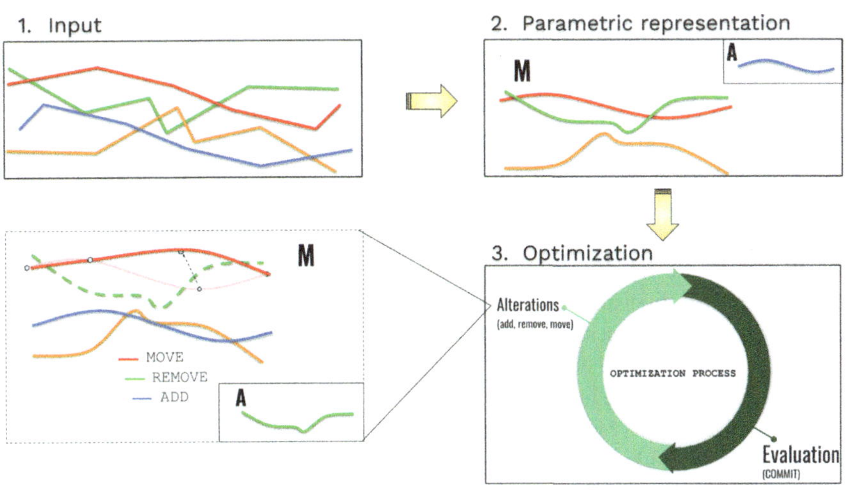

Fig. 1 Schematic representation of the 3 main steps of the proposed approach

2.2 Parametric Representation of the Streamlines

MCMC-based algorithms have introduced different parameterization approaches. The most famous comprehend the adoption of small segments covering the whole white matter volume, as in [11, 12], or the use of spin glass model as in [3]. These approaches, however, need a large number of parameters and fine tuning to drive the spatial optimization. Among the different parameterizations that have been proposed, B-splines has proven to be a valuable alternative and they have been exploited in different ways, starting from Jbabdi et al. [4] until the more recent approach of Lemkaddem et al.

We recall that a B-spline of degree d is a piece-wise concatenation of polynomial curves joining in correspondence of $n+1$ control points, called knots, $\{\boldsymbol{Q}_i\}_{i=0}^{n}$ as follow:

$$X(t) = \sum_{i=0}^{n} N_{i,d}(t)\boldsymbol{Q}_i \quad (1)$$

where $N_{i,d}(t)$ are B-spline basis functions computed recursively using a sequence of scalars t_i along the curve such that:

$$N_{i,n}(t) = \begin{cases} 0 & \text{if } \ t \leq t_i \text{ or } t \geq t_{i+n} \\ \neq 0 & \text{otherwise} \end{cases}. \quad (2)$$

One of the main advantages of this formulation is the possibility to *reduce the number of points* needed to represent the original input set of streamlines. Indeed, instead of using all the original points, we can represent our streamlines by setting an a-priori number of knots (equal for all the streamlines) and using Eq. 1 to uniquely determine each curve. To compute the sequence of knots for each streamline, we employed the Douglas-Peucker reduction algorithm, which minimizes the number of knots needed to represent the curve. In particular, this minimization is efficiently performed by recursively checking that the distance between the original curve and the simplified one is below a certain imposed threshold defined as "smoothing error". Among the variety of different streamline functions, we chose to adopt cubic B-splines , named Catmull-Rom. This class of splines is characterized by the fact that the knots are a subset of the points composing the original streamline and, in particular, the first and the last knots coincide with the first and last point of the streamline. The fact that the control points are chosen along the streamline, combined with the use of cubic polynomial curves for the interpolation, ensures that the corresponding spline has a similar shape of the original streamline. This, in turn, permits to discard curves that lie outside the WM or that do not connect gray matter (GM) regions. Besides being easy to handle, the Catmull-Rom's splines presents also other convenient characteristics: they are able to approximate a large variety of smooth trajectories, making them particularly suitable to represent anatomically plausible WM pathways and, at the same time, they are fast to compute.

2.3 Optimization

The optimization step is driven by a Bayesian approach which allows to exploit global information to guide the streamlines adaptation. We consider the model M as the spatial configuration of the streamlines and the set of its control points as the parameters of the model. According to the Bayesian theory, we can infer on the posterior distributions of these parameters based on the observed data D and the data generated by M. The inference, then, can be used to influence the parameters by back-propagation. The goal of our method is to find the model that better explains the data, i.e., the one that maximizes the posterior probability $P(M|D)$:

$$P(M|D) = \frac{P(D|M)P(M)}{P(D)}. \tag{3}$$

In order to draw samples from the posterior distribution of the parameters, we used MCMC as an approximate inference algorithm. Based on the expectation-maximization approach, the idea is that by alternatively computing the expected values of the unobserved variables while maximizing the posterior, the process converges to the maximum likelihood values of the parameters. In practice, this implies the minimization of a loss function measuring the error between the model and the observed data. In our case, the error related to the model $E(M, D)$ is defined as:

$$E(M, D) = \mathrm{RMSE}(M, D) \tag{4}$$

where

$$\mathrm{RMSE}(M, D) = \sqrt{\left(\frac{1}{n}\right)\sum_{i=1}^{n}(M_i - D_i)^2} \tag{5}$$

is the voxel-wise error between the reconstructed signal obtained using the proposed model M and the observed data D. It is worth to notice that the representation of the streamlines using control points does not affect the computation of the voxel-wise error because each smoothed streamline covers continuously the voxel traversed.

Finally, a configuration is accepted or rejected based on the Metropolis Hasting criterion combined with a simulated annealing approach [6]. This implies that a configuration M_t is automatically accepted if $E(M_t, D) < E(M_{t-1}, D)$, while an eventual increase is admitted proportionally to its magnitude and a "system temperature" factor T, accordingly to the Green's ratio:

$$R = \exp^{-E(M',D)/T}. \tag{6}$$

Based on the simulated annealing approach, at the beginning, the system is characterized by a high temperature (T), which decreases as the process advances. High values of T imply that "bad" configurations are accepted, allowing the system to explore a wider range of configurations. In previous works, it has been shown how a geometric lowering schedule of the temperature ensures the convergence [8] and, at the same time, improves the chances to sample from maxima of $P(M|D)$. Starting from a model M we explore the space of its parameters by altering the configuration.

Adapting Shape and Position of Streamlines

We implemented three different proposals to alter the configuration of streamlines:

1. Move streamline's control points
2. Add a streamline to the set M from the set A
3. Remove a streamline from the set M and add it to A.

With the first proposal we randomly choose a streamline from the set M and we alter its trajectory. This is performed in two possible ways: either by moving one control point or by translating the whole streamline (i.e. simultaneously translate all the control points). In the first case, to find the new spatial position of the control point we sampled the new coordinates from a Gaussian distribution centered around the original position of that point. To reduce the degrees of freedom of this step, we fixed the magnitude of the movement to half of the voxel dimension.

The second proposal consists in randomly choosing a streamline from the auxiliary set A and append it to M. Vice versa, the third proposal removes a random streamline from M and add it to A.

2.4 Data and Experiments

To show the effectiveness of the method, we tested its performances on two different synthetic configurations created with Phantomas,[1] an open source software that allows manually defining 3D geometries of fiber bundles as well as generating the corresponding DW-MRI signal. In both cases, we simulated an acquisition protocol with 64 directions at b-value = 3000 s/mm^2, 1 mm isotropic voxel and signal-to-noise ratio of 30.

Dataset 1. As a proof of concept, we created a simple configuration consisting of a single straight bundle without noise. The input tractogram was composed by 10000 streamlines, keeping only those covering the superior part of the WM volume (see Fig. 2 left).

Dataset 2. We then assessed the performance on a more challenging bundle configuration represented by the dataset provided for the IEEE International Symposium on Biomedical Imaging (ISBI) 2013 Reconstruction Challenge . This dataset consists of 27 bundles arranged in a configuration mimicking the majority of the troublesome scenarios that can be found in the brain. These include branching, kissing, crossing

[1] http://www.emmanuelcaruyer.com/phantomas.php.

Fig. 2 Single bundle dataset: on the left is shown the input tractogram. The tractogram is composed by a set of packed streamlines covering only the superior part of the bundle. The underlying map represents the RMSE computed between the corresponding reconstructed diffusion signal and the observed one. On the right we show the resulting configuration after optimization with out method and the corresponding RMSE map

structures with various diameters and at different angles. From the signal reconstruction point of view, the phantom reproduces both partial volume effects, given by the presence of multiple fiber compartments within the same voxel and cerebrospinal fluid (CSF) contamination. As for the previous dataset, the input tractogram was reconstructed using SD_STREAM with standard parameters, discarding streamlines shorter than 5 mm. To quantitatively evaluate the performances of our method on this phantom, we carried out also the connectivity analysis, comparing the configurations against the known ground-truth (GT); in particular, we assess the number of valid bundles (VBs), i.e., bundles connecting regions known to be connected, as well as the invalid bundles (IBs), i.e., those connecting regions known to be disconnected. We also report the WM volume coverage based on the valid streamlines for each input tractogram and the associated RMSE.

3 Results and Discussion

The single bundle configuration allows to highlight the potential of the method in adapting the position of the streamlines in order to better cover the WM volume. Starting from a set of streamlines covering only the superior part of the created phantom (see Fig. 2 left), COMMIT finds the contributions for each one, discarding some of the redundant that are not necessary to explain the diffusion signal. However, this does not improve the RMSE because the streamline spatial configuration remains fixed, which, in this case, implies that the bundle remains under-represented, as shown by the underlying RMSE map. The result of the optimization performed by our method is reported in Fig. 2 right. The output set of streamlines has been moved and adapted to better cover the entire volume of the bundle which, in turn, decreases the differences between the observed and the reconstructed signals shown by the RMSE map below.

Results of the optimization on the second dataset are reported in Fig. 3. At the top of the first column, we show the geometry of the dataset and, at the bottom, the plot of the RMSE values across the optimization process. The descending trend of the error shows that the streamlines adaptation leads to a configuration that better explain the observed signal. In the second column we compare the RMSE map of the input configuration filtered by COMMIT, with the one computed from the optimized trac-

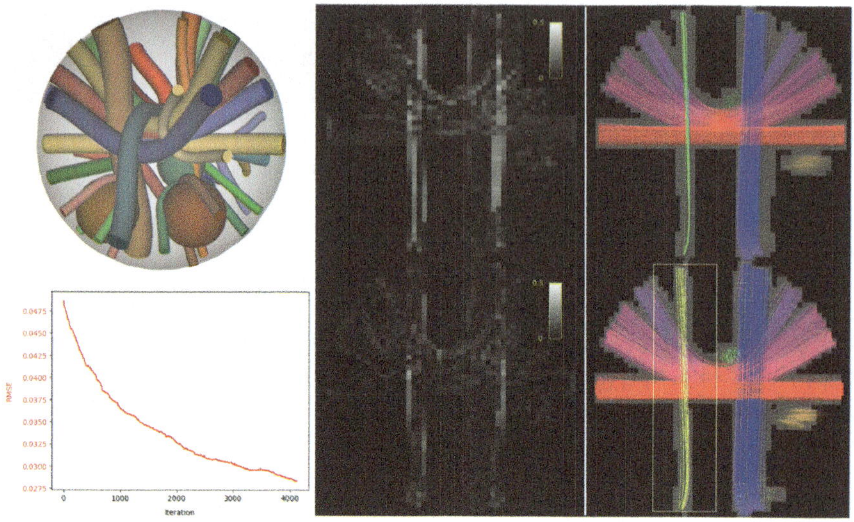

Fig. 3 ISBI 2013 dataset: at the top of the first column is shown the geometry of the phantom and at the bottom the plot of the RMSE values across the optimization process. The second and the third columns shows, respectively the RMSE maps and streamline configurations corresponding to the input tractogram filtered by COMMIT (top) and to the optimized configuration (bottom)

togram. In the third column we report the corresponding configurations, highlighting the impact of our approach on the framed vertical bundle in particular. Starting from an underrepresented bundle, the method is able to better distribute the streamlines inside the WM volume, reducing the difference between the reconstructed and the observed diffusion signal.

Table 1 shows the potential of the presented method with respect to connectivity analysis and compared the quality of the input tractogram, after filtering it with COMMIT and with our method. Our results show that our method is able to simultaneously reduce the complexity of the entire tractogram while improving the reconstruction quality. In fact, by increasing the quality of the reconstructions, we were also able to better discriminate between valid and invalid connections, consequently promoting the removal of more false positives. Also, besides decreasing the fitting error, the optimized configuration has a better WM volume coverage, less IBs, stable VBs with 75% less streamlines.

Finally, it is worth to notice that our formulation is completely independent on how the signal reconstruction is performed, meaning that also other state-of-the-art filtering approaches, e.g., SIFT, SIFT2, LiFE and COMMIT2c, may benefit from the proposed hybrid procedure; future work will be devoted to testing this hypothesis. Also, we will enhance the optimization of configurations characterized by completely unrepresented bundles which, at the moment, are more difficult to reconstruct.

Table 1 Quantitative comparison between the input tractogram (first row), after filtering it with COMMIT (middle) and with the proposed method (bottom row)

Tractograms	VBs	IBs	WM coverage (%)	# streamlines	RMSE
Raw	27	94	73.4	50,205	
COMMIT	27	88	72.7	20,366	0.049 ± 0.056
COMMIT + dynamic filtering	27	**83**	**92.8**	**12,444**	**0.028** ± 0.031

4 Conclusions

Nowadays, the ability of state-of-the-art filtering techniques to improve the accuracy of the tractograms heavily depends on the quality of input tractograms themselves. In this work, we showed how the possibility of adjusting the streamlines configuration during the filtering allows improving further the quality of the tractography reconstruction, both from the qualitative and the quantitative point of view.

We believe that our method could support microstructure-informed techniques by increasing the quality of the reconstruction and helping in the characterization of the brain structural connectivity.

References

1. Daducci, A., Dal Palù, A., Lemkaddem, A., Thiran, J.P.: COMMIT: convex optimization modeling for microstructure informed tractography. IEEE Trans. Med. Imaging **34**(1), 246–257 (2015)
2. Daducci, A., Dal Palu, A., Descoteaux, M., Thiran, J.P.: Microstructure informed tractography: pitfalls and open challenges. Front. Neurosci. **10**(8), 247 (2016)
3. Fillard, P., Poupon, C., Mangin, J.F.: A novel global tractography algorithm based on an adaptive spin glass model. In: International Conference on Medical Image Computing and Computer-Assisted Intervention, vol. 12, pp. 927–934 (2009)
4. Jbabdi, S., Woolrich, M., Andersson, J., Behrens, T.: A Bayesian framework for global tractography. Neuroimage **37**(1), 116–129 (2007)
5. Jbabdi, S., Johansen-Berg, H.: Tractography: where do we go from here? Brain Connect. **1**(3), 169–183 (2011)
6. Kirkpatrick, S., Gelatt, C.D., Vecchi, M.P.: Optimization by simulated annealing. Science **220**(4598), 671–680 (1983)
7. Lemkaddem, A., Sköldebrand, D., Dal Palú, A., Thiran, J.P., Daducci, A.: Global tractography with embedded anatomical priors for quantitative connectivity analysis. Front. Neurol. **5**, 232 (2014)
8. Lieshout, M.N.M.V.: Stochastic annealing for nearest-neighbour point processes with application to object recognition. Adv. Appl. Probab. **26**(2), 281–300 (1994)
9. Maier-Hein, K.H., Neher, P., Stieltjes, B., Descoteaux, M.: The challenge of mapping the human connectome based on diffusion tractography. Nat. Commun. **5** (2017)

10. Pestilli, F., Yeatman, J.D., Rokem, A., Kay, K.N., Wandell, B.A.: Evaluation and statistical inference for human connectomes. Nat. Methods **11**, 1058 (2014)
11. Reisert, M., Mader, I., Anastasopoulos, C., Weigel, M., Schnell, S., Kiselev, V.: Global fiber reconstruction becomes practical. Neuroimage **54**(2), 955–962 (2011)
12. Reisert, M., Kiselev, V., Dihtal, B., Kellner, E., Novikov, D.: MesoFT: unifying diffusion modelling and fiber tracking. Med. Image Comput. Comput. Assist. Interv. **17**, 201–208 (2014)
13. Schiavi, S., Ocampo-Pineda, M., Barakovic, M., Petit, L., Descoteaux, M., Thiran, J.P., Daducci, A.: A new method for accurate in vivo mapping of human brain connections using microstructural and anatomical information. Sci. Adv. **6** (2020)
14. Sherbondy, A.J., Rowe, M.C., Alexander, D.C.: Microtrack: an algorithm for concurrent projectome and microstructure estimation. In: Jiang, T., Navab, N., Pluim, J.P.W., Viergever, M.A. (eds.) Medical Image Computing and Computer-Assisted Intervention, pp. 183–190 (2010)
15. Smith, R.E., Tournier, J.D., Calamante, F., Connelly, A.: SIFT: spherical-deconvolution informed filtering of tractograms. Neuroimage **67**, 298–312 (2013)
16. Smith, R.E., Tournier, J.D., Calamante, F., Connelly, A.: SIFT2: enabling dense quantitative assessment of brain white matter connectivity using streamlines tractography. Neuroimage **119**, 338–351 (2015)
17. Smith, R.E., Tournier, J.D., Calamante, F., Connelly, A.: The effects of SIFT on the reproducibility and biological accuracy of the structural connectome. Neuroimage **104**, 253–265 (2015)
18. Tournier, J.D., Calamante, F., Connelly, A.: Mrtrix: diffusion tractography in crossing fiber regions. Int. J. Imaging Syst. Technol. **22**(1), 53–66 (2012)

Diffeomorphic Alignment of Along-Tract Diffusion Profiles from Tractography

David S. Lee, Ashish Sahib, Antoni Kubicki, Katherine L. Narr, Roger P. Woods, and Shantanu H. Joshi

Abstract Along-tract diffusion measure profiles from white matter fiber tracts have been widely analyzed in population studies of neurodevelopment and disease. An implicit assumption in performing inter-subject comparisons is a one-to-one correspondence of the profiles across subjects. Further, the profiles may also exhibit noise and spatial variability due to misregistration, tractography algorithms, interpolation methods, or inter-subject differences in the population. We present an approach to minimize the variability in the shape of along-tract diffusion profiles by performing diffeomorphic alignment across the tracts as well as across the population. The method represents the tract profiles as configuration functions and defines an objective function to align configurations by minimizing over both global and local warping functions. Following alignment, we show decrease in variability using the measures of coefficient of variability and intraclass correlation coefficients. We also introduce a new measure of inter-tract correlation, obtained by correlation of diffusion profiles between a pair of tracts of interest. We demonstrate our methods in a population of healthy individuals as well as show reliability analysis in subjects scanned twice.

1 Introduction

Techniques in diffusion tractography have facilitated the quantification and analysis of the white matter fibers and anatomical connectivity within the brain. Often, fiber bundles are collapsed into characteristic fibers that conveniently allow representation and analysis of diffusion properties along them. Then, along-tract metrics are

D. S. Lee (✉) · A. Sahib · A. Kubicki · K. L. Narr · R. P. Woods · S. H. Joshi
Department of Neurology, Ahmanson-Lovelace Brain Mapping Center, UCLA, Los Angeles, CA, USA
e-mail: dalee@mednet.ucla.edu

K. L. Narr · R. P. Woods
Department of Psychiatry and Biobehavioral Sciences, UCLA, Los Angeles, CA, USA

D. S. Lee · S. H. Joshi
Department of Bioengineering, UCLA, Los Angeles, CA, USA

N. Gyori et al. (eds.), *Computational Diffusion MRI*, Mathematics and Visualization,
https://doi.org/10.1007/978-3-030-73018-5_5

obtained by sampling the voxelwise diffusion measures along the length of the fibers. These along-tract diffusion profiles have complemented both the conventional voxel-based statistical analysis and network connectivity mapping from tractography by offering means of investigating the local diffusion properties along anatomically relevant fiber tracts. Based on a number of well-established tools for automatically segmenting established fiber tracts [5, 15, 17] and computing the along-tract diffusion profiles [16], many studies have demonstrated the clinical potential of along-tract diffusion profiles in characterizing cognition, healthy aging, and neuropsychiatric conditions [8, 11, 13]. In order to perform inter-subject comparisons, a representative along-tract diffusion profile is computed for each tract of interest. The underlying assumption of the inter-subject comparison of the along-tract diffusion profiles is one-to-one correspondence among the diffusion profiles across a population of subjects. While this is a valid approach, especially when all diffusion images are registered to a common template, there may remain inter-subject variability in the extracted diffusion profiles that is not accounted for with voxelwise spatial registration. This problem of correspondence of diffusion profiles has only been recently posed by researchers, who have proposed two methods to account for such variability. The approach by St-Jean et al. achieves diffusion profile realignment based on maximizing the Fourier-transform based cross-correlation among along-tract profiles across subjects [12]. Another elegant approach introduces a novel measure of Fiber-Flux Diffusion Density that integrates the diffusion and geometric properties along the length of the fiber, and performs alignment by minimizing the dissimilarity among the fibers using a fast marching method [3].

Contributions

In this paper we propose a new method where we seek invariance to the spatial parameterization of the diffusion profiles by aligning the along-tract diffusion profiles based on their shapes. Our approach relies on explicit representations of diffusion profiles and is fully invariant to the spatial parameterization of the profiles. Additionally, we propose to represent the fiber profiles by a vector configuration function for the entire brain. This allows both global (subject-wise) and local (tract-wise) alignment of diffusion profiles. We summarize our contributions as follows. (1) Our method enables a global and local joint alignment of diffusion profiles. (2) The representation is fully invariant to translation and scale, and the diffeomorphic alignment scheme achieves invariance to spatial parameterization. (3) The joint objective function is solved using alternating minimization over global and local parameterizations and uses a derivative-free coordinate-descent approach for optimization. (4) We introduce a new measure for subjects-wise inter-tract correlations that computes spatial correlations between pairs of diffusion profiles. This results in a subjects-wise correlation matrix that reflects the relationships between pairs of white matter fiber tractsbased on the along-tract diffusion profiles. We present results on a population of healthy subjects as well as show reliability analysis on a limited sample of subjects scanned twice.

2 Alignment of Along-Tract Diffusion Measure Profiles

In this section, we describe the representation scheme of diffusion profiles and propose an objective function for their tract-wise, across-subject alignment. The fiber bundles are collapsed to a representative characteristic path usually computed as a minimizer of geometric information, or diffusion information or both [9, 16]. In this paper, we use the method by Yeatman et al. [16] to compute the characteristic paths and extract fractional anisotropy (FA) as the diffusion measure along the tracts.

2.1 *Representation*

For N tracts per subject and K subjects, we use a vector function to represent the subject-wise configuration of characteristic FA profiles. Let $X_1, X_2, \ldots, X_K$ be the configuration functions for K subjects, where $X_i \in \mathbb{L}^2([0, 1], \mathbb{R}^N)$. We resample all FA profiles to unit length. Then the configuration function for N tracts is given by

$$X_i \equiv (x_1^i, x_2^i, \ldots, x_j^i, \ldots, x_N^i)^T, \tag{1}$$

such that $x_j(s) : [0, 1] \to \mathbb{R}, \forall s \in [0, 1]$. For the purpose of matching, we enforce constraints on scale and translation. To achieve invariance to translation, we center the configuration X_i as

$$\tilde{X_i} = X_i - \overline{X_i}, \tag{2}$$

where $\overline{X_i}$ is the mean of the coordinates of X_i. To achieve invariance to scale, we divide $\tilde{X_i}$ by its norm such that $\tilde{X}_{i\,SC} = \frac{\tilde{X_i}}{\|\tilde{X_i}\|}$, where $\|\cdot\|$ is the standard $\mathbb{L}^2$ norm given by $\|X\| = \sqrt{\langle X, X\rangle_{\mathbb{L}^2}} = \sqrt{\int_0^1 \sum_{j=1}^n x_j^2(s)ds}$. With a slight abuse of notation, we refer to the scale and translation invariant diffusion profile configurations $\tilde{X}_{i\,SC}$ as X_i throughout the paper.

We define a smooth vector-valued function $\Phi \in \mathbb{L}^2([0, 1], \mathbb{R}^N)$ given by $\Phi \equiv (\phi_1, \phi_2, \ldots, \phi_j, \ldots, \phi_N)^T$, where ϕ_j is a one-dimensional function. We also enforce a differentiability constraint on the inverse of ϕ_j. Thus $\phi_j \in \text{Diff}^+([0, 1])$ is an element of a set of diffeomorphisms on the unit interval. Next, we define an element-wise composition operator $(\cdot)$ such that

$$X \cdot \Phi \equiv (x_1 \circ \phi_1, x_2 \circ \phi_2, \ldots, x_j \circ \phi_j, \ldots, x_N \circ \phi_N)^T. \tag{3}$$

The function Φ encodes local warping functions in the form of ϕ_j. The application of Φ to X results in a local independent update to each of the tracts in the configuration profile. However, if $\phi_1 = \phi_2 = \ldots = \phi_N$, then Φ can also be thought of as a global warping function where all tracts are transformed in the same manner.

Finally, we introduce a smooth global warping function $\Gamma \in \mathrm{Diff}^+([0, 1])$, such that $\dot{\Gamma} > 0, \forall s \in [0, 1]$. This global warping function can be applied to Φ by a composition operator as $(\Phi, \Gamma) \mapsto \Phi \circ \Gamma$. This application of the warping function Γ results in a global update of all the tracts in the configuration profile.

2.2 *Objective Function for Joint Alignment*

Our goal is to register the diffusion profiles across subjects both locally and globally by aligning features corresponding to peaks and valleys along the diffusion profiles. Further, we prefer that this alignment be invariant to the set of diffeomorphic transformations applied along the domain of parameterization $\forall s \in [0, 1]$.

Thus given a pair of diffusion profiles for two subjects X_1 and X_2, we propose a joint objective function that encodes non-linear mappings of diffusion profiles both at the tract-level and the subject level. This objective function is defined as

$$E = \inf_{\Phi, \Gamma} \|X_1 - X_2 \cdot \Phi \circ \Gamma\|^2. \tag{4}$$

We note that the operators $\cdot$ and $\circ$ do not commute and in general will yield different solutions if the order of Φ and Γ is interchanged for the objective function E. However, we suggest that tracts of the same type across subjects will exhibit lower discrepancy than tracts of different types in a single subject. Thus we suppose that the misalignment errors due to the local warping function Φ will in general be proportionately lower than the errors due to the global warping function Γ. Therefore we fix the order of the operators and the warping function as shown in Eq. 4.

2.3 *Alternating Minimization for Subject-Level and Tract-Level Alignment*

Instead of writing the full gradient of Eq. 4 and using gradient-descent we use dynamic programming to obtain an approximate solution. If one assumes Γ in Eq. 4 to be identity ($\Gamma = s$), then Φ can be obtained by solving

$$\widehat{\Phi} = \underset{\Phi}{\text{argmin}} \int_0^1 \sum_{j=1}^{N} \left(x_j^1(s) - x_j^2(\phi_j(s))\right)^2 ds. \tag{5}$$

Alternatively, if we assume Φ to be comprised of identity functions ($\phi_1 = \phi_2 = \ldots = \phi_N = s$), then Γ can be obtained by solving

$$\widehat{\Gamma} = \underset{\Gamma}{\text{argmin}} \int_0^1 \sum_{j=1}^{n} \left(x_j^1(s) - x_j^2(\Gamma(s))\right)^2 ds. \tag{6}$$

We discretize Eq. 6 by restricting to a finite sampling of the diffusion profiles denoted by T. We then define an uniform $T \times T$ grid with samples $0 \leq p, q \leq T - 1$, and compute the discretized version of the cumulative energy (independently for Eqs. 5 and 6) at each point (p, q) while enforcing a positivity constraint on the derivatives $\dot{\Phi} > 0$ and $\dot{\Gamma} > 0$ respectively. Since the cost functions given in Eqs. 5 and 6 are additive on the possible paths ϕ_j and Γ over s, there are several algorithms including recursive approaches to solve this problem. In this paper, we solve both equations by employing a two-pass update scheme, where the first pass updates a cumulative energy function over each node (p, q), and uses this cumulative energy to construct the paths ϕ_j and Γ by interpolating along the valleys of the cumulative energy during the second pass. Since the cumulative energy cost can be recursively defined at each node, we obtain a global solution for $\widehat{\Phi}$ and $\widehat{\Gamma}$ from Eqs. 5 and 6 respectively.

To solve Eq. 4, we use a combination of alternating minimization over the variables Φ and γ to solve the underlying sub-problems given by Eqs. 5 and 6. Further, to solve Eq. 5, we employ a coordinate descent minimization scheme over the sub-variables $\{\phi_1, \ldots, \phi_N\}$. Both these procedures involve solving for 1−dimensional local warping functions $\{\phi_1, \ldots, \phi_N\}$ at each step and combining the results with the solutions obtained for a 1−dimensional global warping function Γ. The coordinate-descent approach suitably lends itself to a derivative-free minimization as long as we can solve for the minimizer of each coordinate function. Algorithm 1 outlines the detailed steps for the minimization procedure. A MATLAB® implementation of Algorithm 1 on a 3.1 GHz Intel Core i7 platform for a discretization $T = 100$ efficiently executes in less than 15 seconds for aligning a pair of subjects.

3 Results

3.1 Data

We apply our algorithm to a subset of $n = 43$ healthy adult subjects (Age: 32.2 ± 11.8, Sex: 21M/22F). To demonstrate the reliability of the algorithm, we also show

Algorithm 1: Alternating Minimization for global Subject-level and local Tract-level Diffusion Profile Alignment

Input: $X_1, X_2, \ldots, X_K$ as the configuration profiles for K subjects

Output: Aligned $\tilde{X}_1, \tilde{X}_2, \ldots, \tilde{X}_K$ configuration profiles for K subjects

$\overline{X} \leftarrow \frac{1}{K}\sum_i X_i$, s.t. $\overline{x}_j \leftarrow \frac{1}{K}\sum_{i=1}^{K} x_j^i$
$\Gamma \leftarrow s$
for $i \leftarrow 1$ **to** K **do**
    $\tilde{X}_i \leftarrow X_i$
    $\Delta_e \leftarrow \left\|\overline{X} - X_i\right\|$
    $\Phi \leftarrow (s, s, \ldots, s)^T$, where $s : [0, 1] \rightarrow [0, 1]$ is an identity function.
    while $\Delta_e > \epsilon$ **do**
        for $k \leftarrow 1$ **to** N **do**
            $\widehat{\Phi} \leftarrow \text{argmin}_{\phi_k}\, bigintssss_0^1 \sum_{j=1}^{n} \left(\overline{x}_j(s) - x_j^i(\phi_j(s))\right)^2 dt$
            $\tilde{X}_i \leftarrow X_i \cdot \widehat{\Phi}$
            $\widehat{\Gamma} \leftarrow \text{argmin}_{\Gamma} \int_0^1 \sum_{j=1}^{n} \left(\overline{x}_j(s) - \tilde{x}_j^i(\Gamma(s))\right)^2 dt$
            $\tilde{X}_i \leftarrow \tilde{X}_i \circ \widehat{\Gamma}$
        end
        $\Delta_e \leftarrow \left\|\overline{X} - \tilde{X}_i\right\|$
    end
end

results on a healthy population of $n = 12$ adult subjects scanned at two time points (Age: 28.9 ± 7.0, Sex: 7M/5F).

Image Acquisition and Preprocessing

We obtained diffusion-weighted and structural images using a 3T Siemens Prisma scanner and a 32-channel head coil. Diffusion-weighted images were acquired using a spin-echo echo planar sequence (EPI), which included 14 reference images ($b = 0\,\text{s/mm}^2$), and multishell images ($b = 1500, 3000\ \text{s/mm}^2$) with 92 gradient directions for each shell (repetition time (TR)/echo time (TE)=3230ms/89.2ms, voxel size $= 1.5\,\text{mm}^3$), with two runs for each of anterior-posterior and posterior-anterior gradient directions. The structural image was acquired using a T1-weighted multi-echo magnetization prepared rapid gradient echo sequence (TR/TE=2500ms/ 1.81, 1.79, 7.18 ms, inversion time (TI) = 1000 ms, voxel size $= 0.8\,\text{mm}^3$). We followed the Human Connectome Project minimal preprocessing pipeline to process the imaging data [6]. The diffusion weighted images were corrected for EPI susceptibility and eddy current distortions using FSL's topup tool, and eddy tool, respectively [1]. T1-weighted images were first processed with Freesurfer's recon-all pipeline, producing cortical parcellations and subcortical segmentations [4]. T1-weighted structural images were registered to Montreal Neurological Institute (MNI) 152 T1 standard space for anterior commissure—posterior commissure (AC-PC) alignment, and each subject's diffusion weighted image was registered to the structural image [6].

Tractography and Along-Tract Diffusion Measure Extraction

Whole brain tractography was performed with MRtrix3, using multi-tissue multi-shell anatomically-constrained spherical deconvolution from the estimated fiber orientation distribution functions (step size 0.75 mm, angle threshold 45°, amplitude threshold 0.05, min/max fiber length 3/250 mm), followed by spherical-deconvolution informed filtering of tractograms to obtain more biologically relevant fibers, ultimately producing 10 million fibers for each subject [7]. FA maps were obtained by tensor construction using FSL's DTIFIT tool [2].

Whole brain fiber tracts were segmented into 20 fiber groups; Left (L) and Right (R) Thalamic Radiation (Th Rad), Corticospinal Tract (CST), Cingulate Cingulum (CnCn), Cingulate Hippocampus (CnHp), Inferior Fronto-Occipital Fasciculus (IFOF), Superior/Inferior Longitudinal Fasciulus (SLF/ILF), Uncinate (Unc), and Arcuate (Arc), and Corpus Callosum Forceps Major/Minor (CC F Maj/Min) using Automated Fiber Quantification (AFQ) [16]. Pairs of ROIs predefined on a standard white matter atlas were warped to the subject's native space, and the fibers that pass through the pair of ROIs were classified into a fiber group. The spurious fibers in the resulting fiber groups were filtered, the FA values along the fibers of each fiber group were sampled across 100 points, and the along-tract FA profile was computed across the length of a representative *superfiber* of each fiber group, based on the average of the along-tract FA profile for all fibers in the fiber group weighted by the inverse of each fiber's distance from the superfiber [16]. An example of the 20 fiber groups and their representative superfibers are shown in Fig. 1. The along-tract FA values are overlaid in color, reflecting the local variability of the FA measure along the length of the fibers.

Fig. 1 Example of 20 superfibers overlaid with color-coded FA

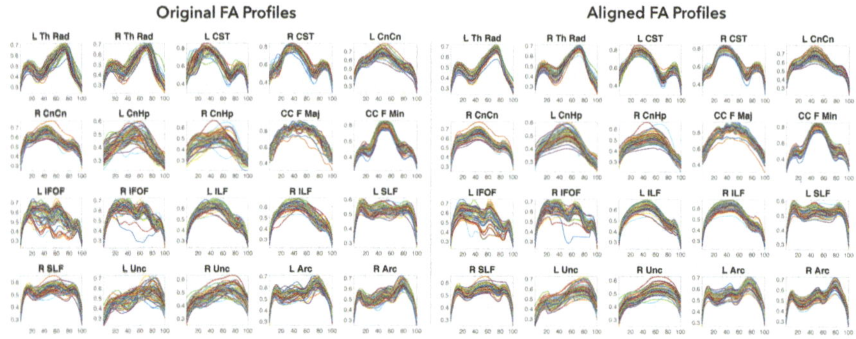

Fig. 2 FA profiles before and after alignment

3.2 Along-Tract FA Profiles Before and After Joint Alignment

Figure 2 shows the along-tract FA profiles across all healthy control subjects ($n = 43$) before and after alignment for the 20 fiber groups. Visually, the aligned outputs show enhancements in prominent features along the tracts. These improved alignments across peaks and valleys of the FA measures are increasingly visible in tracts with very low variability such as the L/R Th Rad. It is also observed that there may be common underlying patterns for the FA profiles along the tract, which are highlighted after alignment.

3.3 Reduced Coefficient of Variation

To quantitatively analyze the effect of the alignment of along-tract diffusion profiles, we computed the coefficient of variation (CV) along the length of each tract before and after alignment (Fig. 3). CV is also equivalent to the inverse of the temporal signal to noise ratio (tSNR) commonly used in fMRI literature to evaluate data quality. Defined as $\frac{\mu}{\sigma}$, where μ and σ are the mean and standard deviation of the diffusion profiles at each point on the fiber for each tract across the population respectively, CV quantifies the variability of the along-tract diffusion profiles across the population. In accordance with the alignment results in Fig. 2 with the visual inspection of the along-tract profiles in Fig. 3, we observe lower CVs following alignment in many locations along several of the 20 superfibers (also overlaid on the tracts in Fig. 4).

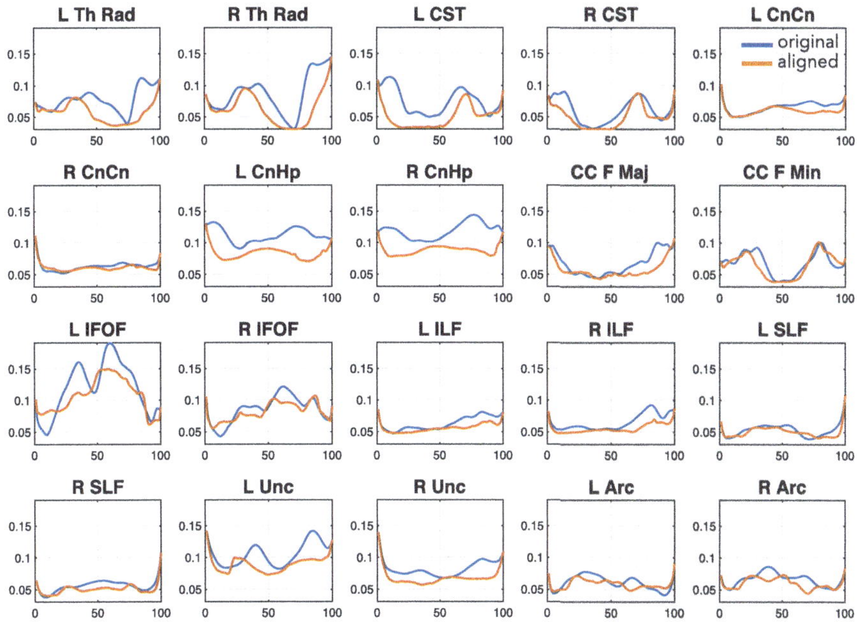

Fig. 3 Reduction in the along-tract spatial coefficient of variation after alignment

3.4 Subject-Wise Inter-tract Correlations

Here, we introduce a new subject-wise measure of inter-tract correlation. Different from the population-level inter-tract correlation by Wahl et al. [14], which correlates mean FA values of a pair of tracts at the population-level, we propose the computation of the spatial correlation of the full along-tract FA profile between a pair of tracts for each individual subject. This yields a subject-specific correlation matrix that could be potentially used for investigating relationships between the fiber tracts for each individual, and for performing statistical analyses across groups of subjects.

To evaluate the significance of the inter-tract correlations within a group of subjects, we performed a one-sample t-test on the correlation value for each tract-pair node against the null hypothesis of zero. To statistically compare the inter-tract correlations between two groups of subjects, we performed a paired t-test between the unaligned and the aligned profiles along the length of each tract. All results, shown in Fig. 5, were corrected for multiple comparison by controlling for false discovery rate (FDR) at $q = 0.05$. Although the within-group inter-tract correlations were significant both in the original and the aligned data (first and second columns), we observed an increase in the inter-tract correlations after alignment (third column). We also noted that since the profiles of the Th Rad and the CST demonstrate mirrored shapes, we observed a significant negative inter-tract correlation between them.

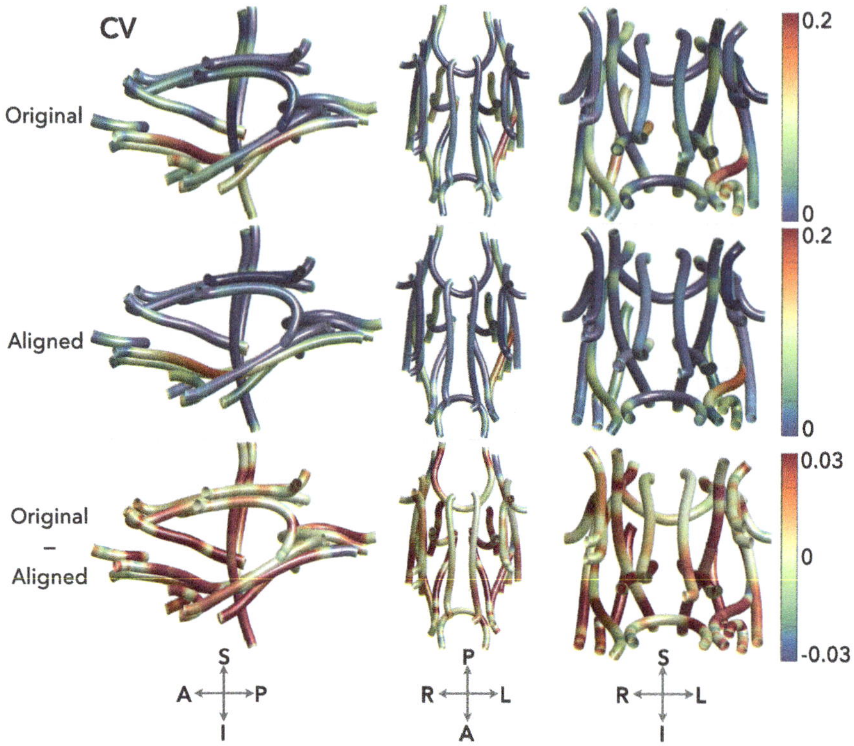

Fig. 4 Spatial coefficient of variation overlaid on the tract profiles

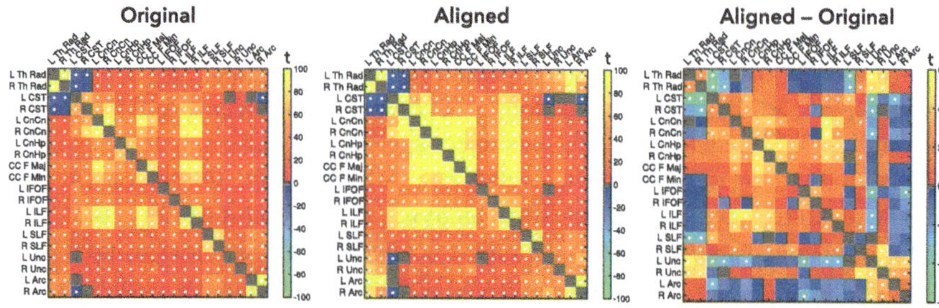

Fig. 5 Inter-tract correlations for healthy subjects for original data (left) and aligned data (middle). The right panel shows inter-tract correlation differences between the original and aligned data

3.5 *Intraclass Correlation Coefficient for Reliability Across Time Points*

To evaluate the reliability of diffusion profiles for a subset of subjects ($N = 12$) scanned twice, we registered the diffusion profiles across the two time points to the extrinsic (Euclidean mean) of the configurations and computed the intraclass correlation coefficient (ICC) before and after alignment [10]. Figure 6 shows a large improvement in ICC in parts of fiber groups (L CST, L/R CnHp, R ILF, L Unc), whereas only marginal improvement in other groups. We also note that several fiber groups already showed high ICCs in the original data. Although tested on a small sample, this finding provides additional support for the potential utility of along-tract profiles in diffusion imaging studies of health and disease. We also computed ICC to evaluate the reliability of inter-tract correlations across two timepoints. As seen in Fig. 7, ICC of the inter-tract correlations significantly increases throughout the fiber groups after alignment. The ICC for the inter-tract correlations with L IFOF remains low across almost all fiber groups. This may be due to the presence of high variability or noise in FA for L IFOF. This is also confirmed from Fig. 2, which shows noisy FA profiles for L IFOF as well as Fig. 3, where its CV is the highest among all fiber groups.

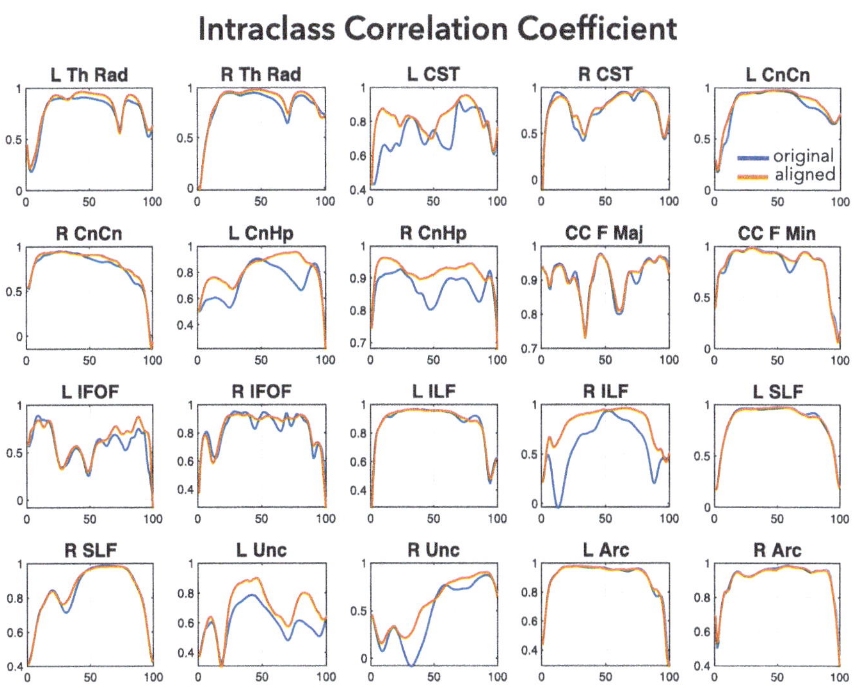

Fig. 6 Pointwise intraclass correlations along the tract profiles before and after alignment

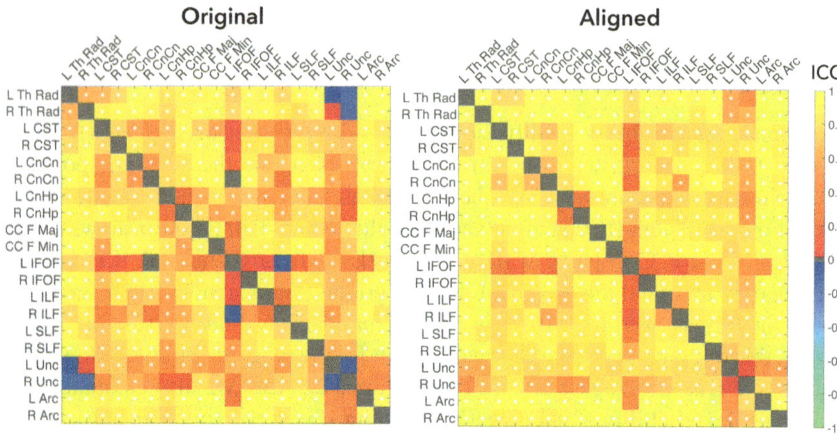

Fig. 7 Intraclass correlations performed on intertract correlations before and after alignment

4 Discussion

We presented a framework for the diffeomorphic alignment of along-tract diffusion measure profiles, by tract-wise registration, followed by a global reparameterization across all tracts between a pair of subjects. This joint alignment provides means to reduce the variability of diffusion profiles across the population by matching the shape of individual profiles to that of a template by minimizing over the spatial parameterization of the domain along the fiber tract. Reduction in the coefficient of variability (CV), and improvements in intraclass correlation coefficients (ICC) following alignment suggests that the consistency between the diffusion profiles across subjects has been improved. Improvement in CV is particularly evident in regions where fibers are passing near each other, for example, intersections between CST and the tracts oriented anterior-posterior-wise, such as cingulum, SLF, ILF, IFOF. This may suggest that each tract's diffusion profiles in the vicinity of crossroads may be affected by its neighboring fibers, and the overall variability may be reduced by performing both tract-wise and subject-wise alignment. We further introduced the subject-wise inter-tract correlation metric, by computing the correlation of diffusion profiles between pairs of tracts. The correlations between fiber tracts were enhanced after the diffeomorphic alignment, which may be attributed to improved coherence of the tract profiles across the subjects. The increase in the ICC of inter-tract correlations further supports the consistency of the diffusion profiles obtained with the diffeomorphic alignment. Our future work will be focused on validating alignments in noisy data and to determine whether the alignment reflects a true reduction of biological variability.

Acknowledgements This research was partially supported by the NIH/NIAAA awards K25AA0 24192 and R01-AA026834, NIH/NIMH grant MH110008 and MH102743, and NIH/NIGMS grant T32-GM008042.

References

1. Andersson, J.L., Sotiropoulos, S.N.: An integrated approach to correction for off-resonance effects and subject movement in diffusion MR imaging. Neuroimage **125**, 1063–1078 (2016)
2. Basser, P.J., Mattiello, J., LeBihan, D.: Estimation of the effective self-diffusion tensor from the NMR spin echo. J. Magn. Reson. Ser. B **103**(3), 247–254 (1994)
3. Benou, I., Veksler, R., Friedman, A., Raviv, T.R.: Combining white matter diffusion and geometry for tract-specific alignment and variability analysis. Neuroimage **200**, 674–689 (2019)
4. Dale, A.M., Fischl, B., Sereno, M.I.: Cortical surface-based analysis: I. Segmentation and surface reconstruction. NeuroImage **9**(2), 179–194 (1999)
5. Garyfallidis, E., Côté, M.A., Rheault, F., Sidhu, J., Hau, J., Petit, L., Fortin, D., Cunanne, S., Descoteaux, M.: Recognition of white matter bundles using local and global streamline-based registration and clustering. Neuroimage **170**, 283–295 (2018)
6. Glasser, M.F., Sotiropoulos, S.N., Wilson, J.A., Coalson, T.S., Fischl, B., Andersson, J.L., Xu, J., Jbabdi, S., Webster, M., Polimeni, J.R., et al.: The minimal preprocessing pipelines for the Human Connectome Project. Neuroimage **80**, 105–124 (2013)
7. Jeurissen, B., Tournier, J.D., Dhollander, T., Connelly, A., Sijbers, J.: Multi-tissue constrained spherical deconvolution for improved analysis of multi-shell diffusion MRI data. Neuroimage **103**, 411–426 (2014)
8. Karahan, E., Costigan, A.G., Graham, K.S., Lawrence, A.D., Zhang, J.: Cognitive and white-matter compartment models reveal selective relations between corticospinal tract microstructure and simple reaction time. J. Neurosci. **39**(30), 5910–5921 (2019)
9. Prasad, G., Joshi, S.H., Jahanshad, N., Villalon-Reina, J., Aganj, I., Lenglet, C., Sapiro, G., McMahon, K.L., de Zubicaray, G.I., Martin, N.G., et al.: Automatic clustering and population analysis of white matter tracts using maximum density paths. Neuroimage **97**, 284–295 (2014)
10. Shrout, P., Fleiss, J.: Intraclass correlations: uses in assessing rater reliability. Psychol. Bull. **86**(2), 420–428 (1979)
11. Slater, D.A., Melie-Garcia, L., Preisig, M., Kherif, F., Lutti, A., Draganski, B.: Evolution of white matter tract microstructure across the life span. Hum. Brain Mapp. **40**(7), 2252–2268 (2019)
12. St-Jean, S., Chamberland, M., Viergever, M.A., Leemans, A.: Reducing variability in along-tract analysis with diffusion profile realignment. Neuroimage **199**, 663–679 (2019)
13. Versace, A., Graur, S., Greenberg, T., Santos, J.L., Chase, H., Bonar, L., Stiffler, R., Hudak, R., Kim, T., Yendiki, A., Greenberg, B., Rasmussen, S., Liu, H., Haber, S., Phillips, M.: Reduced focal fiber collinearity in the cingulum bundle in adults with obsessive-compulsive disorder. Neuropsychopharmacology **44**(7), 1182–1188 (2019)
14. Wahl, M., Li, Y.O., Ng, J., LaHue, S.C., Cooper, S.R., Sherr, E.H., Mukherjee, P.: Microstructural correlations of white matter tracts in the human brain. Neuroimage **51**(2), 531–541 (2010)
15. Wasserthal, J., Neher, P., Maier-Hein, K.H.: Tractseg-fast and accurate white matter tract segmentation. Neuroimage **183**, 239–253 (2018)
16. Yeatman, J.D., Dougherty, R.F., Myall, N.J., Wandell, B.A., Feldman, H.M.: Tract profiles of white matter properties: automating fiber-tract quantification. PLOS ONE **7**(11), 1–15 (2012)
17. Yendiki, A., Panneck, P., Srinivasan, P., Stevens, A., Zöllei, L., Augustinack, J., Wang, R., Salat, D., Ehrlich, S., Behrens, T., Jbabdi, S., Gollub, R., Fischl, B.: Automated probabilistic reconstruction of white-matter pathways in health and disease using an atlas of the underlying anatomy. Front. Neuroinform. **5**, 23 (2011)

Direct Reconstruction of Crossing Muscle Fibers in the Human Tongue Using a Deep Neural Network

Muhan Shao, Aaron Carass, Arnold D. Gomez, Jiachen Zhuo, Xiao Liang, Maureen Stone, and Jerry L. Prince

Abstract The human tongue is made entirely of muscle fibers that either group in a single direction or cross orthogonally in pairs. Reconstructing the muscle fiber orientations throughout the tongue can be beneficial for better understanding the tongue's function in speaking, swallowing, and breathing. Diffusion weighted imaging (DWI) can quantify the anisotropic diffusion of water in muscle and has been used to image tongue muscles. To resolve crossing muscle fibers in the tongue, high angular resolution diffusion imaging (HARDI) can be used to capture the complex fiber configuration. However, existing fiber reconstruction methods, which primarily focus on white matter tracts in the human brain, do not account for the orthogonal nature of the crossing muscle fibers in the tongue. In this paper, we propose a deep convolutional neural network to directly reconstruct the crossing muscle fiber orientations in the tongue from HARDI. The network takes the spherical harmonics (SH) coefficients of the HARDI signals as input and estimates both the SH coefficients of the fiber orientation distribution function (fODF) and the fiber orientations as outputs. Signals from neighboring voxels are incorporated in each estimate to encourage spatial consistency and a novel separation loss is used to encourage orthogonality of the crossing fibers. The network predicts the fiber orientations in a fully automatic manner, without setting a threshold to extract peaks. The proposed method provides

M. Shao (✉) · A. Carass · J. L. Prince
Department of Electrical and Computer Engineering, Johns Hopkins University, Baltimore, MD, USA
e-mail: muhan@jhu.edu

A. Carass · J. L. Prince
Department of Computer Science, Johns Hopkins University, Baltimore, MD, USA

A. D. Gomez
Department of Neurology, Johns Hopkins University School of Medicine, Baltimore, MD, USA

J. Zhuo · X. Liang
Department of Diagnostic Radiology and Nuclear Medicine, University of Maryland School of Medicine, Baltimore, MD, USA

M. Stone
Department of Neural and Pain Sciences, University of Maryland School of Dentistry, Baltimore, MD, USA

N. Gyori et al. (eds.), *Computational Diffusion MRI*, Mathematics and Visualization,
https://doi.org/10.1007/978-3-030-73018-5_6

superior quantitative performance compared to two state-of-the-art methods when evaluated on synthetic tongue data with different noise levels. Application to post-mortem human tongue data revealed the complex muscle fibers of the human tongue and showed qualitative improvements over the competing methods.

1 Introduction

The tongue plays an important role in multiple vital human functions including breathing, swallowing, and speaking [17]. These functions are possible because the tongue can produce complex deformations with its muscle architecture. This muscle architecture consists of intrinsic muscles including the transversalis (T), verticalis (V), superior (SL) and inferior longitudinalis (IL), and extrinsic muscles including the genioglossus (GG), hyoglossus (HG), styloglossus (SG), and geniohyoid (GH). The tongue muscle fibers are extensively interdigitated and orthogonal in three dimensions [9]. Some regions contain a single fiber orientation, but most regions are interdigitated with two fiber orientations crossing at approximately 90°. Figure 1 shows a T2-weighted magnetic resonance imaging (MRI) of a human tongue and idealized tongue muscle fiber orientations from a fiber orientation model [7].

There are compelling scientific reasons to obtain a better understanding of the relationship between individual tongue muscles and the complex functions [21]. Diffusion weighted MRI (DWI) has been used to image tongue muscle architecture and interpret its function [5, 16, 21]. To discriminate crossing fibers within the same voxel, high angular resolution diffusion imaging (HARDI) [20] can be used. Numerous methods have been proposed to reconstruct fibers from HARDI data [2, 3, 18, 19], including some recently proposed deep learning based methods

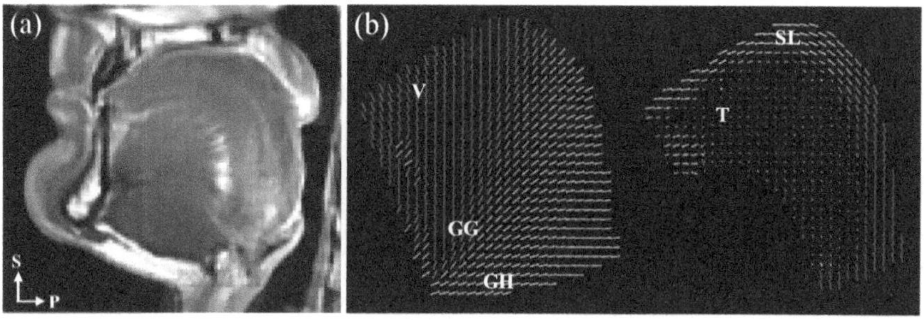

Fig. 1 **a** Mid-sagittal slice of a T2-w MRI of a human tongue. S: superior; P: posterior. **b** Tongue muscle fiber orientations generated based on a mechanical tongue model [7]. The muscle groups in **b** are: verticalis (V), genioglossus (GG), geniohyoid (GH), superior longitudinalis (SL), transversalis (T). The fiber orientations are conventionally color-coded (red: right-left, green: anterior-posterior, blue: inferior-superior)

[11–14]. These methods were designed for brain tissue, and do not account for the orthogonal nature of the crossing muscle fibers in the tongue.

In this paper, we propose a deep convolutional neural network (CNN) to directly reconstruct the fiber orientations. Both the HARDI signal and the fiber orientation distribution function (fODF) can be completely represented by the spherical harmonic (SH) coefficients. Neighboring information is incorporated using convolutional layers to make the fibers spatially consistent. The major contributions of this paper can be summarized as:

- We propose an algorithm that can reconstruct both the fODF and the fiber orientations from HARDI tongue data in a fully automatic manner.
- We introduce a separation loss during training to encourage orthogonality of the predicted crossing fibers, which is critical in tongue muscle fiber orientation reconstruction.

2 Methods

2.1 Training Data and Ground Truth

We used data from [22] as training data for the proposed CNN. The ground truth fiber orientations and the corresponding volume fractions were generated by applying the multi-fiber ball-and-stick method [2] on the brain of a healthy subject. In our experiments, up to two fibers per voxel were chosen to synthesize the training data, which is consistent with the tongue anatomy. The synthetic DW images were created using a multi-tensor model [20] with the diffusion-weighting $b = 1000\,\text{s/mm}^2$ applied in 60, 90, or 120 diffusion directions in different images. In each image, Rician noise was added to the synthesized signal. The signal-to-noise ratio (SNR) of the synthetic DWI was set to 9, 18, or 36 in different images. The fiber orientations and fractions formed the fODF and can be represented as SH coefficients.

2.2 Fiber Estimation Network

Given HARDI data of a human tongue, our goal is to estimate the number of muscle fibers and the corresponding fiber orientations at each voxel. To ensure that the network can be applied to HARDI data with different numbers of diffusion directions, the input to the network is defined as the SH coefficients of the HARDI signal, which are calculated by least-squares linear fitting to the HARDI signal, as in [18]. The output at each voxel is divided into two parts. The first part is the SH coefficients of the fODF. The second part contains two binary terms indicating if the 1st and 2nd fibers exist and two vectors representing the fiber orientations.

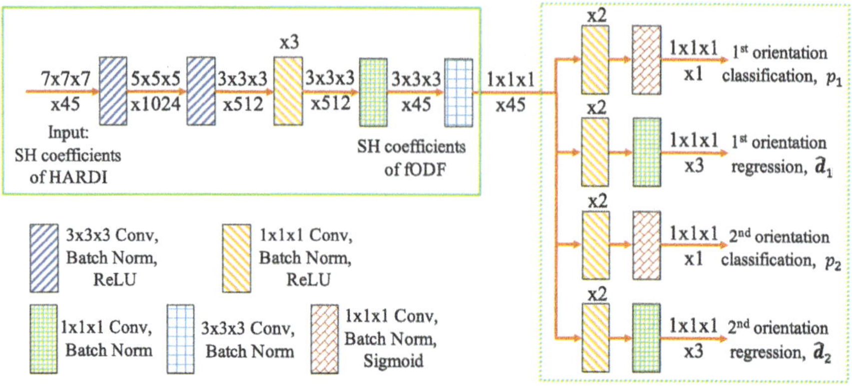

Fig. 2 Architecture of the proposed fiber estimation network. The first stage (solid box) of the network predicts the SH coefficients of the fODF. The second stage (dots and dashes box) detects if the 1st and 2nd fibers exist and predicts the orientations. The numbers indicate the shape of the tensor and the number of features

The proposed network architecture (Fig. 2) is composed of two stages. The first stage (solid green box) predicts the SH coefficients of the fODF, and the second one (dots and dashes green box) predicts the fiber orientations, which simulates the peak finding process in HARDI reconstruction methods. In our experiments, both input HARDI signal and fODF were fitted to 8th order SH (45 SH coefficients).

In the first stage, the input $7 \times 7 \times 7 \times 45$ patch is passed through two $3 \times 3 \times 3$ convolutional blocks and four $1 \times 1 \times 1$ convolutional blocks with the number of features per block being: 1024, 512, 512, 256, 512, and 45 (output). The $1 \times 1 \times 1$ convolutional blocks are used to add more features to the network without decreasing the size the patches. Each convolutional block has a convolutional layer, followed by a batch normalization layer and a rectified linear unit (ReLU) layer. ReLU is not used in the last block, since the SH coefficients can be negative. The output is a $3 \times 3 \times 3 \times 45$ patch representing the SH coefficients of the fODF. This patch is passed through a $3 \times 3 \times 3$ convolutional layer, a batch normalization layer and becomes the input to the second stage.

The second stage consists of four parallel branches, including two classifiers that predict the existence of the 1st and 2nd fibers in the central voxel and two regressors that predict the fiber orientations. Each classifier consists of three $1 \times 1 \times 1$ convolutional layers. The final layer is activated by sigmoid function to obtain a probability, which is used to classify the fiber as exist or not. Each regressor consists of six $1 \times 1 \times 1$ convolutional layers. The final layer is not activated to introduce negativity. Non-zero output vector is normalized as a unit vector to represent fiber orientation.

2.3 Fiber Estimation Loss

The loss function is a combination of the mean squared error in the first stage (s1) and the fiber orientation error (FOE) in the second stage (s2). The FOE in s2 is given by:

$$l_{s2} = l_{\mathrm{CE}}(p_1, \mathbb{1}_1) + l_{\mathrm{FOE}}(\hat{\boldsymbol{d}}_1, \boldsymbol{d}_1) + l_{\mathrm{CE}}(p_2, \mathbb{1}_2) + l_{\mathrm{FOE}}(\hat{\boldsymbol{d}}_2, \boldsymbol{d}_2) + l_{\mathrm{sep}}(\hat{\boldsymbol{d}}_1, \hat{\boldsymbol{d}}_2) \quad (1)$$

where $l_{\mathrm{CE}}(p_i, \mathbb{1}_i)$ is the cross entropy loss for the ith fiber between the predicted probability map and the ground truth and is defined as: $l_{\mathrm{CE}}(p_i, \mathbb{1}_i) = y_i \log p_i + (1 - y_i) \log (1 - p_i)$, where y_i is the ground truth label indicating if the ith fiber exists. $l_{\mathrm{FOE}}(\hat{\boldsymbol{d}}_i, \boldsymbol{d}_i)$ is the fiber orientation error loss between the predicted ith fiber orientation $\hat{\boldsymbol{d}}_i$ and the truth orientation $\boldsymbol{d}_i$, which is modified from the cosine similarity loss and is defined as:

$$l_{\mathrm{FOE}}(\hat{\boldsymbol{d}}_i, \boldsymbol{d}_i) = -\frac{|\hat{\boldsymbol{d}}_i \cdot \boldsymbol{d}_i| + \epsilon}{\|\hat{\boldsymbol{d}}_i\| + \|\boldsymbol{d}_i\| + \epsilon} \quad (2)$$

where ϵ is a small positive number to avoid a zero denominator. To minimize the angular error loss, we enforce that $\hat{\boldsymbol{d}}_i$ should point to the truth direction $\boldsymbol{d}_i$ as close as possible. We take the absolute value of the inner product to account for opposite orientations being equivalent.

To encourage the orthogonality of the crossing fibers in the tongue [9], we propose a separation loss l_{sep} between the predicted crossing fiber orientations $\hat{\boldsymbol{d}}_1$ and $\hat{\boldsymbol{d}}_2$, which is defined as:

$$l_{\mathrm{sep}}(\hat{\boldsymbol{d}}_1, \hat{\boldsymbol{d}}_2) = |\hat{\boldsymbol{d}}_1 \cdot \hat{\boldsymbol{d}}_2| \quad (3)$$

This separation loss at each voxel will be minimized when the crossing angle of the two predicted orientations is 90°. Therefore, the network is encouraged to predict orthogonal crossing fibers, which is an important characteristic in tongue muscles.

2.4 Training Procedure

The fiber estimation network was trained on 5 synthetic HARDI data sets. Each image contains around 80k patches and the batch size was 1024. Another synthetic data set was used for validation. The network was trained using the Adam optimizer [10] with initial learning rate of 0.0001. The performance of the trained network was evaluated at 150 epochs as the validation loss plateaued.

3 Experiments and Results

The performance of the proposed CNN method was compared with two state-of-the-art fiber reconstruction methods: constrained spherical deconvolution (CSD) [18], and Bayesian estimation of diffusion parameters obtained using sampling techniques (BEDPOSTX) [2, 3]. CSD was implemented in Dipy [6], and BEDPOSTX was implemented in FMRIB Software Library (FSL, version 6.0) [8]. When running CSD, a response function for a single fiber bundle is required. We manually selected the region of interest in the GG muscle in a known non-crossing area to estimate the response function. The output of CSD is the fODF, where a peak finding algorithm [18] was performed to identify distinct fiber orientations. A relative peak threshold was used to filter out some small peaks introduced by noise. In this study, we set this threshold to 0.2, as recommended in [15]. When running BEDPOSTX, we thresholded the anisotropic volume fraction at 0.01 to remove spurious peaks. Default threshold in FSL is 0.05. However, BEDPOSTX identified the number of voxels contain crossing fibers as 4% in a synthetic HARDI tongue data, using the default threshold, when the true number of crossing fiber voxels is 67%. Therefore, we lower the volume fraction threshold to reduce the false negative results.

We also report the results of our proposed network but without the separation loss (Eq. 3) during training. This experiment was designed to highlight the importance of the separation loss in improving the accuracy of the predicted fiber orientations. We report the performance of all the algorithms on three synthetic tongue HARDI data and one post-mortem human tongue data.

3.1 *Quantitative Evaluation on Synthetic Tongue HARDI Data*

In order to quantitatively evaluate the proposed method, we simulated three single-shell HARDI data sets with the diffusion weighting of $b = 1000s/mm^2$ and 60 diffusion directions. The ground truth of the fiber orientations were generated based on a mechanical tongue model [7] (see Fig. 1b). In total, there were 10,888 voxels containing fibers. In each voxel, there were at most two crossing fibers. The HARDI data were synthesized using a multi-tensor model [20]. Each fiber's diffusion tensor was computed by rotating a default single tensor with the desired fiber orientation. The default single-fiber tensor was set with diffusivities in orthogonal directions as $\{\lambda_1, \lambda_2, \lambda_3\} = \{0.002, 0.001, 0.001\}$ s/mm^2. The diffusivities were determined by fitting a diffusion tensor model on real HARDI tongue data. Although the diffusivity of the tongue muscle tissue is different from that of the brain tissue, the SH coefficients of the tongue and brain HARDI data, which are the input to the proposed network, have similar distributions. We thus believe that training this network with brain data is a reasonable surrogate, in the absence of tongue data. In voxels with crossing fibers, the fraction of the first fiber f_1 was sampled from a Gaussian distribution with mean of 0.6 and standard deviation of 0.05, and the fraction of the second fiber $f_2 = 1 - f_1$. We added three different levels of Rician noise to the synthesized signal; corresponding to SNR in the synthetic DWI of 10, 15, and 20.

The performance of the different methods were compared using the following criteria: (1) angular correlation coefficient (ACC) [1] between the predicted and truth SH coefficients of the fODF; (2) correct assessment of the number of fiber orientations in each voxel; (3) angular accuracy of the predicted fiber orientations. ACC is a similarity measure that ranges from -1 to 1, where 1 indicates a perfect match. To assess the correctness of the fiber number estimation, we computed the success rate (SR) [4], which is defined as the proportion of voxels where the reconstruction algorithm correctly estimated the number of the fiber orientations within a small angular tolerance. In this study, the tolerance was set to 20°, as in [4]. To quantify the incorrect assessment, we also computed the false positive rate (FP), the false negative rate (FN), and the Dice score for the crossing fiber regions, which is an overlap measurement between the crossing fiber regions in the ground truth and the regions where the automatic algorithm predicted two fibers. The angular accuracy of the estimated orientations was evaluated by calculating the angular error (AE) between the estimated and true orientations. AE was calculated only in the voxels where the reconstruction algorithm correctly predicted the number of muscle fibers. In voxels with crossing fibers, the two estimated fibers were first matched to the ground-truth to minimize the error.

The quantitative results are presented in Figs. 3, 4 and 5, and Table 1. In Fig. 3, ACC was not computed for BEDPOSTX since it did not output the fODF. We see that both the proposed CNN methods (with and without the separation loss) provided higher ACC than the CSD method, especially when the SNR is low. Table 1 presents the performance of each method on the estimation of the number of fiber orientations. The proposed CNN method with the separation loss outperforms all the other methods at each noise level in terms of SR and FN. The overestimated metric FP of the proposed method is higher than the best method, but remains at a low level. The mean Dice scores on the crossing fiber regions for each method are 0.57, 0.23, 0.38, and 0.64, respectively. The proposed method shows better accuracy estimating the number of fibers in crossing fiber regions. Figure 4 shows the success rate and the mean angular error simultaneously on single and crossing fiber regions. The proposed method shows competitive results in single fiber voxels and better results in crossing fiber voxels.

From the AE boxplots in Fig. 5, we can observe that in single fiber regions, the CNN methods produced slightly more accurate fiber orientation when the noise level was high, which is closer to real tongue DWI. In crossing fiber regions, the proposed CNN with the separation loss shows the lowest AE compared to all the other methods. It is worth noting that when removing the separation loss, the same network architecture did not show much improvement compared to CSD and BEDPOSTX in crossing fiber regions. This clearly demonstrates that the separation loss contributes to improving the accuracy of the predicted fiber orientations.

Table 1 Success rate (SR), false positive rate (FP), and false negative rate (FN) on three synthetic HARDI tongue data with different noise levels. The total number of voxels is 10,888. Bold: best performance among the four reconstruction algorithms

	SNR = 10			SNR = 15			SNR = 20		
	SR	FP	FN	SR	FP	FN	SR	FP	FN
CSD	0.451	0.036	0.513	0.616	0.007	0.376	0.715	**0.001**	0.284
BEDPOSTX	0.328	0.016	0.656	0.389	0.020	0.591	0.517	0.028	0.455
CNN w/o l_{sep}	0.439	**0.006**	0.555	0.495	**0.003**	0.502	0.534	0.002	0.464
CNN w/ l_{sep}	**0.546**	0.012	**0.442**	**0.674**	0.007	**0.319**	**0.717**	0.006	**0.277**

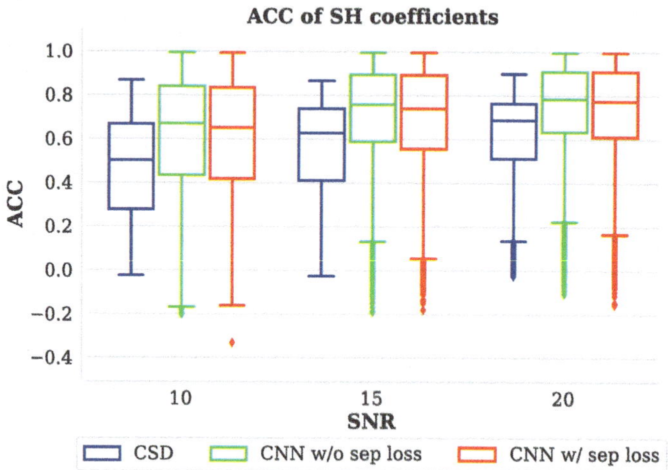

Fig. 3 Boxplots of the angular correlation coefficients (ACC) for the spherical harmonic (SH) coefficients of the fiber orientation distribution function (fODF). ACC was not computed for BEDPOSTX since it did not output the fODF

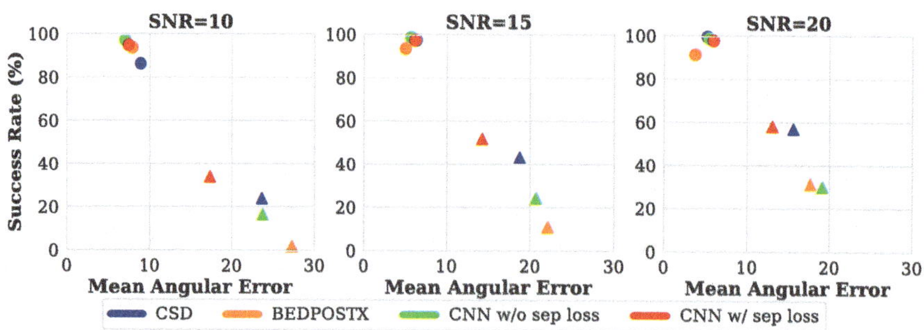

Fig. 4 Success rate versus mean angular error in single (circle) and crossing (triangle) fiber regions

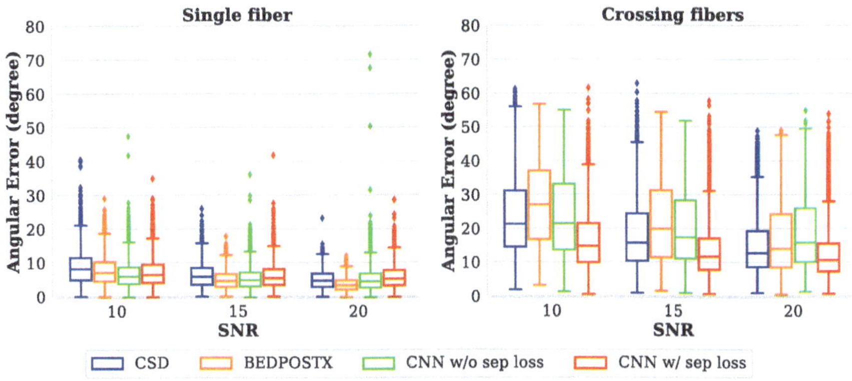

Fig. 5 Angular error in voxels with single (left) and crossing (right) fibers

3.2 Qualitative Results on Post-mortem Human Tongue Data

Post-mortem human tongue data were obtained on an 86-year-old female, around 48 h post death. Use of post-morten specimen was approved by the Institutional Biosafety Committee. MRI was acquired on a Siemens Prisma scanner using a Readout-Segmented Echo-Planar diffusion sequence with the following parameters: readout segment size = 3, TE = 70 ms, TR = 6230 ms, in-plane resolution = 2.5 mm, slice thickness = 2.5 mm, sampling bandwidth = 1645 Hz/Px, and Echo spacing = 0.34 ms. 200 diffusion directions were acquired with $b = 2000\,\text{s/mm}^2$ along with 14 b0 images. A tongue mask was manually drawn to restrict the processing region. A visual comparison of the fiber orientation reconstruction results produced by the four methods is shown in Fig. 6. The predicted fiber orientations are overlapped on a b0 image. We observe that all the methods provided similar results on the first fiber orientation. However, CSD and BEDPOSTX provided more false positive predictions on the second fiber orientation in areas where only GG muscle should exist (see the yellow arrows in Fig. 6a, b). Compared to the other three methods, CNN with the separation loss provided cleaner and more consistent fiber predictions. Comparing Fig. 6c, d reveals the effect of the separation loss. In regions where GG and T muscles are crossing, CNN with the separation loss produced more accurate T muscle orientations, which are orthogonal to GG muscle and should be perpendicular to the sagittal plane. We can also observe that in some regions where T and V, or T and GH muscles are crossing, the T muscle orientation appeared as the first orientation instead of the second one. The reason could be that the strength of the T muscle is higher in these regions.

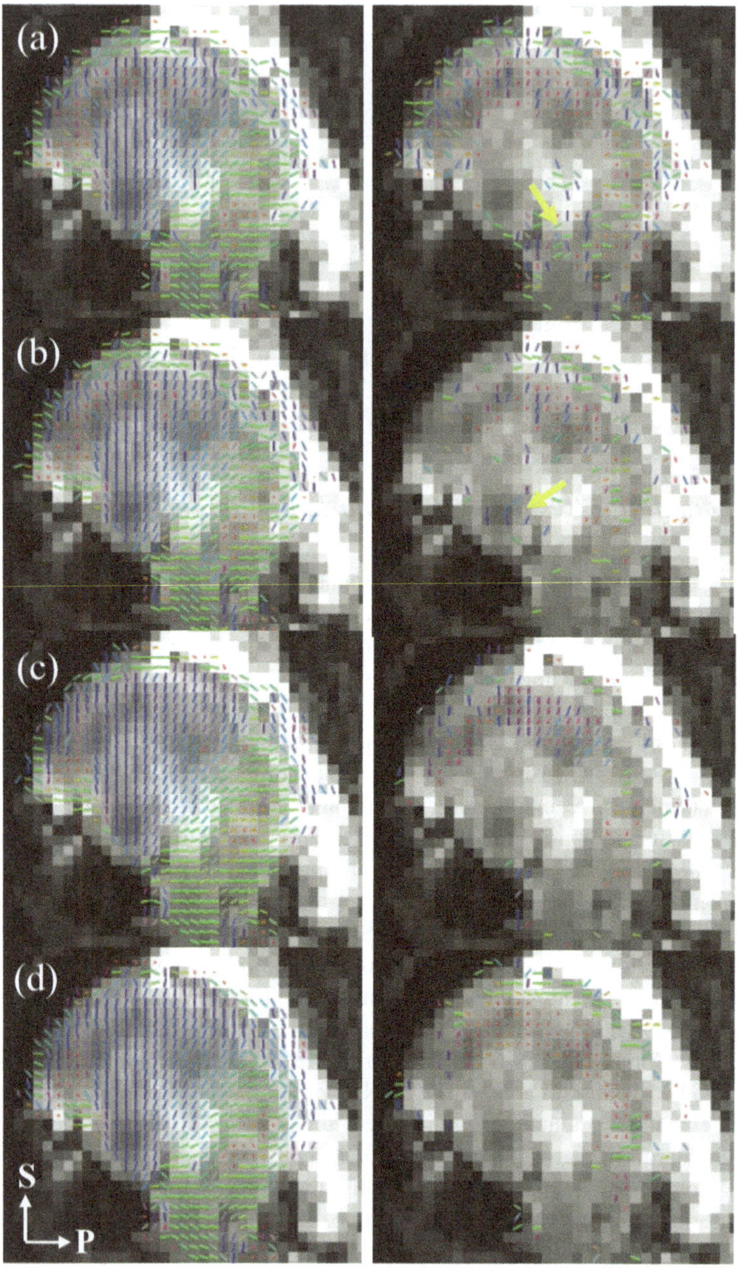

Fig. 6 Sagittal view of the fiber orientations produced by the four fiber reconstruction methods on a post-mortem human tongue data: **a** CSD; **b** BEDPOSTX; **c** CNN without the separation loss; **d** CNN with the separation loss. S: superior; P: posterior. Left column shows the first orientation and right column shows the second orientation. The yellow arrows in **a** and **b** point to some false positive prediction on the second orientation. The fiber orientations are conventionally color-coded (red: right-left, green: anterior-posterior, blue: inferior-superior)

4 Discussion and Conclusions

In this paper, we proposed a convolutional neural network that can reconstruct the fODF and fiber orientations directly from tongue HARDI data. A separation loss was added to the network to encourage the predicted crossing fibers to be orthogonal, which is an important characteristic of tongue muscles. The proposed method was evaluated on three synthetic tongue HARDI data and one post-mortem human tongue data, showing that our method can successfully reconstruct the complex muscle fibers in the tongue. The proposed method outperformed two state-of-the-art fiber reconstruction methods, CSD and BEDPOSTX. Furthermore, our method can be run in an automatic manner. While in CSD, we need to manually select the region of non-crossing area to create the response function. When extracting peak fiber orientations, both CSD and BEDPOSTX require a threshold to remove suspicious fibers. We also showed that the proposed network architecture without the separation loss produced worse reconstruction results, demonstrating the significance of the proposed loss function.

References

1. Anderson, A.W.: Measurement of fiber orientation distributions using high angular resolution diffusion imaging. Magn. Reson. Med. **54**(5), 1194–1206 (2005)
2. Behrens, T.E., Woolrich, M.W., Jenkinson, M., Johansen-Berg, H., Nunes, R.G., Clare, S., Matthews, P.M., Brady, J.M., Smith, S.M.: Characterization and propagation of uncertainty in diffusion-weighted MR imaging. Magn. Reson. Med. **50**(5), 1077–1088 (2003)
3. Behrens, T.E., Berg, H.J., Jbabdi, S., Rushworth, M.F., Woolrich, M.W.: Probabilistic diffusion tractography with multiple fibre orientations: what can we gain? Neuroimage **34**(1), 144–155 (2007)
4. Daducci, A., Canales-Rodrı, E.J., Descoteaux, M., Garyfallidis, E., Gur, Y., Lin, Y.C., Mani, M., Merlet, S., Paquette, M., Ramirez-Manzanares, A., et al.: Quantitative comparison of reconstruction methods for intra-voxel fiber recovery from diffusion MRI. IEEE Trans. Med. Imaging **33**(2), 384–399 (2013)
5. Gaige, T.A., Benner, T., Wang, R., Wedeen, V.J., Gilbert, R.J.: Three dimensional myoarchitecture of the human tongue determined in vivo by diffusion tensor imaging with tractography. J. Magn. Reson. Imaging **26**(3), 654–661 (2007)
6. Garyfallidis, E., Brett, M., Amirbekian, B., Rokem, A., Van Der Walt, S., Descoteaux, M., Nimmo-Smith, I.: Dipy, a library for the analysis of diffusion MRI data. Front. Neuroinform. **8**, 8 (2014)
7. Gomez, A.D., Elsaid, N., Stone, M.L., Zhuo, J., Prince, J.L.: Laplace-based modeling of fiber orientation in the tongue. Biomech. Model. Mechanobiol. **17**(4), 1119–1130 (2018)
8. Jenkinson, M., Beckmann, C., Behrens, T., Woolrich, M., Smith, S.: FSL. NeuroImage **62**, 782–90 (2012)
9. Kier, W.M., Smith, K.K.: Tongues, tentacles and trunks: the biomechanics of movement in muscular-hydrostats. Zool. J. Linn. Soc. **83**(4), 307–324 (1985)
10. Kingma, D.P., Ba, J.: Adam: a method for stochastic optimization (2014). arXiv:1412.6980
11. Koppers, S., Merhof, D.: Direct estimation of fiber orientations using deep learning in diffusion imaging. In: International Workshop on Machine Learning in Medical Imaging, pp. 53–60. Springer (2016)

12. Koppers, S., Friedrichs, M., Merhof, D.: Reconstruction of diffusion anisotropies using 3D deep convolutional neural networks in diffusion imaging. In: Modeling, Analysis, and Visualization of Anisotropy, pp. 393–404. Springer (2017)
13. Lin, Z., Gong, T., Wang, K., Li, Z., He, H., Tong, Q., Yu, F., Zhong, J.: Fast learning of fiber orientation distribution function for MR tractography using convolutional neural network. Med. Phys. **46**(7), 3101–3116 (2019)
14. Nath, V., Schilling, K.G., Parvathaneni, P., Hansen, C.B., Hainline, A.E., Huo, Y., Blaber, J.A., Lyu, I., Janve, V., Gao, Y., et al.: Deep learning reveals untapped information for local white-matter fiber reconstruction in diffusion-weighted MRI. Magn. Reson. Imaging **62**, 220–227 (2019)
15. Schilling, K.G., Janve, V., Gao, Y., Stepniewska, I., Landman, B.A., Anderson, A.W.: Histological validation of diffusion MRI fiber orientation distributions and dispersion. Neuroimage **165**, 200–221 (2018)
16. Shinagawa, H., Murano, E.Z., Zhuo, J., Landman, B., Gullapalli, R.P., Prince, J.L., Stone, M.: Tongue muscle fiber tracking during rest and tongue protrusion with oral appliances: a preliminary study with diffusion tensor imaging. Acoust. Sci. Technol. **29**(4), 291–294 (2008)
17. Stone, M., Woo, J., Lee, J., Poole, T., Seagraves, A., Chung, M., Kim, E., Murano, E.Z., Prince, J.L., Blemker, S.S.: Structure and variability in human tongue muscle anatomy. Comput. Methods Biomech. Biomed. Eng.: Imaging Vis. **6**(5), 499–507 (2018)
18. Tournier, J.D., Calamante, F., Connelly, A.: Robust determination of the fibre orientation distribution in diffusion MRI: non-negativity constrained super-resolved spherical deconvolution. Neuroimage **35**(4), 1459–1472 (2007)
19. Tuch, D.S.: Q-ball imaging. Magn. Reson. Med. **52**(6), 1358–1372 (2004)
20. Tuch, D.S., Reese, T.G., Wiegell, M.R., Makris, N., Belliveau, J.W., Wedeen, V.J.: High angular resolution diffusion imaging reveals intravoxel white matter fiber heterogeneity. Magn. Reson. Med. **48**(4), 577–582 (2002)
21. Voskuilen, L., Mazzoli, V., Oudeman, J., Balm, A.J., van der Heijden, F., Froeling, M., de Win, M.M., Strijkers, G.J., Smeele, L.E., Nederveen, A.J.: Crossing muscle fibers of the human tongue resolved in vivo using constrained spherical deconvolution. J. Magn. Reson. Imaging **50**(1), 96–105 (2019)
22. Wilkins, B., Lee, N., Gajawelli, N., Law, M., Leporé, N.: Fiber estimation and tractography in diffusion MRI: development of simulated brain images and comparison of multi-fiber analysis methods at clinical b-values. Neuroimage **109**, 341–356 (2015)

Learning Anatomical Segmentations for Tractography from Diffusion MRI

Christian Ewert, David Kügler, Anastasia Yendiki, and Martin Reuter

Abstract Deep learning approaches for diffusion MRI have so far focused primarily on voxel-based segmentation of lesions or white-matter fiber tracts. A drawback of representing tracts as volumetric labels, rather than sets of streamlines, is that it precludes point-wise analyses of microstructural or geometric features along a tract. Traditional tractography pipelines, which do allow such analyses, can benefit from detailed whole-brain segmentations to guide tract reconstruction. Here, we introduce fast, deep learning-based segmentation of 170 anatomical regions directly on diffusion-weighted MR images, removing the dependency of conventional segmentation methods on T1-weighted images and slow pre-processing pipelines. Working natively in diffusion space avoids non-linear distortions and registration errors across modalities, as well as interpolation artifacts. We demonstrate consistent segmentation results between 0.70 and 0.87 Dice depending on the tissue type. We investigate various combinations of diffusion-derived inputs and show generalization across different numbers of gradient directions. Finally, integrating our approach to provide anatomical priors for tractography pipelines, such as TRACULA, removes hours of pre-processing time and permits processing even in the absence of high-quality T1-weighted scans, without degrading the quality of the resulting tract estimates.

Christian Ewert and David Kügler—These authors contributed equally.

C. Ewert · D. Kügler · M. Reuter (✉)
German Center for Neurodegenerative Diseases (DZNE), Bonn, Germany
e-mail: martin.reuter@dzne.de

A. Yendiki · M. Reuter
Athinoula A. Martinos Center for Biomedical Imaging, Massachusetts General Hospital, Boston, MA, USA

Department of Radiology, Harvard Medical School, Boston, MA, USA

N. Gyori et al. (eds.), *Computational Diffusion MRI*, Mathematics and Visualization,
https://doi.org/10.1007/978-3-030-73018-5_7

1 Introduction

Tractography has significantly advanced clinical applications [5, 6, 25] and has enabled neuroscientists to study developmental and pathological effects on the human connectome [14]. Traditional tractography pipelines often use anatomical segmentations to obtain priors for reconstructing tracts from diffusion-weighted MRI (dMRI). This introduces a dependency on T1-weighted (T1w) images, which are required for anatomical segmentation by neuroimaging suites such as FreeSurfer [9]. For dMRI microstructural analyses, accurate segmentations of the grey/white matter (GM/WM) boundary are particularly important as different biophysical models have been proposed for each tissue type [22]. However, segmenting in T1w, rather than diffusion image space is problematic due to non-linear distortions between modalities, as well as potential registration inaccuracies and interpolation artifacts when mapping segmentation labels from anatomical to diffusion image space. Furthermore, enabled by acquisitions with high angular resolution and multiple b-values, dMRI-derived cytoarchitectonic boundaries may in the future complement or supersede T1w-derived segmentations for morphometric analyses. Addressing this need for fast, accurate, dMRI-based segmentation, we present a framework for segmenting 170 GM, WM, and subcortical regions in native diffusion space, without requiring high-quality T1w images.

Methods such as SLANT and FastSurfer [11, 12] introduce deep learning for neuromorphometry, yet still rely on T1w images. Traditional [13, 30, 31, 35] and deep learning-based [15, 19] methods extend segmentation to diffusion-weighted images (DWIs) for various acquisition protocols and dMRI representations. Applications of segmentation based on the inherently multi-channel dMRI signals include whole-brain GM/WM/Cerebrospinal fluid [31, 35], WM regions [19, 30], nuclei (cerebellar [15] and thalamic [13]), organs [3, 8, 26, 39], tumors [28] and stroke lesions [4, 20]. As an alternative approach to traditional tractography, neural networks can directly segment WM tracts based on dMRI [16, 17, 21, 23] or diffusion orientations [32, 33, 37, 38], from clinical [21] or high-quality [32] datasets. Unfortunately, segmenting WM tracts as volumetric labels does not provide an along-the-tract parameterization which is useful for point-wise analyses of microstructural and geometric features of the tracts. In contrast to direct, volumetric tract segmentation approaches, the present work aims to provide the information that guides traditional tractography methods. No work to date has addressed whole-brain segmentation, including subcortical structures, cortical regions, and WM regions underlying the cortex, directly from DWIs.

Here, we introduce deep learning-based segmentation of 170 distinct regions (cortical, subcortical, and WM) from DWIs (see Fig. 1). Thus, we provide a fast, deep learning alternative for a critical step in dMRI pre-processing streams, which can facilitate the use of classical tractography methods, without limiting their outputs and subsequent analysis options. Basing segmentations purely on dMRI data removes the dependency on high-quality T1w images, the potentially error-prone, cross-modal co-registration and interpolation, and avoids confounding non-linearities between

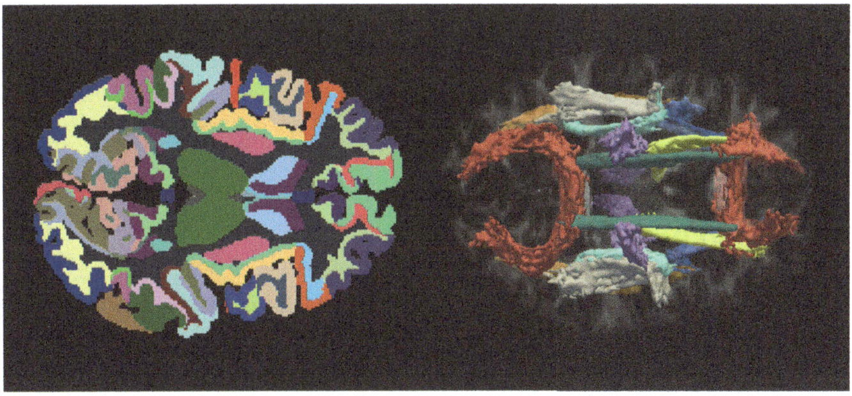

Fig. 1 Whole-brain segmentation of 170 regions (cortical, sub-cortical and white matter) directly from diffusion MRI (left), and probabilistic white matter tracts generated by TRACULA based on anatomical priors from dMRI-based segmentations (right)

anatomical and diffusion spaces. To develop effective segmentation of dMRI data, we compile a dataset of DWIs and reference segmentations generating the latter by mapping FreeSurfer segmentations [9] to diffusion space. We adopt the anatomy-targeted FastSurfer architecture [11], which already supports the segmentation of a large number of regions. Moreover, expanding on the work of Li et al. [17] for WM tract segmentation, we explore suitable dMRI data representations for learning-based segmentation. We compare inputs consisting of images without diffusion-weighting, diffusion tensor components, or DWIs and vary the number of DWIs from which tensors are generated.

When compared against FreeSurfer, our method achieves performance comparable to the state-of-the-art at orders of magnitude faster processing times. As a use case, we integrate dMRI-based segmentations into the tractography package TRACULA (TRActs Constrained by UnderLying Anatomy) [36], which performs global probabilistic tractography with anatomical priors. Differences of TRACULA tracts based on deep learning versus traditional anatomical initialization (see Fig. 4) are in the order of previously reported differences between automated TRACULA reconstructions and manual labels while reducing the TRACULA pre-processing run-time by 4 h. With our approach, the segmentation of 170 regions in dMRI space can be achieved in 32 s on a GPU.

2 Materials and Methods

The tract posterior probabilities computed by TRACULA involve anatomical priors that are calculated from an anatomical segmentation. We perform this segmentation directly in diffusion space and remove the requirement for T1w images by replacing the corresponding component of the TRACULA pipeline with a deep learning network.

2.1 Data

Diffusion MRI Data

We use pre-processed DWIs from the WU-Minn Human Connectome Project (HCP) [7, 18, 24, 27, 29, 34], which are already corrected for eddy-currents and subject motion. These images are acquired with a 3-shell protocol at b-values of 1000, 2000, and 3000 s/mm^2. Each shell is composed of 90 diffusion-encoding gradient directions, approximately uniformly distributed on the sphere. In addition, 18 images without diffusion-weighting are interleaved with the DWIs. For training, validation, and test, we create gender-balanced non-intersecting subsets of 250, 50, and 100 subjects, respectively.

Segmentation Labels

Anatomical segmentations of T1-weighted images of the same subjects are obtained with FreeSurfer 6.0 [9]. We project cortical parcellations from the surface models up to 2 mm deep into the WM, as required by TRACULA. The registration to the diffusion space is performed with the boundary-based rigid registration method *bbregister* [10].

2.2 Data Representations

Since q-space sampling schemes may comprise anywhere from six to several hundred measurements, a general segmentation approach should be independent of the exact choice of diffusion-encoding directions and b-values. Instead, a suitable representation has to abstract from acquisition details yet contain sufficient relevant information. A parsimonious model that is often fitted to DWIs acquired on shells is the diffusion tensor [1], which models local diffusion as a single (uni-modal) Gaussian distribution. The symmetric 3×3 diffusion tensor can be understood as a condensed *summary* of the local diffusion behavior at a given voxel and can be reconstructed from any q-shell acquisition scheme that includes at least 6 directions. To explore how the number of DWIs used to fit the tensor affects its performance for the segmentation task, we extract multiple subsets of gradient directions on the same shell (approximately uniformly distributed). For each of these subsets, the diffusion tensor is fitted to the data with *FSL*'s *dtifit* function and the six unique tensor components are stacked and used as input.

2.3 Architecture

FastSurferCNN [11] is a U-Net-based neuroimage segmentation network validated extensively on anatomical MRI datasets. Three fully convolutional networks are trained independently on axial, coronal, and sagittal slices of MR images. The predictions from the three views are then combined into the final prediction volume by means of a weighted average (view-aggregation). The network uses skip-connections between encoder- and decoder-blocks. In the decoder-blocks, information from the previous decoder-block and the corresponding encoder-block are combined. Instead of simply concatenating these feature maps, FastSurferCNN employs competitive dense blocks to reduce the number of parameters. Competitive dense blocks rely on max-out activations to encourage the network to learn which parts of the provided feature maps are relevant for the segmentation task.

All networks are trained with a combined loss-function containing a median frequency balanced logistic loss with edge-focus and a Dice loss:

$$\mathcal{L} = \underbrace{-\sum_{x} \omega(x) g_c(x) log(p_c(x))}_{\text{Logistic Loss}} - \underbrace{\frac{2\sum_x p_c(x) g_c(x)}{\sum_x p_c^2(x) + \sum_x g_c^2(x)}}_{\text{Dice Loss}}$$

with $\omega(x) = \omega_F(x) + \omega_E(x)$, median frequency balanced weights $\omega_F(x)$, edge-weighting $\omega_E(x)$ at voxel x, references g, prediction p and class c. In order to provide 3D spatial context, FastSurferCNN's input consists not only of the slice of interest but also a sequence of neighboring slices.

2.4 Training

We train all networks with an initial learning rate of 0.01, which is reduced every 10 epochs (multiplied by 0.2). Early stopping is applied when the loss on the validation set does not improve for 15 epochs.

2.5 Tracts

To calculate priors for tractography, TRACULA needs registrations between anatomical, diffusion and MNI spaces. Since we provide the segmentation natively in diffusion space, a registration to anatomical T1w space is obsolete and a single registration from diffusion to MNI space suffices. This mapping can easily be established without a T1w image, yet the two different registration-methods (one includes the anatomical image whereas the other does not) introduce an additional variance that prevents a consistent evaluation of tract similarity. We, therefore, re-use the mapping from diffusion to MNI space that was established to generate the reference tracts. Furthermore, to mitigate artifacts from nearest-neighbor interpolation to large voxel sizes, we instead predict and map probability distributions over classes (soft-labels) via tri-linear interpolation, taking the argmax over classes in the target space.

2.6 *Evaluation Criteria*

Segmentation Quality

We measure the similarity of the segmentation resulting from our method and the segmentation of FreeSurfer mapped to diffusion space, with two evaluation criteria.

The *Dice score* $\mathcal{D}_c$ for region c measures the relative overlap between the binary labels of the prediction $P_c = \{p_{ic} \mid i = 1, \ldots, N\}$ and the reference $R_c = \{r_{ic} \mid i = 1, \ldots, N\}$ segmentation:

$$\mathcal{D}_c(P, R) = \frac{2\sum_{i=1}^{N} p_{ic} r_{ic}}{\sum_{i=1}^{N} p_{ic} + \sum_{i=1}^{N} r_{ic}}.$$

The voxel-based mean *Hausdorff distance* $\mathcal{H}_c$ for region c measures the difference between prediction P_c and reference labels R_c via

$$\mathcal{H}_c(P, R) = \frac{1}{|R_c|} \sum_{r \in R_c} \min_{p \in P_c} ||p - r||_2 + \frac{1}{|P_c|} \sum_{p \in P_c} \min_{r \in R_c} ||p - r||_2.$$

Tractography

Similarly to the segmentation case, we quantify the similarity between pairs of tracts reconstructed with anatomical priors from either segmentation via voxel-based mean *Hausdorff distance*. Since TRACULA relies on a Markov-Chain Monte-Carlo method, two different sets of tracts are not per se comparable. Thus, in accordance with previous tract evaluations [36, 40], we threshold the tracts at 20% of their maximum intensity.

3 Results and Discussion

To determine the impact on segmentation quality, we evaluate networks with respect to different data subsets and input representations, keeping the network architecture (e.g. number of filters) fixed.

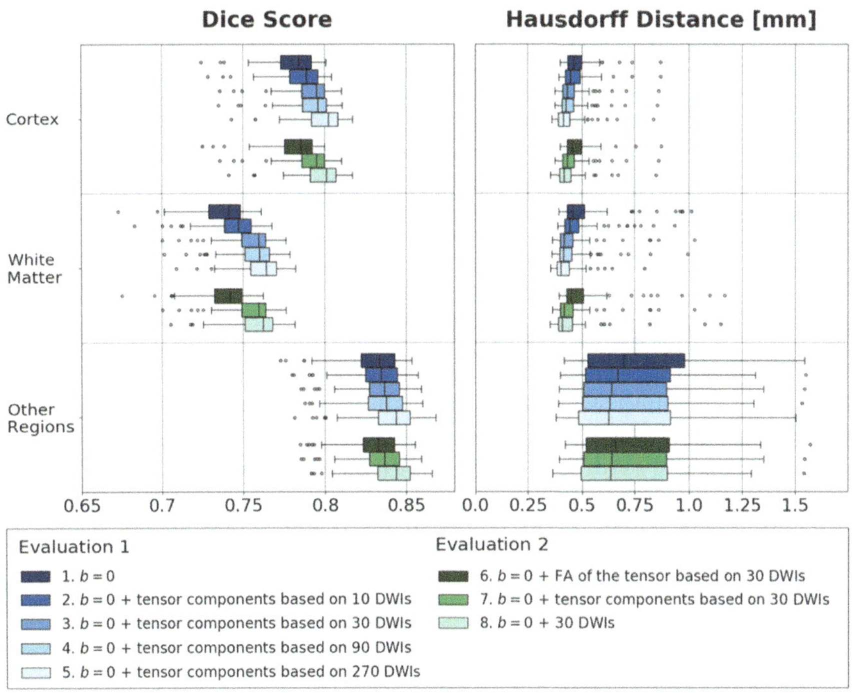

Fig. 2 Ablation of neural network input in comparison to FreeSurfer reference segmentation. **Evaluation 1**: Q-Space Sampling Density (top group of blue bars): 1. Only $b = 0$ image, 2.–5. $b = 0$ + diffusion tensors fitted with varying sampling density (2.–4. only on the first shell with $b = 1000\,\mathrm{s/mm}^2$, 5. on three shells); **Evaluation 2**: dMRI data representation (bottom group of green bars) of a fixed set of 30 DWIs on the first shell: $b = 0$ plus 6. an FA map, 7. the diffusion tensor, and 8. DWIs directly

3.1 Evaluation 1: *Q*-Space Sampling Density

When q-space is sampled more densely, we expect the diffusion tensor to be more accurate due to the improved signal-to-noise ratio (SNR). To explore how different single-shell q-space samplings influence the segmentation quality, we compare FreeSurfer segmentations against our network's prediction for several scenarios which are displayed in Fig. 2.

The Dice score seems to correlate with the compactness of the shape of anatomical regions. Scores are high for sub-cortical regions, where shapes feature large volume-to-surface ratios, lower in the folded cortical areas, and lowest for the thin cortical projections into the WM. The slim, 2 mm projection on a coarse voxel grid with 1.25 mm isotropic size could be another reason for the smaller Dice scores in WM regions. In fact, the ratio of voxels on the boundary is significantly higher for white matter regions compared to cortical regions. The Hausdorff distances paint a very similar picture in terms of ranking methods while they are more consistent than Dice scores across WM- and cortical regions.

As expected, the segmentation quality increases when the diffusion tensor is fitted with more DWIs (1.–5.). The segmentation based solely on the image without diffusion-weighting ($b = 0$) provides a strong baseline, potentially due to the higher SNR at $b = 0$. Yet, the inclusion of additional diffusion information increases segmentation performance further.

3.2 Evaluation 2: Input Representations

We also assess the effect of different input data on the segmentation quality (see Fig. 2, second, green bar group: 6.–8.). *Fractional Anisotropy (FA)* measures the coherence of water diffusion in a voxel and is frequently used in tract-based analyses. Since *FA* is a scalar measure computed from the eigenvalues of the tensor, it only contains a subset of the tensor information. As a result, the performance is worse with *FA* (6.) than with the full tensor (7.). More broadly, the diffusion tensor is a simple model that fails to accurately describe water diffusion in full detail. Thus, it is not surprising that a segmentation directly based on the DWIs (8.) yields better results than one based on the diffusion tensor. However, for the task of segmentation, the diffusion tensor seems to capture most of the relevant information present in the DWIs.

3.3 Evaluation 3: Generalization

In the previous evaluation, networks were trained separately for each set of inputs. This evaluation explores how a network *trained* on tensor components based on n DWIs performs when it is *evaluated* on tensor components based on m DWIs (with $n \neq m$). This kind of generalizability is a critical property when accommodating data with a variety of acquisition details at test time. Notably, while neither network generalizes perfectly to the different input format, the stable results (Fig. 3) suggest that generalizability can be asserted to a large extent.

3.4 Evaluation 4: Tract Similarity

Finally, we assess the stability of WM tract generation when switching from traditional T1w image segmentation to our dMRI-based, deep learning approach: For a gender-balanced subset of 20 subjects from the HCP, we reconstruct 18 different WM tracts with TRACULA, using either the proposed, dMRI-based segmentations or the FreeSurfer T1w-based segmentations (see Fig. 4). The deviation is within the margin

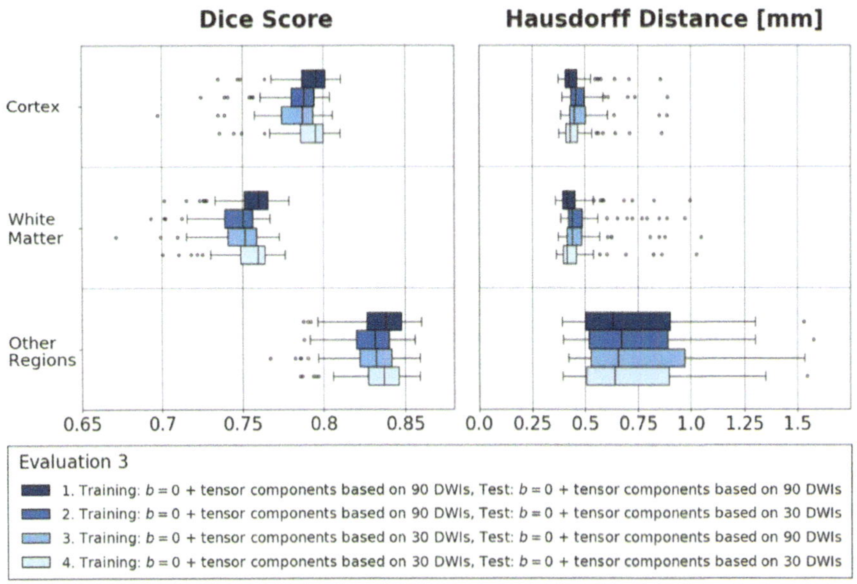

Fig. 3 Network generalization when tensors are fitted with differently many DWIs

of the deviation of TRACULA's tracts from manually-annotated bundles [36], which is around 2 mm for most tracts.

We time both applications on five representative cases (Evaluation 1 and 4). Our method takes 32 s for the anatomical segmentation in diffusion space compared to 4 h with FreeSurfer (parallelization of hemispheres and 4 threads, 7 h sequentially). For the tractography pipeline, our work accelerates the total run time from 283 to 56 min.

4 Conclusion

Our work presents and analyzes the application of deep learning for anatomical segmentation directly on DWIs. Applied to probabilistic tractography with anatomical priors, our method enables processing without the requirement of T1w images and thus avoids errors from non-linear distortions, registration inaccuracies, or interpolation artifacts. As a consequence, dMRI-based anatomical segmentation achieves results similar to corresponding state-of-the-art T1w-based segmentation and speeds up pre-processing in the TRACULA pipeline by hours. Accelerating heavy processing pipelines is essential especially for large cohort studies such as HCP [29] or the Rhineland study [2], where large diffusion datasets from thousands of participants require efficient methods.

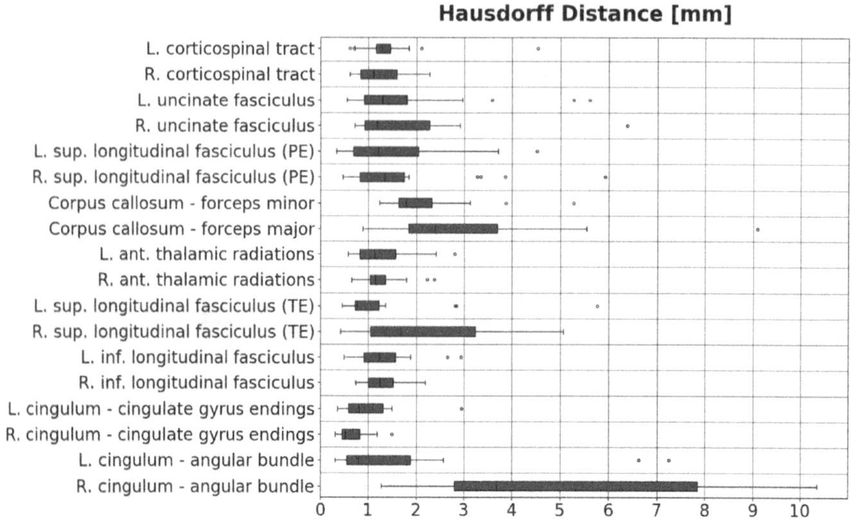

Fig. 4 Similarity of tracts based on DWI segmentations versus FreeSurfer's segmentation. L. = Left, R. = Right, sup. = superior, ant. = anterior, inf. = inferior, TE = temporal endings, PE = parietal endings

Furthermore, we analyze how the choice of input images for the neural network affects segmentation performance. We confirm that segmentation quality increases as more q-space samples are included when fitting diffusion tensors—likely due to increased SNR. Skipping the tensor fit and directly learning from DWIs increases segmentation performance further. However, simply increasing q-space samples is not an option due to memory limitations and reliance on the availability of the same set of q-space samples for future input cases. Tensor-based inputs, on the other hand, provide a widely applicable alternative [17] and yield results that remain relatively stable and close to the direct DWI performance. In our opinion, tensors fitted to 30 DWIs, a number that is feasible in clinical studies, offer a good balance. Future work will explore diffusion models other than the tensor as segmentation inputs and assess generalizability across a large variety of different dMRI datasets.

While we illustrate the use of our deep learning-based segmentation in a pipeline for probabilistic tractography with anatomical priors, it can be useful in a wide range of other applications. These include improved WM/GM and pallidum-putamen segmentation, seed-based tractography, network analysis, or ROI-based analysis of microstructural measures, to name a few.

Acknowledgements Data were provided in part by the Human Connectome Project, WU-Minn Consortium (Principal Investigators: David Van Essen and Kamil Ugurbil; 1U54MH091657) funded by the 16 NIH Institutes and Centers that support the NIH Blueprint for Neuroscience Research; and by the McDonnell Center for Systems Neuroscience at Washington University. AY was partly supported by NIH awards R01-EB021265 and U01-EB026996.

References

1. Basser, P.J., Mattiello, J., LeBihan, D.: MR diffusion tensor spectroscopy and imaging. Biophys. J. **66**, 259–67 (1994)
2. Breteler, M.M., Stöcker, T., Pracht, E., Brenner, D., Stirnberg, R.: MRI in the Rhineland study: a novel protocol for population neuroimaging. Alzheimer's & Dement. **10**, P92–P92 (2014)
3. Clark, T., Wong, A., Haider, M.A., Khalvati, F.: Fully deep convolutional neural networks for segmentation of the prostate gland in diffusion-weighted MR images. In: Image Analysis and Recognition, pp. 97–104. Springer (2017)
4. Clèrigues, A., Valverde, S., Bernal, J., Freixenet, J., Oliver, A., Lladó, X.: Acute and sub-acute stroke lesion segmentation from multimodal MRI. Comput. Methods Prog. Biomed. **194** (2020)
5. Costabile, J.D., Alaswad, E., D'Souza, S., Thompson, J.A., Ormond, D.R.: Current applications of diffusion tensor imaging and tractography in intracranial tumor resection. Front. Oncol. **9**, 426 (2019)
6. Essayed, W.I., Zhang, F., Unadkat, P., Cosgrove, G.R., Golby, A.J., O'Donnell, L.J.: White matter tractography for neurosurgical planning: a topography-based review of the current state of the art. NeuroImage. Clin. **15**, 659–672 (2017)
7. Feinberg, D.A., Moeller, S., Smith, S.M., Auerbach, E., Ramanna, S., Gunther, M., Glasser, M.F., Miller, K.L., Ugurbil, K., Yacoub, E.: Correction: multiplexed echo planar imaging for sub-second whole brain FMRI and fast diffusion imaging. PLoS ONE **6**(9) (2011)
8. Ferreira, P.F., Martin, R.R., Scott, A.D., Khalique, Z., Yang, G., Nielles-Vallespin, S., Pennell, D.J., Firmin, D.N.: Automating in vivo cardiac diffusion tensor postprocessing with deep learning-based segmentation. Magn. Reson. Med. (2020)
9. Fischl, B.: FreeSurfer. NeuroImage **62**(2), 774–781 (2012)
10. Greve, D., Fischl, B.: Accurate and robust brain image alignment using boundary-based registration. Neuroimage **48**, 63–72 (2009)
11. Henschel, L., Conjeti, S., Estrada, S., Diers, K., Fischl, B., Reuter, M.: FastSurfer - a fast and accurate deep learning based neuroimaging pipeline. NeuroImage **219**, 117012 (2020). http://www.sciencedirect.com/science/article/pii/S1053811920304985
12. Huo, Y., Xu, Z., Xiong, Y., Aboud, K., Parvathaneni, P., Bao, S., Bermudez, C., Resnick, S.M., Cutting, L.E., Landman, B.A.: 3D whole brain segmentation using spatially localized atlas network tiles. Neuroimage **194**, 105–119 (2019)
13. Iglesias, J.E., van Leemput, K., Golland, P., Yendiki, A.: Joint inference on structural and diffusion MRI for sequence-adaptive Bayesian segmentation of thalamic nuclei with probabilistic atlases. In: Information Processing in Medical Imaging, pp. 767–779. Springer (2019)
14. Jeurissen, B., Descoteaux, M., Mori, S., Leemans, A.: Diffusion MRI fiber tractography of the brain. NMR Biomed. **32**(4) (2019)
15. Kim, J., Patriat, R., Kaplan, J., Solomon, O., Harel, N.: Deep cerebellar nuclei segmentation via semi-supervised deep context-aware learning from 7T diffusion MRI. IEEE Access **8**, 101550–101568 (2020)
16. Li, B., Niessen, W.J., Klein, S., de Groot, M., Ikram, M.A., Vernooij, M.W., Bron, E.E.: A hybrid deep learning framework for integrated segmentation and registration: evaluation on longitudinal white matter tract changes. In: Medical Image Computing and Computer Assisted Intervention, pp. 645–653. Springer (2019)
17. Li, B., de Groot, M., Steketee, R.M.E., Meijboom, R., Smits, M., Vernooij, M.W., Ikram, M.A., Liu, J., Niessen, W.J., Bron, E.E.: Neuro4Neuro: a neural network approach for neural tract segmentation using large-scale population-based diffusion imaging. Neuroimage **218** (2020)
18. Moeller, S., Yacoub, E., Olman, C.A., Auerbach, E., Strupp, J., Harel, N., Uğurbil, K.: Multiband multislice GE-EPI at 7 tesla, with 16-fold acceleration using partial parallel imaging with application to high spatial and temporal whole-brain fMRI. Magn. Reson. Med. **63**(5), 1144–1153 (2010)
19. Mu, Y., Li, Q., Zhang, Y.: White matter segmentation algorithm for DTI images based on super-pixel full convolutional network. J. Med. Syst. **43**(9), 303 (2019)

20. Nazari-Farsani, S., Nyman, M., Karjalainen, T., Bucci, M., Isojärvi, J., Nummenmaa, L.: Automated segmentation of acute stroke lesions using a data-driven anomaly detection on diffusion weighted MRI. J. Neurosci. Methods **333** (2020)
21. Nelkenbaum, I., Tsarfaty, G., Kiryati, N., Konen, E., Mayer, A.: Automatic segmentation of white matter tracts using multiple brain MRI sequences. In: International Symposium on Biomedical Imaging, pp. 368–371. IEEE (2020)
22. Palombo, M., Ianus, A., Guerreri, M., Nunes, D., Alexander, D.C., Shemesh, N., Zhang, H.: SANDI: a compartment-based model for non-invasive apparent soma and neurite imaging by diffusion MRI. Neuroimage **215** (2020)
23. Pomiecko, K., Sestili, C., Fissell, K., Pathak, S., Okonkwo, D., Schneider, W.: 3D Convolutional neural network segmentation of white matter tract masks from MR diffusion anisotropy maps. In: 16th International Symposium on Biomedical Imaging, pp. 1–5. IEEE (2019)
24. Setsompop, K., Gagoski, B.A., Polimeni, J.R., Witzel, T., van Wedeen, J., Wald, L.L.: Blipped-controlled aliasing in parallel imaging for simultaneous multislice echo planar imaging with reduced g-factor penalty. Magn. Reson. Med. **67**(5), 1210–1224 (2012)
25. Shapey, J., Vos, S.B., Vercauteren, T., Bradford, R., Saeed, S.R., Bisdas, S., Ourselin, S.: Clinical applications for diffusion MRI and tractography of cranial nerves within the posterior fossa: a systematic review. Front. Neurosci. **13**, 23 (2019)
26. Shehata, M., Khalifa, F., Soliman, A., Ghazal, M., Taher, F., El-Ghar, M.A., Dwyer, A.C., Gimel'farb, G., Keynton, R.S., El-Baz, A.: Computer-aided diagnostic system for early detection of acute renal transplant rejection using diffusion-weighted MRI. IEEE Trans. Biomed. Eng. **66**(2), 539–552 (2019)
27. Sotiropoulos, S.N., Moeller, S., Jbabdi, S., Xu, J., Andersson, J.L., Auerbach, E.J., Yacoub, E., Feinberg, D., Setsompop, K., Wald, L.L., Behrens, T.E.J., Ugurbil, K., Lenglet, C.: Effects of image reconstruction on fiber orientation mapping from multichannel diffusion MRI: reducing the noise floor using SENSE. Magn. Reson. Med. **70**(6), 1682–1689 (2013)
28. Trebeschi, S., van Griethuysen, J.J.M., Lambregts, D.M.J., Lahaye, M.J., Parmar, C., Bakers, F.C.H., Peters, N.H.G.M., Beets-Tan, R.G.H., Aerts, H.J.W.L.: Deep learning for fully-automated localization and segmentation of rectal cancer on multiparametric MR. Sci. Rep. **7**(1), 5301 (2017)
29. van Essen, D.C., Ugurbil, K., Auerbach, E., Barch, D., Behrens, T.E.J., Bucholz, R., Chang, A., Chen, L., Corbetta, M., Curtiss, S.W., Della Penna, S., Feinberg, D., Glasser, M.F., Harel, N., Heath, A.C., Larson-Prior, L., Marcus, D., Michalareas, G., Moeller, S., Oostenveld, R., Petersen, S.E., Prior, F., Schlaggar, B.L., Smith, S.M., Snyder, A.Z., Xu, J., Yacoub, E.: The Human Connectome Project: a data acquisition perspective. Neuroimage **62**(4), 2222–2231 (2012)
30. Wang, Y., Zhao, Y., Guo, Z., Qi, M., Fan, Y., Meng, H.: Diffusion tensor image segmentation based on multi-atlas active shape model. Multimedia Tools Appl. **78**(24), 34231–34246 (2019)
31. Wang, J., Cheng, H., Newman, S.D.: Sparse representation of DWI images for fully automated brain tissue segmentation. Journal of neuroscience methods **343**, (2020)
32. Wasserthal, J., Neher, P., Maier-Hein, K.H.: TractSeg - fast and accurate white matter tract segmentation. Neuroimage **183**, 239–253 (2018)
33. Wasserthal, J., Neher, P.F., Hirjak, D., Maier-Hein, K.H.: Combined tract segmentation and orientation mapping for bundle-specific tractography. Med. Image Anal. **58** (2019)
34. Xu, J., Moeller, S., Strupp, J., Auerbach, E., Feinberg, D.A., Ugurbil, K., Yacoub, E.: Highly accelerated whole brain imaging using aligned-blipped-controlled-aliasing multiband EPI. In: Proceedings of the 20th Annual Meeting of ISMRM (2012)
35. Yap, P.T., Zhang, Y., Shen, D.: Brain tissue segmentation based on diffusion MRI using $\ell 0$ sparse-group representation classification. In: Medical Image Computing and Computer-Assisted Intervention, pp. 132–139. Springer (2015)
36. Yendiki, A., Panneck, P., Srinivasan, P., Stevens, A., Zöllei, L., Augustinack, J., Wang, R., Salat, D., Ehrlich, S., Behrens, T., Jbabdi, S., Gollub, R., Fischl, B.: Automated probabilistic reconstruction of white-matter pathways in health and disease using an atlas of the underlying anatomy. Front. Neuroinformatics **5**, 23 (2011)

37. Zhang, F., Hoffmann, N., Karayumak, S.C., Rathi, Y., Golby, A.J., O'Donnell, L.J.: Deep white matter analysis: fast, consistent tractography segmentation across populations and dMRI acquisitions. In: Medical Image Computing and Computer Assisted Intervention, pp. 599–608. Springer (2019)
38. Zhang, F., Cetin Karayumak, S., Hoffmann, N., Rathi, Y., Golby, A.J., O'Donnell, L.J.: Deep white matter analysis (DeepWMA): fast and consistent tractography segmentation. Med. Image Anal. (2020)
39. Zhang, L., Mohamed, A.A., Chai, R., Guo, Y., Zheng, B., Wu, S.: Automated deep learning method for whole-breast segmentation in diffusion-weighted breast MRI. J. Magn. Reson. Imaging **51**(2), 635–643 (2020)
40. Zöllei, L., Jaimes, C.E., Saliba, E., Grant, P.E., Yendiki, A.: Tracts constrained by underlying infant anatomy (traculina): an automated probabilistic tractography tool with anatomical priors for use in the newborn brain. Neuroimage **199**, 1–17 (2019)

Diffusion MRI Fiber Orientation Distribution Function Estimation Using Voxel-Wise Spherical U-Net

Sara Sedlar, Théodore Papadopoulo, Rachid Deriche, and Samuel Deslauriers-Gauthier

Abstract Diffusion Magnetic Resonance Imaging (dMRI) is an imaging technique which enables analysis of the brain tissue at a microscopic scale, particularly the analysis of white matter. Given a high enough angular resolution, a common way to explain the measured signal is via fiber orientation distribution function (fODF). This function describes the orientation and volume fraction of axon bundles within each voxel and is an essential ingredient of tractography. In this work, we have investigated a deep learning approach for the fODF estimation. U-nets enable fast and high resolution inference by combining multi-scale features from contracting and expanding parts of the network. As dMRI signals are most commonly acquired on spheres, we propose a spherical U-net which is adjusted to the properties of the dMRI data, namely its real nature, antipodal symmetry, uniform sampling and axial symmetry of the signals corresponding to individual fibers. We compared our model with another deep learning approach based on a 3D convolutional neural network and a state-of-the-art approach—multi-shell multi-tissue constrained spherical deconvolution, on real data from Human Connectome Project and synthetic data generated using ball and stick model. The methods are compared in terms of mean square error and mean angular error for dMRI signals of different angular resolutions. Provided quantitative analyses show improved performance with our approach even with significantly reduced number of parameters and results obtained on synthetic data indicate its robustness with respect to noise. Qualitative results illustrating the performance of the methods are also presented.

S. Sedlar (✉) · T. Papadopoulo · R. Deriche · S. Deslauriers-Gauthier
Inria, Université Côte d'Azur, Nice, France
e-mail: sara.sedlar@inria.fr

T. Papadopoulo
e-mail: theodore.papadopoulo@inria.fr

R. Deriche
e-mail: rachid.deriche@inria.fr

S. Deslauriers-Gauthier
e-mail: samuel.deslauriers-gauthier@inria.fr

N. Gyori et al. (eds.), *Computational Diffusion MRI*, Mathematics and Visualization,
https://doi.org/10.1007/978-3-030-73018-5_8

1 Introduction

Diffusion MRI is an imaging modality tailored to capture interactions of diffusing water molecules with surrounding micro-structures within examined tissue. As such, it has shown importance in neuroimaging, particularly in the analysis of white matter micro-structures. It opened the possibility to examine properties of axon bundles such as orientation, volume fraction, dispersion, etc. Models proposed to explain the dMRI signals have evolved with the improvement of the acquisition process. Initially, in Diffusion Tensor Imaging [1], axon bundles were described via diffusion tensors [2]. With the increase of dMRI angular resolution, more informative models have been proposed, specifically in voxels containing crossing or kissing fibers, fiber fanning or bending. A number of these models include estimation of probability density functions (PDF) such as Ensemble Average Propagator (EAP) [3, 4] describing average relative spin displacements, diffusion Orientation Distribution Function (dODF) [5, 6] and fiber Orientation Distribution Function (fODF) [7–9]. These voxel-wise quantities opened the possibility of tracking white matter pathways—tractography [10], a process of a great potential in the analysis of brain structural connectivity [11].

The fODF is a spherical PDF that reveals orientations and volumes of the underlying axon bundles. Traditional methods include estimation of a single fiber response function that is deconvolved from the dMRI signal in order to obtain the fODF [7–9].

Recently, a 3D convolutional neural network (3DCNN) directly applied on spherical harmonic (SH) coefficients has been proposed for the fast estimation of fODFs [12]. In [13], for the same problem, residual CNN (ResCNN) and dense neural network (ResDNN) have been investigated. In both works, potential of the models has been demonstrated for significantly downsampled acquisition sampling schemes, what is often a requirement in clinical applications.

U-nets have shown potential in high resolution inference from planar data by combining multi-scale features from contracting and expanding parts of the network [14]. As sampling of dMRI signals is most commonly performed on spheres, all building blocks of U-net need to be adjusted to the properties of spherical signals. Recently, in [15], a spherical U-net has been proposed for the cortical surface parcellation and prediction of attribute maps, with convolutions, pooling and transposed convolutions adjusted to the spherical space. A neural network model, similar to the planar CNN, for the analysis of spherical data—spherical CNN (S^2CNN) has been introduced in [16], where, contrary to [15], in order to avoid computationally expensive interpolations, convolutions of signals and kernels are performed in spectral domain. Similar approach has been developed in [17], where a significant speed up of convolutions has been achieved with zonal kernels.

In this work, we have addressed the problem of the fODF estimation from dMRI data acquired with significantly downsampled acquisition schemes. Exploiting the properties of the *U-net* and S^2CNN, we propose a voxel-wise spherical U-net, that is tailored to the properties of dMRI signals acquired on spheres, namely real nature, uniform distribution of samples, antipodal and axial symmetry of the signals generated by individual fibers.

2 Background and Method

The main operations in U-nets are convolutions, pooling, and transposed convolutions. While the convolution of equidistantly discretized planar signals with kernels is well defined, convolution between S^2 signals and kernels faces two challenges. First of all, the operation analogue to the translation in Euclidean space during convolution is not a rotation in S^2 space, but in the $SO(3)$ manifold. Secondly, the discretization of signals in Euclidean space is usually done in an equidistant manner, what cannot be achieved in S^2 domain. An interpolation must therefore be performed for each step of convolution. These problems are addressed in the work presented in [16] where the spherical CNN—S^2CNN has been introduced. In this framework, to avoid the computationally expensive interpolations, convolutions of S^2 and $SO(3)$ signals and kernels are performed in spectral domain, and as in standard CNNs, activation function is applied in signal domain. Furthermore, to achieve the same effect as pooling, in each new layer, the bandwidth of the input signal is reduced and the support of the kernel is spread. In [17], additional speed up has been achieved by constraining kernels to be zonal. As a consequence, the convolution can be more efficiently performed in S^2. In this work, we propose a spherical U-net with convolutional building blocks from [16, 17] adjusted to the properties of dMRI data.

Given the antipodal symmetry of the dMRI signals, we use only the SH basis of even degree for their representation. A signal $s : (\theta, \phi) \rightarrow \mathbb{R}$ can be written as

$$s(\theta, \phi) = \sum_{l=0}^{L_{max}} \sum_{m=-l}^{m=l} \hat{s}_l^m Y_l^m(\theta, \phi), \quad \text{for } l \in \{0, 2, \ldots, L_{max}\} \tag{1}$$

where θ and ϕ are inclination and azimuth angles, $Y_l^m(\theta, \phi)$ are SH basis of order m and degree l, and $\hat{s}_l^m$ are the corresponding SH coefficients. L_{max} is the signal's bandwidth determined as

$$N \geq (L_{max} + 1)(L_{max} + 2)/2 \tag{2}$$

where N is the number of sampling points. In addition, as dMRI signals are real, we reduce computational complexity by using the real SH basis

$$Y_{lm} = \begin{cases} \sqrt{2}(-1)^m Im[Y_l^{|m|}] & \text{if } m < 0 \\ Y_l^0 & \text{if } m = 0 \ . \\ \sqrt{2}(-1)^m Re[Y_l^m] & \text{if } m > 0 \end{cases} \tag{3}$$

Consequently, spherical kernels are also real and antipodally symmetric. Another important property of dMRI signals is the axial symmetry of the signals coming from individual axon bundles. This motivated us to use kernels that are axially symmetric around z axis—zonal kernels and in this way promote an axon bundle-wise feature extraction. Zonal kernels have been introduced in [17] in order to decrease the computational complexity imposed by performing convolutions of $SO(3)$ signals and kernels as in [16]. Given this, kernel $h : (\theta) \rightarrow \mathbb{R}$ can be represented as a linear combination of zonal harmonics (ZH) as

$$h(\theta) = \sum_{l=0}^{L_{max}} \hat{h}_l Y_l^0(\theta, 0), \quad \text{for } l \in \{0, 2, \ldots, L_{max}\}. \tag{4}$$

This significantly simplifies convolution between signals and kernels, as the resulting signal is no longer in $SO(3)$ manifold, but in S^2 space. In addition, since the number of ZHs necessary to represent such kernels is rather small—$L_{max}/2 + 1$, we used directly ZH coefficients as trainable parameters as it was initially introduced in [17]. Convolution between a signal $s : (\theta, \phi) \rightarrow \mathbb{R}$ and an axially symmetric kernel $h : (\theta) \rightarrow \mathbb{R}$, represented with ZH coefficients $\hat{h}_l$ can be written as

$$c(\theta, \phi) = \sum_{l=0}^{L_{max}} \hat{h}_l \sum_{m=-l}^{m=l} \hat{s}_l^m Y_l^m(\theta, \phi), \quad \text{for } l \in \{0, 2, \ldots, L_{max}\}. \tag{5}$$

As we are dealing with discrete signals, Eq. 1 can be simply written as matrix-vector product as $\boldsymbol{s} = Y\hat{\boldsymbol{s}}$, where Y contains SH basis $Y_l^m(\theta_k, \phi_k)$ in columns, sampled at the angles (θ_k, ϕ_k), $\hat{\boldsymbol{s}}$ are corresponding SH coefficients and $\boldsymbol{s}$ is a discrete spherical signal. Although the discretization of the band-limited planar signals without information loss is well defined with Nyquist–Shannon sampling theorem and their transformation to spectral domain is trivial, discretization of the spherical signals and calculation/estimation of SH coefficients is a challenging task. Sampling theorem for band-limited spherical signals has been introduced in the work of Driscoll and Healy [18], where they have defined an equiangular sampling grid that guarantees information preservation and calculation of SH coefficients. The total number of required samples is $N = 4(L_{max} + 1)^2$. Given a signal sampled on Driscoll–Heally grid, $\boldsymbol{s} : (\theta_k, \phi_k) \rightarrow \mathbb{R}$ and SH basis discretized in the same way in a matrix Y, calculation of SH coefficients can be simply written as $\hat{\boldsymbol{s}} = WY^H\boldsymbol{s}$, where H refers to conjugate transpose and W are quadrature weights necessary to account for the basis orthogonality loss due to discretization. This sampling is quite excessive and given a real world situation where a signal is not completely band-limited and is affected by noise, signal segments around poles that are oversampled would be more accurately represented. Due to this, sampling on a sphere is, in general, application dependent and dMRI signals are usually sampled uniformly over multiple shells in a way that an optimal angular coverage is achieved [19]. As a consequence, some information can be lost and several methods for the estimation of SH coefficients have been proposed [20–22]. In this work, we have used Gram–Schmidt orthonormalization process to estimate the basis Y' for the transformation of S^2 signals into spectral domain, similarly as introduced by Yeo [20]. This is performed in an iterative manner, if y_i and y_i' correspond to $i - th$ columns of Y and Y', respectively, y_i' are determined as follows:

$$y_i' = y_i - \sum_{j=0}^{i-1} \frac{\langle y_i, y_j' \rangle}{\langle y_j', y_j' \rangle} y_j', \quad y_i' = \frac{y_i'}{||y_i'||_2}. \tag{6}$$

where $y_0' = y_0$. In this iterative process, as we start from basis that correspond to lower frequencies, more importance is given to them. This is convenient as we know that aliasing affects higher frequencies. In order to avoid bias due to ordering of the basis, Gram–Schmidt process is repeated multiple times, each time randomly shuffling the order of the basis of the same degree, which are at the end averaged. SH coefficients are simply estimated as $\hat{s} = Y'^T s$.

2.1 Voxel-Wise Spherical U-Net

Figure 1 depicts an illustration of the proposed spherical U-net. Input to the U-net is composed of $n^3 \cdot n_{shells}$ discrete S^2 channels, where n is the size of neighbourhood and n_{shells} is the number of dMRI shells. Output corresponds to the SH coefficients of the estimated fODF. We refer to the results of (transposed) convolution of input S^2 signals and zonal kernels, followed by activation function, as *feature maps*, which are sampled at uniformly distributed points on sphere, generated using *Q-sampling tool* [19]. The network is composed of contracting and expanding parts. Each layer of the contracting part extracts feature maps that are of the same bandwidth as its input (that is used as a part of the input to the parallel layer in the expanding part, black horizontal arrows in Fig. 1) and corresponding feature maps with decreased bandwidth that serve as the input to the following layer of the contracting part (pink arrows oriented down in Fig. 1). The decrease in bandwidth imitates pooling of the planar CNNs. Feature maps of the same bandwidth are computed as convolution of signals/feature maps transformed into spectral domain and kernels, represented with ZH coefficients, as in Eq. 5, followed by Rectified Linear Unit (ReLU) activation function. These feature maps are further transformed into spectral domain with decreased bandwidth and serves as the input to the following layer of the contracting part. Each layer of the expanding part learns up-sampling of the feature maps which serve as the input to the following layer in the expanding chain or as the final inference. In general, as input, it receives the feature maps from the parallel layer of the contracting part, if such layer exists (black horizontal arrows in Fig. 1) and the feature maps estimated by the previous layer of the expanding part (turquoise arrows oriented up in Fig. 1). Transposed convolution in planar CNN simply corresponds to the insertion of zeros between points and convolution with kernels. We have implemented the transposed convolution as follows

- Let N_i be the number of sampling points of the input feature maps of layer i with bandwidth L^i_{max} determined according to inequality 2.
- To up-sample the feature maps from layer i to layer $i-1$ to have bandwidth $L^{(i-1)}_{max}$, we first generate N_{i-1} sampling points using *Q-space sampling tool* [19] and compute the corresponding basis Y' as in Eq. 6.

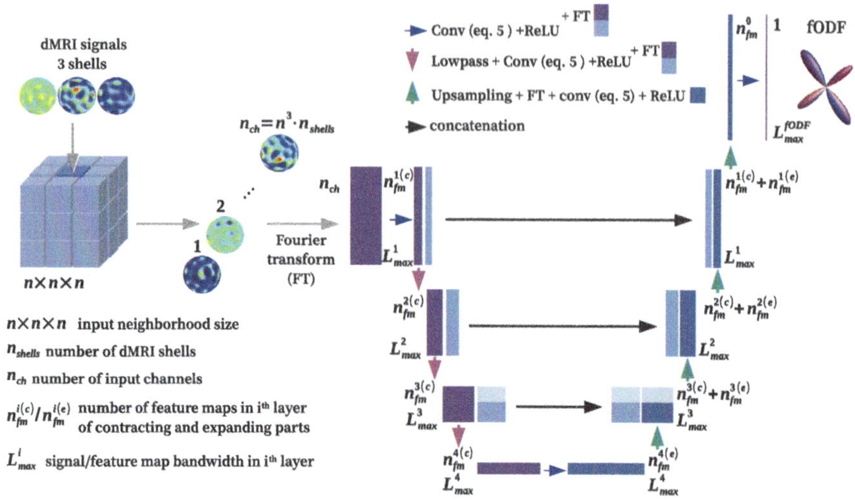

Fig. 1 Illustration of a spherical U-net architecture with corresponding convolutional operations in contracting and expanding parts

- Since Q-space sampling points are generated incrementally, positions of the points of the layer i correspond to the first N_i points of the sampling scheme of the layer $i-1$, so inserted zeros correspond to the last $N_{i-1} - N_i$ points.
- Up-sampled SH coefficients are computed as $\hat{s}_{i-1} = Y'^{T}_{:,1:N_i} s_i$, where $:, 1 : N_i$ refers to the cropping of the matrix Y'^{T} to N_i columns.
- Convolution of the up-sampled signals and kernels is performed as in Eq. 5 followed by an activation function.

3 Dataset

We used in our experiments two types of datasets, real data from Human Connectome Project (HCP) [23] (referred to as *Real dataset*) and synthetic data generated from the same real HCP scans using multi-fiber ball and stick biophysical model [24] following the procedure described in [25]. Real data was acquired on Siemens 3T Skyra system with 100 mT/m gradient, over three shells with b-values of 1000, 2000 and 3000 s/mm^2, each with 90 gradient directions and 18 b $=0$ images at resolution $1.25 \times 1.25 \times 1.25$ mm^3. To generate synthetic data, first, up to three fiber orientations and corresponding volume fractions were estimated per voxel using the *bedpostx* tool from the *FSL* library [26]. These parameters were then used to generate synthetic data using the multi-fiber ball and stick model as in [25] for each shell independently. In the generation process, the free diffusivity coefficients are set to $\{0.68, 0.96, 2.25\} \cdot 10^{-3}$ s/mm^2 for the white matter, gray mat-

Table 1 Sizes of $3DCNNs$ and $S^2U\text{-}nets$ (MB) for N_p sampling points

Model/N_p	20	30	40	60	90	120
$3DCNN$	18.12	18.12	18.12	18.96	20.18	20.18
$S^2U\text{-}net^{1\times1\times1}$	15.65	15.65	15.65	19.30	20.52	20.52
$S^2U\text{-}net_s^{3\times3\times3}$	3.99	3.99	3.99	4.89	5.17	5.17
$S^2U\text{-}net^{3\times3\times3}$	15.80	15.80	15.80	19.42	20.60	20.60

ter and cerebrospinal fluid, respectively [25]. Single-fiber tensor's eigenvalues are set to $\{\lambda_1, \lambda_2, \lambda_3\} = \{1.7, 0.17, 0.17\} \cdot 10^{-3}$ s/mm^2 [25]. To simulate more realistic dMRI data, Rician noise with SNR=18 was added to the synthesized data. In addition, in order to investigate the robustness of the compared methods, one synthetic dataset is generated with the constant diffusion single-fiber tensor eigenvalues (*Synthetic dataset 1*) as in [25] and another one with the eigenvalues taken from the uniform distribution around these values (values taken from the range of $\pm10\%$) (*Synthetic dataset 2*). Experiments are conducted on *Real dataset*, *Synthetic dataset 1* and *Synthetic dataset 2* with downsampled acquisition schemes. To select relevant white matter voxels, we used brain tissue segmentation computed from T1w images using the *FAST* algorithm [27] implemented in the *mrtrix* library [28]. Gold standard fODFs, of SH degree 8, were estimated using the multi-shell multi-tissue constrained spherical deconvolution (MSMT-CSD) approach [9], on signals acquired on full sampling scheme, using *mrtrix* library [28]. In the case of synthetic data, fODFs were estimated on the noise-less data. We used 50 subjects in total, 30 for training, 10 for validation and 10 for testing.

4 Experiments and Implementation Details

In order to evaluate our method on data similar to those used in clinical practice, experiments are performed on data with significantly reduced number of sampling points N_p (20, 30, 40, 60, 90 and 120 in total for the three shells). We compared our method with another deep learning approach—3DCNN [12] and with MSMT-CSD [9]. To investigate importance of neighbourhood information, our model is trained with single voxel multi-shell ($S^2U\text{-}net^{1\times1\times1}$) signals and with $3 \times 3 \times 3$ neighbourhood multi-shell input ($S^2U\text{-}net^{3\times3\times3}$), what is also the case with the 3DCNN model. In addition, to investigate potential of our approach, we trained one model with significantly lower number of trainable parameters—$S^2U\text{-}net_s^{3\times3\times3}$. Sizes of the deep learning networks are given in Table 1.

Both deep learning approaches are implemented using the *tensorflow* library [29]. Models are trained over 100 epochs. In each epoch, 3 dMRI samples are randomly selected from 30 training samples. For both models loss function is defined as mean square error (MSE) between estimated and gold standard fODFs represented in spectral domain. Initial learning rate is 0.001 and after 50 epochs it is reduced to 0.0001. Model weights updates are computed using the Adam optimization algorithm [30].

5 Results and Conclusions

Results are compared quantitatively in terms of MSE and mean angular error (MAE) for single fiber voxels and voxels containing two crossing fibers. To compute peaks of the estimated and gold standard fODFs we used the *mrtrix* library [28] and the threshold of 0.1 of the highest peak is used to eliminate spurious fibers. In Fig. 2 we can see that our models $S^2U\text{-}net^{3\times3\times3}$ achieve significantly lower MSE compared to the models that do not use neighbouring information and slightly, but consistently lower MSE compared to $3DCNN$. In addition, almost equal performance can be achieved with a more compact model—$S^2U\text{-}net_s^{3\times3\times3}$. In Fig. 3 we can notice that for single fiber voxels and real dataset, MAE is almost equal to the one achieved with MSMT-CSD, however the results obtained on synthetic data indicate that our approach is more robust to noise. As depicted in Fig. 3, $S^2U\text{-}net^{3\times3\times3}$ and $S^2U\text{-}net_s^{3\times3\times3}$ achieve lower MAE in voxels with crossing fibers. Qualitative comparison of MSMT-CSD, 3DCNN and $S^2U\text{-}net^{3\times3\times3}$ is provided in Fig. 4 for 60 sampling points. We can notice that MSMT-CSD compared to the 3DCNN and $S^2U\text{-}net^{3\times3\times3}$ is more prone to produce spurious fibers, while these deep learning approaches are more likely to omit some.

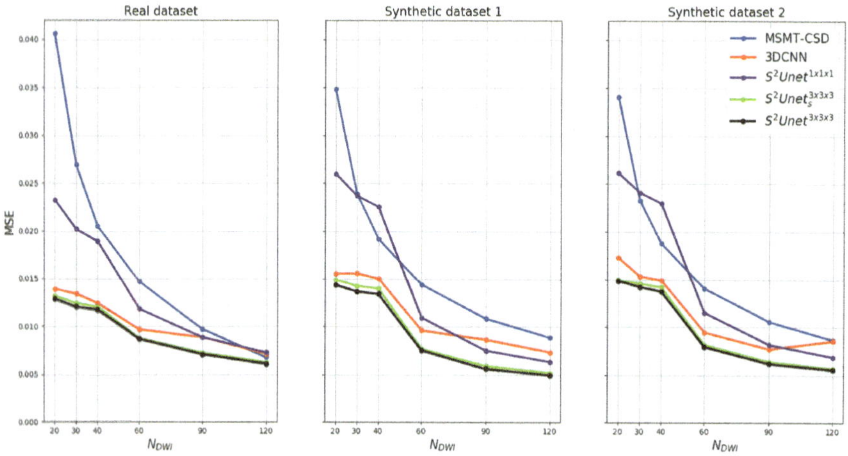

Fig. 2 Comparison of MSE averaged over 10 testing subjects for *real HCP dataset, Synthetic dataset 1* and *Synthetic dataset 2* for different number of sampling points

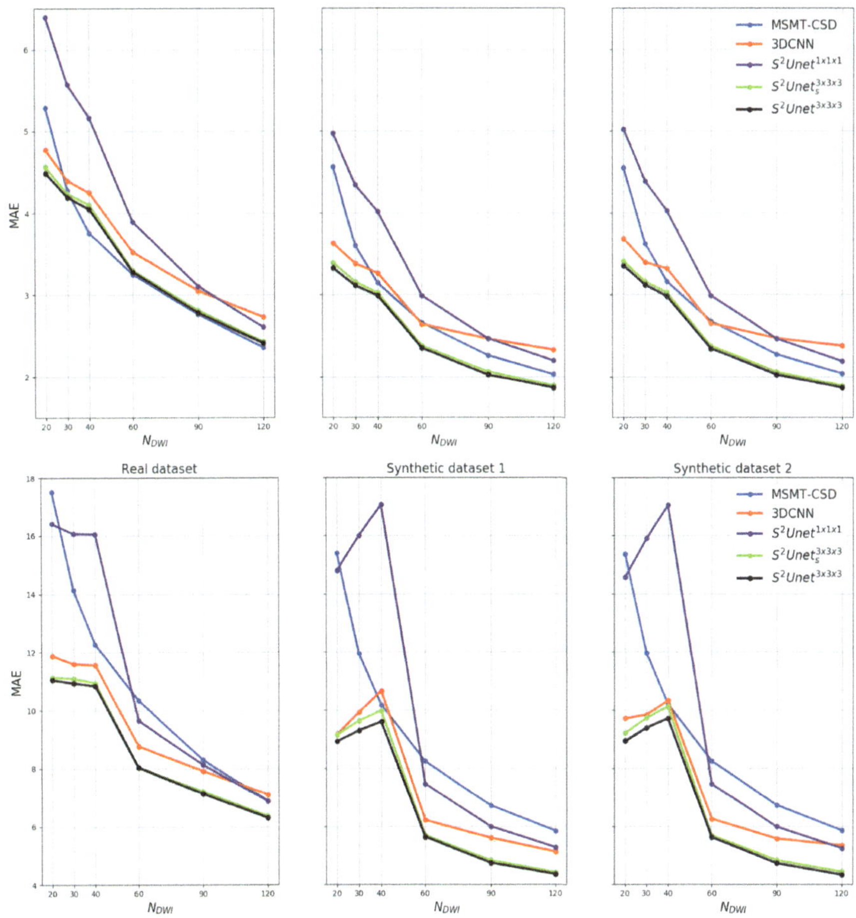

Fig. 3 Comparison of MAE averaged over 10 testing subjects for *real HCP dataset*, *Synthetic dataset 1* and *Synthetic dataset 2* for different number of sampling points for voxels containing single fibers (upper three sub-figures) and voxels containing two crossing fibers (lower three sub-figures)

In this work we have proposed a deep learning method that is adjusted to the properties of dMRI signals, namely real and spherical nature of the signals, antipodal symmetry, random distribution of the sampling points and axial symmetry of signals coming from individual fibers. We have demonstrated that the proposed method is suitable for high resolution inference such as the estimation of the fODFs and can successfully incorporate neighbouring information to boost its performance. Compared with the 3DCNN, the method is capable to produce better fODF estimates even with a significantly reduced number of parameters. Results obtained on synthetic data indicate a better robustness with respect to noise.

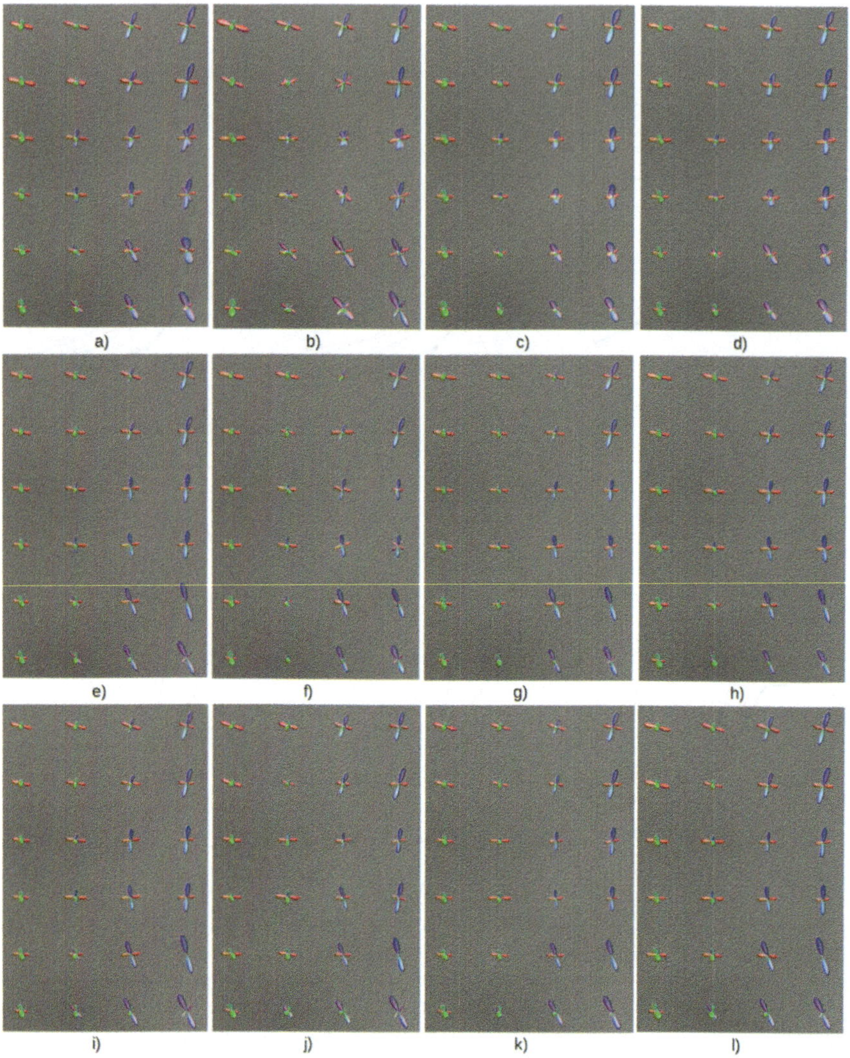

Fig. 4 Illustration of fODF gold standard and estimates obtained using MSMT-CSD, 3DCNN and $S^2U\text{-}net^{3\times3\times3}$ with angular resolution decreased to 60 points in total for the three shells. Sub-figures **a**, **e** and **i** correspond to the gold standard fODFs for *real HCP dataset*, *Synthetic dataset 1* and *Synthetic dataset 2*, respectively. Sub-figures **b**, **f** and **j** correspond to the fODF estimates obtained using MSMT-CSD; sub-figures **c**, **g** and **k** using 3DCNN and sub-figures **d**, **h** and **l** correspond to the fODF estimation with $S^2U\text{-}net^{3\times3\times3}$

Acknowledgements This work was supported by the ERC under the European Union's Horizon 2020 research and innovation program (ERC Advanced Grant agreement No 694665: CoBCoM: Computational Brain Connectivity Mapping). This work has been partly supported by the French government, through the 3IA Côte d'Azur Investments in the Future project managed by the National Research Agency (ANR) with the reference number ANR-19-P3IA-0002. Data were provided [in part] by the Human Connectome Project, WU-Minn Consortium (Principal Investigators: David Van Essen and Kamil Ugurbil; 1U54MH091657) funded by the 16 NIH Institutes and Centers that support the NIH Blueprint for Neuroscience Research; and by the McDonnell Center for Systems Neuroscience at Washington University.

The authors are grateful to Inria Sophia Antipolis—Méditerranée https://wiki.inria.fr/Clusters Sophia/Clusters_Home"Nef" computation cluster for providing resources and support.

References

1. Le Bihan, D., et al.: Diffusion tensor imaging: concepts and applications. J. Magn. Reson. Imaging: Off. J. Int. Soc. Magn. Reson. Med. **13**(4), 534–546 (2001)
2. Basser, P.J., Mattiello, J., LeBihan, D.: MR diffusion tensor spectroscopy and imaging. Biophys. J. **66**(1), 259–267 (1994)
3. Van Wedeen, J., et al.: Mapping complex tissue architecture with diffusion spectrum magnetic resonance imaging. Magn. Reson. Med. **54**(6), 1377–1386 (2005)
4. Merlet, S.L., Deriche, R.: Continuous diffusion signal, EAP and ODF estimation via compressive sensing in diffusion MRI. Med. Image Anal. **17**(5), 556–572 (2013)
5. Tuch, D.S.: Q-ball imaging. Magn. Reson. Med.: Off. J. Int. Soc. Magn. Reson. Med. **52**(6), 1358–1372 (2004)
6. Descoteaux, M., et al.: Regularized, fast, and robust analytical Q-ball imaging. Magn. Reson. Med.: Off. J. Int. Soc. Magn. Reson. Med. **58**(3), 497–510 (2007)
7. Tournier, J.-D., et al.: Direct estimation of the fiber orientation density function from diffusion-weighted MRI data using spherical deconvolution. Neuroimage **23**(3), 1176–1185 (2004)
8. Tournier, J.-D., Calamante, F., Connelly, A.: Robust determination of the fibre orientation distribution in diffusion MRI: non-negativity constrained super-resolved spherical deconvolution. Neuroimage **35**(4), 1459–1472 (2007)
9. Jeurissen, B., et al.: Multi-tissue constrained spherical deconvolution for improved analysis of multi-shell diffusion MRI data. NeuroImage **103**, 411–426 (2014)
10. Basser, P.J., et al.: In vivo fiber tractography using DT-MRI data. Magn. Reson. Med. **44**(4), 625–632 (2000)
11. Jbabdi, S., et al.: Measuring macroscopic brain connections in vivo. Nat. Neurosci. **18**(11), 1546 (2015)
12. Lin, Z., et al.: Fast learning of fiber orientation distribution function for MR tractography using convolutional neural network. Med. Phys. **46**(7), 3101–3116 (2019)
13. Nath, V., et al.: Deep learning estimation of multi-tissue constrained spherical deconvolution with limited single shell DW-MRI (2020). arXiv:2002.08820
14. Ronneberger, O., Fischer, P., Brox, T.: U-net: convolutional networks for biomedical image segmentation. In: International Conference on Medical Image Computing and Computer-Assisted Intervention. Springer, Cham (2015)
15. Zhao, F., et al.: Spherical U-Net on cortical surfaces: methods and applications. In: International Conference on Information Processing in Medical Imaging. Springer, Cham (2019)
16. Cohen, T.S., et al.: Spherical CNNS (2018). arXiv:1801.10130
17. Esteves, C., et al.: Learning so (3) equivariant representations with spherical CNNS. In: Proceedings of the European Conference on Computer Vision (ECCV) (2018)
18. Driscoll, J.R., Healy, D.M.: Computing Fourier transforms and convolutions on the 2-sphere. Adv. Appl. Math. **15**(2), 202–250 (1994)

19. Caruyer, E., Lenglet, C., Sapiro, G., Deriche, R.: Design of multishell sampling schemes with uniform coverage in diffusion MRI. Magn. Reson. Med. (Wiley) **69**(6), 1534–1540 (2013). https://doi.org/10.1002/mrm.24736
20. Yeo, B.T.T.: Computing spherical transform and convolution on the 2-sphere. Manuscript, MIT (2005)
21. Descoteaux, M., et al.: Regularized, fast, and robust analytical Q'ball imaging. Magn. Reson. Med.: Off. J. Int. Soc. Magn. Reson. Med. **58**(3), 497–510 (2007)
22. Rauhut, H., Ward, R.: Sparse recovery for spherical harmonic expansions (2011). arXiv:1102.4097
23. Van Essen, D.C., et al.: The WU-Minn human connectome project: an overview. Neuroimage **80**, 62–79 (2013)
24. Behrens, T.E.J., et al.: Probabilistic diffusion tractography with multiple fibre orientations: What can we gain? Neuroimage **34**(11), 144–155 (2007)
25. Wilkins, B., et al.: Fiber estimation and tractography in diffusion MRI: development of simulated brain images and comparison of multi-fiber analysis methods at clinical b-values. Neuroimage **109**, 341–356 (2015)
26. Smith, S.M., et al.: Advances in functional and structural MR image analysis and implementation as FSL. Neuroimage **23**, S208–S219 (2004)
27. Zhang, Y., Brady, M., Smith, S.: Segmentation of brain MR images through a hidden Markov random field model and the expectation-maximization algorithm. IEEE Trans. Med. Imaging **20**(1), 45–57 (2001)
28. Tournier, J.-D., Smith, R.E., Raffelt, D., Tabbara, R., Dhollander, T., Pietsch, M., Christiaens, D., Jeurissen, B., Yeh, C.-H., Connelly, A.: MRtrix3: a fast, flexible and open software framework for medical image processing and visualisation. NeuroImage **202**, 116–37 (2019)
29. Abadi, M., et al.: Tensorflow: a system for large-scale machine learning. In: 12th USENIX Symposium on Operating Systems Design and Implementation (OSDI 16) (2016)
30. Kingma, D.P., Ba, J.: Adam: a method for stochastic optimization (2014). arXiv:1412.6980

Microstructure Modeling and Representation

Stick Stippling for Joint 3D Visualization of Diffusion MRI Fiber Orientations and Density

Ryan P. Cabeen, David H. Laidlaw, and Arthur W. Toga

Abstract We investigate a stick stippling approach for glyph-based visualization of neural fiber architecture derived from diffusion magnetic resonance imaging. The presence of subvoxel crossing fibers in the brain has prompted the development of advanced modeling techniques; however, there remains a need for improved visualization techniques to more clearly convey their complex structure. While tractography can illustrate large scale anatomy, visualization of diffusion models can provide a more complete picture of local anatomy without the known limitations of tracking. We identify challenges and evaluate techniques for visualizing multi-fiber models and identify benefits of a stick stippling technique relative to existing methods. We conducted experiments to compare these representations and evaluated them with in vivo diffusion MR datasets that vary in voxel resolution and anisotropy. We found that stick rendering as 3D tubes increased legibility of fiber orientation and that encoding fiber density by tube radius reduced clutter and reduced dependence on viewing orientation. Furthermore, we identified techniques to reduce the negative perceptual effects of voxel gridding through a jittering and resampling approach to produce a stippling effect. Looking forward, this approach provides a new way to explore diffusion MRI datasets that may aid in the visual analysis of white matter fiber architecture and microstructure. Our software implementation is available in the Quantitative Imaging Toolkit (QIT).

R. P. Cabeen (✉) · A. W. Toga
Laboratory of Neuro Imaging (LONI), USC Mark and Mary Stevens Neuroimaging and Informatics Institute (INI), Keck School of Medicine of USC, University of Southern California, Los Angeles, CA, USA
e-mail: rcabeen@loni.usc.edu

A. W. Toga
e-mail: toga@loni.usc.edu

D. H. Laidlaw
Department of Computer Science, Brown University, Providence, RI, USA
e-mail: dhl@cs.brown.edu

N. Gyori et al. (eds.), *Computational Diffusion MRI*, Mathematics and Visualization,
https://doi.org/10.1007/978-3-030-73018-5_9

1 Introduction

Diffusion magnetic resonance imaging (dMRI) provides a unique probe of water molecule diffusion that can reveal the complex architecture of neural tissue, including crossing, kissing, and bending fibers [1]. This complexity poses a challenge when visualizing dMRI datasets due to the dense and overlapping nature of fiber bundles in the brain [2, 3]. While tractography provides a powerful tool for understanding large scale neural structures, there is high variability across tracking methods and false positives are difficult to avoid [4] and there are limitations to its anatomical accuracy [5]. To help provide a more complete picture of the underlying anatomy and to understand why tracking algorithms fail, it can be useful to instead visualize the underlying voxel models representing local diffusion properties in each voxel [3, 6, 7].

Glyph rendering is a common technique for visualizing local features of dMRI data [8]; in this process, markers or symbols are used to visually encode the diffusion or microstructure properties at various positions of a volume. Previous work has included a variety of useful glyphs [2], including ellipsoids [6] and superquadratics representing tensors [9, 10]. Glyphs have also been used to represent more complex models, such as orientation distribution functions (ODFs) [11], and spherical harmonic representations of fiber orientation distributions (FODs) [12, 13]. Other work has focused on directly visualizing fiber organization through glyph packing [14], and two-dimensional techniques such as line rendering [15–17] and line stippling [18, 19]. Nevertheless, the visualization of complex fiber configurations remains a major challenge, due to cluttering in high resolution data, dependence on viewing orientation, and the introduction of perceptual artifacts due to the gridding of voxels.

Our primary contribution is the development of a visualization technique that combines several of these existing approaches to better depict crossing fiber configurations with fixels, a term that refers to the combination of fiber orientations and associated fiber density or volume fraction (Fig. 1) [20]. We propose methods that build on related superquadratic and fiber stippling work by using data-modulated 3D stick glyphs and by using a stippling effect to more clearly depict crossing fiber orientation and density. We compare these and other multi-fiber glyphs and show how they are affected by image gridding and viewing orientation. We then evaluate this technique by creating visualizations of in vivo neuroimaging data with varying

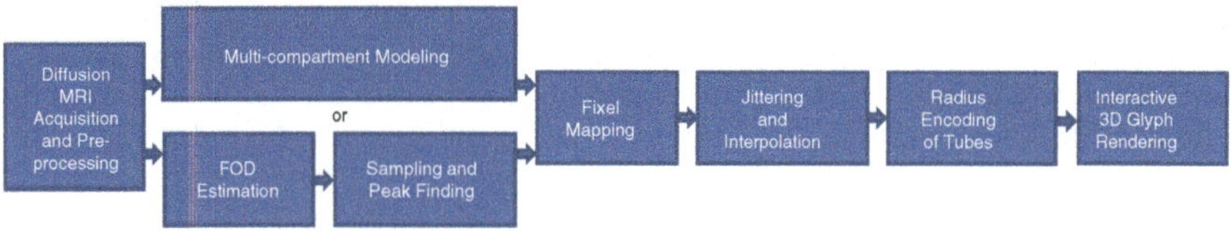

Fig. 1 The proposed computational pipeline for creating stick stippling visualizations from diffusion MRI data. The pipeline applies to both multi-compartment models, such as the ball-and-sticks model, as well as FODs represented by spherical harmonics

voxel resolution and anisotropy to show the relative benefits of this approach for typical data exploration sessions.

2 Methods

2.1 Diffusion Modeling and the Fixel Representation

We focus here on the visualization of multi-fiber models using a fixel representation that can summarize either FODs [21] or multi-compartment models such as the ball-and-sticks model [22]. Each fixel consists of a fiber orientation and its associated fiber density, which are extracted as follows. For FOD modeling, fiber orientations are extracted by sampling the distribution on a four-fold subdivision of an spherical icosahedral mesh. From this, fiber orientations are extracted from the local maxima on the mesh, and fiber density is taken by the peak density. FOD fixels were further processed to remove duplicates at antipodal points and local ridges using hierarchical clustering with a sine angle distance. For multi-compartment modeling, fixels are directly derived from the principal direction and volume fraction of each compartment. While previous work has used fixels primarily for FODs [23], we more generally use fixel density to represent either volume fraction or density, depending on the underlying model.

More formally, the fixels of the ith voxel are given by $M_i = \{(f_{ij}, \mathbf{v}_{ij})\}_{j=1}^{N_i}$, with fixel count N, density f, and orientation v. Our approach also requires interpolation for jittering fixels in a neighborhood around each voxel, and for this, we use a kernel regression framework [24, 25], which estimates the fixels $\hat{M}$ at an arbitrary position p from a neighborhood of voxels $\mathcal{N}$:

$$\hat{M}(p) = \underset{M}{\operatorname{argmin}} \sum_{i \in \mathcal{N}} K(p_i, p) d^2(M_i, M) + \lambda N \quad (1)$$

$$d^2(M, \hat{M}) = \min_{\pi} \sum_{j} f_j \left(1 - (\mathbf{v}_j \cdot \hat{\mathbf{v}}_{\pi(j)})^2\right) \quad (2)$$

$$K(p_i, p) = \exp\left(-\|p_i - p\|^2 / h^2\right) \quad (3)$$

given spatial bandwidth h, and regularization term $\lambda = 0.99$. While previous work has used this with ball-and-sticks modeling, we more generally use this here with fixels obtained from FODs [23] as well. The primary goal here is then to develop improved glyph-based visualization techniques based on this approach, which we describe next.

2.2 Fixel Glyph Visualization

The design of glyph visualizations generally includes a number of technical challenges related to visual perception [26–29], and we identified three major issues specifically related to fixels (Fig. 2). First, the complexity of multi-fiber models often creates cluttered scenes, making it difficult to distinguish important fibers from unimportant ones. Second, the viewing orientation of the scene can reduce the legibility of fixel orientation and density, particularly those parallel to the camera's viewing direction [15]. Third, the discretization of the voxel grid can break gestalt principles of perceptual organization, making some bundles appear more organized than others, even when they only differ in orientation relative to the viewing angle.

We can address these issues by using the 3D shape and layout of fibers to reduce visual ambiguities (Fig. 1). First, we render fixels as tubes (or sticks) to provide a clearer reading of fiber orientation, as shading from the 3D shape can provide depth cues, unlike rendering solid lines. We also color fixels according to the widely used RGB-directional color mapping [30]. Second, we modulate the radius of the tube to reflect the fiber density. The advantage of modulating the radius is that it can be appreciated from any viewing orientation, unlike tube length, which can be illegible when parallel to the viewing direction and can be confused with orientation due to perspective foreshortening. Finally, we render fixels using jittering of 3D glyphs to produce a stippling effect, a process that greatly reduces the effect of gridding by drawing glyphs at uniformly sampled points within each voxel. This approach builds on related work on stippling and superquadratics to visualize fixels. Furthermore. In the following section, we conduct experiments to make comparisons with previously used multi-fiber visualizations and to understand how effectively they address the challenges posed in fixel visualization.

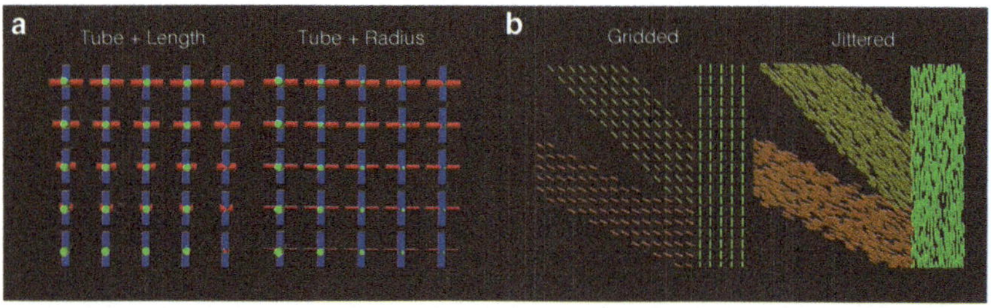

Fig. 2 Perceptual challenges for fixel visualization shown in synthetic data. Panel A shows challenges related to viewing orientation, where length and radius encoding of fiber density are compared. Red fiber density decreases from top to bottom, and green fiber density decreases left to right. This illustrates how the density of fibers parallel to the camera view (green) is only legible with radius encoding. Panel B shows challenges related to voxel discretization, where gridded and jittered glyphs are compared. Three bundles were created with equal thickness but with different orientations. This illustrates how gridding can make axis-aligned bundles look more coherent, while jittering can avoid these issues to give an equally coherent appearance to all bundles

3 Experiments and Results

We conducted experiments with three datasets to understand the strengths and limitations of various fixel rendering techniques with regard to how they represent complex anatomy. These datasets were chosen to include voxels of varied resolution and anisotropy in order to show performance across anatomical scales and imaging parameters. In addition to techniques described in the previous section, we compared the results to solid line rendering, tube length encoding, and spherical harmonic FOD rendering when applicable. Our experiments were implemented using the Quantitative Imaging Toolkit (QIT) [31].

3.1 Clinical Data Experiment

Our first experiment examined data from a typical clinical acquisition. Diffusion-weighted MR image volumes were acquired from a healthy 34 year old volunteer, conducted on a GE 1.5T scanner with voxel size of 2 mm^3, matrix size 128 $\times$ 128, and 72 contiguous axial slices. A total of 71 volumes were acquired, with seven T_2-weighted volumes (b-value 0 s/mm^2) and 64 diffusion-weighted volumes (b-value 1000 s/mm^2) and distinct gradient encoding directions. Ball-and-sticks were estimated using MCMC optimization procedure of Behrens et al. [22].

Eight visualizations were generated to depict a coronal slice showing complex interactions among the corpus callosum (CC), and superior longitudinal fasciculus (SLF), cingulum bundle, and corona radiata (Fig. 3). This figure compares grid-

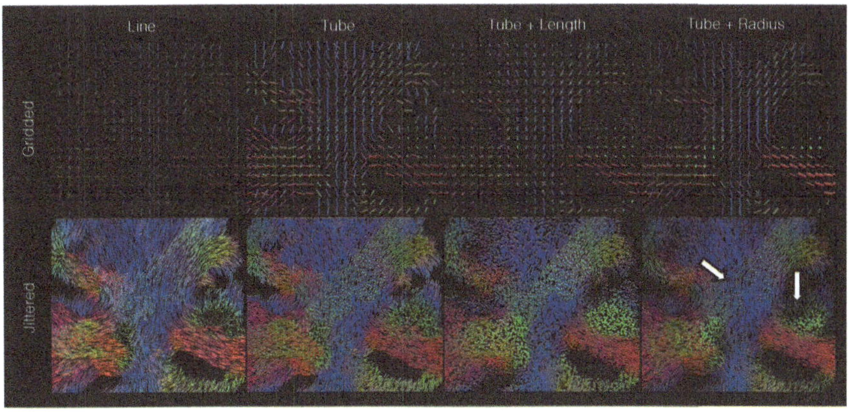

Fig. 3 Eight fixel visualizations with clinically feasible data. A comparison of the first and second columns shows how 3D tube rendering more clearly depicts fiber orientation. A comparison of the third and fourth columns show how radius encoding can illustrate variation in fiber density, e.g. the variation in SLF fibers shown in green (white arrows)

ded and jittered glyph placement, line and 3D tube drawing, and length and radius encodings of fiber density. We found that jittering effectively removed evidence of the underlying voxel grid, and visualization representing fiber density had less clutter. We found that length encoding of volume fraction was only effective in fibers orthogonal to the viewing direction, however, the density of fibers parallel to the view, such as the SLF and cingulum, were only appreciable with radius encoding.

3.2 HCP Experiment

Our second experiment examined state-of-the-art dataset from the Human Connectome Project [32, 33],[1] specifically the single subject dataset with identifier 100307. Diffusion-weighted MR imaging was conducted on a Siemens 3T scanner with voxel size of 1.25 mm^3, matrix size 145 $\times$ 145, and 174 slices. A total of 288 volumes were acquired, with 18 T_2-weighted volumes (b-value 0 s/mm^2) and the remainder distributed among roughly three shells (b-values 1000, 2000, 3000 s/mm^2) with distinct gradient encoding directions. FODs were estimated using the compartmental modeling approach of Tran and Shi [34] using 16 order spherical harmonics. The FODs were discretized using a subdivided icosahedron, and fixels were created from the peak directions and fiber density.

Four visualizations were generated to depict a coronal slice that includes a triple crossing of the corticospinal tract, CC, and SLF (Fig. 4). These visualizations included line rendering, spherical FOD rendering, gridding tubes, and jittered tubes. We found that the high resolution of the data make gridded visualizations more difficult to read, due to the smaller size of fixel glyphs. The tube renderings were more legible than either line or FOD rendering, but jittered tubes provided the greatest legibility. In particular, the jittering more clearly showed triple crossings, density reductions at the white-gray matter interface and the variation in density of the SLF.

3.3 RESOLVE Experiment

Our third experiment examined an advanced presurgical acquisition with highly anisotropic voxels. Diffusion-weighted MR imaging was conducted prior to surgery to treat temporal lobe epilepsy. The RESOLVE DWI sequence was run on a Siemens Prisma 3T scanner with a high resolution coronal in-plane resolution of 0.6 mm $\times$ 0.6 mm with matrix size 272 $\times$ 360. 19 contiguous slices were acquired with thickness 2.1 mm. A total of 56 volumes were acquired, with six T_2-weighted volumes (b-value 0 s/mm^2) and 50 diffusion-weighted volumes (b-values 800 and 2000 s/mm^2) and distinct gradient encoding directions. Ball-and-sticks were esti-

[1] http://www.humanconnectome.org.

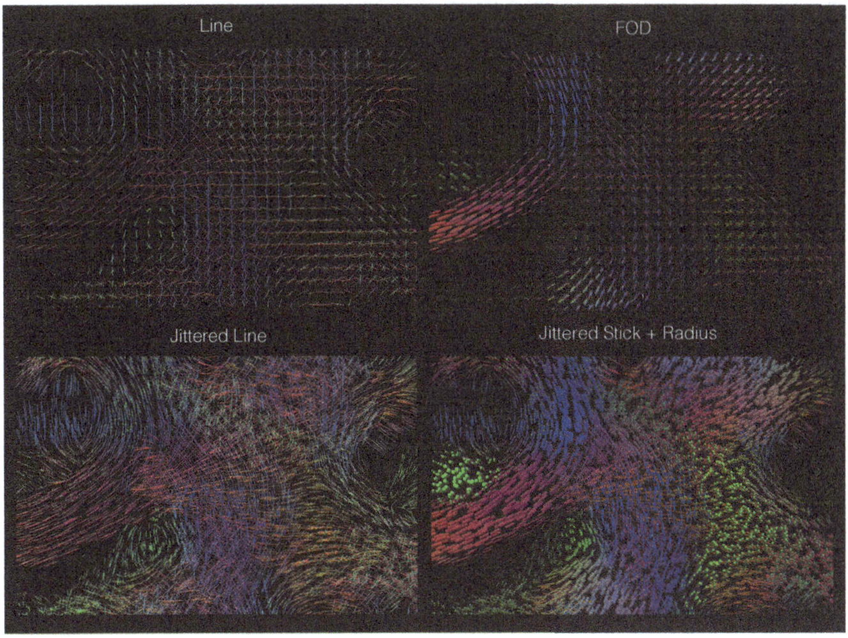

Fig. 4 Visualizations of human connectome project data. The top row shows how high spatial resolution can impact visibility in line and ODF rendering. The bottom right shows how jittering can avoid gridding artifacts and improve visibility

mated using MCMC optimization procedure of Behrens et al. [22] with the continuous exponential model for multi-shell experiments.

Four visualizations were generated to compare fixel visualization of the coronal high in-plane resolution data with an axial slice showing voxel anisotropy (Fig. 5). Each slice was rendered using both gridded and jittered tubes with radius encoding. The coronal slice was found to have fine detail in crossing fibers; however, the axial slice showed perceptual irregularities due to the thick slices. Specifically, the anisotropic axial slice showed greater coherence of SLF fibers than CC fibers, due to the greater separation of endpoints of fixels representing the CC. In contrast, the jittered glyphs showed no discrepancy between coronal and axial views, reducing the effects of both voxel gridding and anisotropy.

4 Discussion and Conclusions

Our results indicate that when visualizing complex fiber architecture, there are limitations caused by clutter, viewing orientation, and voxel gridding; however, a comparison of glyph visualization techniques showed that these can be mitigated. First,

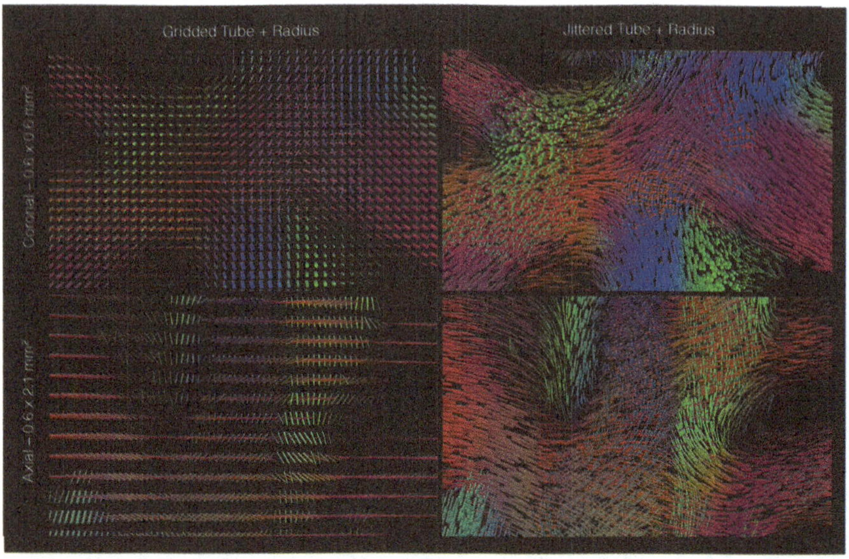

Fig. 5 Visualizations of RESOLVE DWI data. The top left shows how high in-plane resolution can greatly reduce the size and visibility of glyphs. The bottom left shows how this introduces large spacing between glyphs when voxels are anisotropic. By contrast, jittered glyphs were found to greatly reduce these negative effects

rendering fixels as 3D tube-shaped sticks helped to better indicate orientation through shading and depth cues, and modulating the radius by fiber density helped to reduce clutter and ambiguities based on viewing direction. Second, a stippling effect, produced by glyph jittering and resampling, was found to reduce gridding artifacts, which is perhaps similar to anti-aliasing through randomized glyph placement. Nevertheless, a number of open questions remain. Many applications require the visualization of other per-compartment parameters, e.g. statistical metrics, axon diameter, or myelination, and this requires further work to understand how to effectively encode such parameters with glyphs. There may also be cases where fixels are not a sufficient representation, e.g. when there is fanning or dispersion that may be more apparent by visualizing a complete FOD, and a stippling approach for FOD visualization may also be advantageous, given some mechanism for suppressing peaks along the viewing axis. There also remain important open problems related to visualizing glyphs in conjunction with tractography [35] and surfaces [36]. More broadly, an important open challenge is the translation of diffusion MRI visualization tools into use by clinical and basic neuroscience researchers, which may require further investigations with user studies, more intuitive software interfaces, and collaborative research programs. Looking forward, the described stick stippling approach may help to interpret imaging results and aid the development of advanced visualization systems for both human neuroimaging and preclinical imaging [37]. Our software implementation is available online as part of the Quantitative Imaging Toolkit (QIT).[2,3]

Acknowledgements This work was supported by the National Institute of Biomedical Imaging and Bioengineering (NIBIB) under NIH grant number P41EB015922. Data were provided in part by the Human Connectome Project, WU-Minn Consortium (Principal Investigators: David Van Essen and Kamil Ugurbil; 1U54MH091657) funded by the 16 NIH Institutes and Centers that support the NIH Blueprint for Neuroscience Research; and by the McDonnell Center for Systems Neuroscience at Washington University.

References

1. Tuch, D.S., Reese, T.G., Wiegell, M.R., Wedeen, V.J.: Diffusion MRI of complex neural architecture. Neuron 40(5) (2003) 885–895
2. Margulies, D.S., Böttger, J., Watanabe, A., Gorgolewski, K.J.: Visualizing the human connectome. NeuroImage 80 (2013) 445–461
3. Tax, C.M., Chamberland, M., van Stralen, M., Viergever, M.A., Whittingstall, K., Fortin, D., Descoteaux, M., Leemans, A.: Seeing more by showing less: Orientation-dependent transparency rendering for fiber tractography visualization. PLoS ONE **10**(10) (2015)
4. Maier-Hein, K., Neher, P., Houde, J.C., Cote, M.A., Garyfallidis, E., Zhong, J., Chamberland, M., Yeh, F.C., Lin, Y.C., Ji, Q., et al.: Tractography-based connectomes are dominated by false-positive connections (2016). 084137
5. Schilling, K.G., Nath, V., Hansen, C., Parvathaneni, P., Blaber, J., Gao, Y., Neher, P., Aydogan, D.B., Shi, Y., Ocampo-Pineda, M., et al.: Limits to anatomical accuracy of diffusion tractography using modern approaches. NeuroImage 185 (2019) 1–11
6. Laidlaw, D.H., Ahrens, E.T., Kremers, D., Avalos, M.J., Jacobs, R.E., Readhead, C.: Visualizing diffusion tensor images of the mouse spinal cord. In: Visualization'98. Proceedings, pp. 127–134. IEEE (1998)
7. Zhang, S., Laidlaw, D.H., Kindlmann, G.: Diffusion tensor MRI visualization. In: The Visualization Handbook, pp. 327–340 (2004)
8. Vilanova, A., Zhang, S., Kindlmann, G., Laidlaw, D.: An introduction to visualization of diffusion tensor imaging and its applications. In: Visualization and Processing of Tensor Fields, pp. 121–153. Springer (2006)
9. Kindlmann, G.: Superquadric tensor glyphs. In: IEEE Conference on Visualization, pp. 147–154. Eurographics Association (2004)
10. Ennis, D.B., Kindlman, G., Rodriguez, I., Helm, P.A., McVeigh, E.R.: Visualization of tensor fields using superquadric glyphs. Magnetic Resonance in Medicine 53(1) (2005) 169–176
11. Shattuck, D.W., Chiang, M.C., Barysheva, M., McMahon, K.L., De Zubicaray, G.I., Meredith, M., Wright, M.J., Toga, A.W., Thompson, P.M.: Visualization tools for high angular resolution diffusion imaging. In: MICCA I. (2008) 298–305
12. Peeters, T.H., Prckovska, V., van Almsick, M., Vilanova, A., ter Haar Romeny, B.M.: Fast and sleek glyph rendering for interactive hardi data exploration. In: PacificVis 2009, pp. 153–160. IEEE (2009)
13. Vaillancourt, O., Chamberland, M., Houde, J.C., Descoteaux, M.: Visualization of diffusion propagator and multiple parameter diffusion signal. In: Visualization and Processing of Higher Order Descriptors for Multi-Valued Data, pp. 191–212. Springer (2015)

[2] https://cabeen.io/qitwiki.

[3] https://resource.loni.usc.edu/resources/downloads/.

14. Kindlmann, G., Westin, C.F.: Diffusion tensor visualization with glyph packing. IEEE Trans. Vis. Comput. Graph. **12**(5) (2006)
15. Schurade, R., Hlawitschka, M., Hamann, B., Scheuermann, G., Knösche, T.R., Anwander, A.: Visualizing white matter fiber tracts with optimally fitted curved dissection surfaces. In: Proceedings of the 2nd Eurographics Conference on Visual Computing for Biology and Medicine, pp. 41–48 (2010)
16. Vaillancourt, O., Boré, A., Girard, G., Descoteaux, M.: A fiber navigator for neurosurgical planning (neuroplanningnavigator). In: IEEE Visualization, vol. 231 (2010)
17. Höller, M., Otto, K.M., Klose, U., Groeschel, S., Ehricke, H.H.: Fiber visualization with LIC maps using multidirectional anisotropic glyph samples. Journal of Biomedical Imaging **2014**, 9 (2014)
18. Goldau, M., Wiebel, A., Gorbach, N.S., Melzer, C., Hlawitschka, M., Scheuermann, G., Tittgemeyer, M.: Fiber stippling: an illustrative rendering for probabilistic diffusion tractography. In: IEEE BioVis, pp. 23–30. IEEE (2011)
19. Goldau, M., Reichenbach, A., Hlawitschka, M.: Visualizing crossing probabilistic tracts. In: IEEE Scientific Visualization, pp. 147–148. IEEE (2015)
20. Tournier, J., Calamante, F., Connelly, A., et al.: MRtrix: diffusion tractography in crossing fiber regions. International Journal of Imaging Systems and Technology 22(1) (2012) 53–66
21. Tournier, J.D., Yeh, C.H., Calamante, F., Cho, K.H., Connelly, A., Lin, C.P.: Resolving crossing fibres using constrained spherical deconvolution: validation using diffusion-weighted imaging phantom data. NeuroImage 42(2) (2008) 617–625
22. Behrens, T.E., Berg, H.J., Jbabdi, S., Rushworth, M., Woolrich, M.: Probabilistic diffusion tractography with multiple fibre orientations: What can we gain? Neuroimage **34**(1), 144–155 (2007)
23. Raffelt, D.A., Tournier, J.D., Smith, R.E., Vaughan, D.N., Jackson, G., Ridgway, G.R., Connelly, A.: Investigating white matter fibre density and morphology using fixel-based analysis. NeuroImage 144 (2017) 58–73
24. Cabeen, R.P., Bastin, M.E., Laidlaw, D.H.: Kernel regression estimation of fiber orientation mixtures in diffusion MRI. NeuroImage 127 (2016) 158–172
25. Cabeen, R.P., Bastin, M.E., Laidlaw, D.H.: Estimating constrained multi-fiber diffusion mr volumes by orientation clustering. MICCA **I**, 82–89 (2013)
26. Borgo, R., Kehrer, J., Chung, D.H., Maguire, E., Laramee, R.S., Hauser, H., Ward, M., Chen, M.: Glyph-based visualization: foundations, design guidelines, techniques and applications. In: Eurographics (STARs), pp. 39–63 (2013)
27. Wagemans, J., Elder, J.H., Kubovy, M., Palmer, S.E., Peterson, M.A., Singh, M., von der Heydt, R.: A century of gestalt psychology in visual perception: I. Perceptual grouping and figure–ground organization. Psychol. Bull. **138**(6), 1172 (2012)
28. Senay, H., Ignatius, E.: Rules and principles of scientific data visualization. Institute for Information Science and Technology, Department of Electrical Engineering and Computer Science, School of Engineering and Applied Science, George Washington University (1990)
29. Fuchs, J., Isenberg, P., Bezerianos, A., Keim, D.: A systematic review of experimental studies on data glyphs. IEEE Trans. Vis. Comput. Graph. (2016)
30. Pajevic, S., Pierpaoli, C.: Color schemes to represent the orientation of anisotropic tissues from diffusion tensor data: application to white matter fiber tract mapping in the human brain. Magnetic Resonance in Medicine: An Official Journal of the International Society for Magnetic Resonance in Medicine **42**(3), 526–540 (1999)
31. Cabeen, R., Laidlaw, D., Toga, A.: Quantitative Imaging Toolkit: software for interactive 3D visualization, data exploration, and computational analysis of neuroimaging datasets. In: In Proceedings of the International Society for Magnetic Resonance in Medicine (ISMRM). Paris, France, p. 2854 (2018)
32. Van Essen, D.C., Smith, S.M., Barch, D.M., Behrens, T.E., Yacoub, E., Ugurbil, K., Consortium, W.M.H., et al.: The WU-MINN human connectome project: an overview. Neuroimage **80**, 62–79 (2013)

33. Sotiropoulos, S.N., Jbabdi, S., Xu, J., Andersson, J.L., Moeller, S., Auerbach, E.J., Glasser, M.F., Hernandez, M., Sapiro, G., Jenkinson, M., et al.: Advances in diffusion mri acquisition and processing in the human connectome project. Neuroimage **80**, 125–143 (2013)
34. Tran, G., Shi, Y.: Fiber orientation and compartment parameter estimation from multi-shell diffusion imaging. IEEE TMI 34(11) (2015) 2320–2332
35. Vos, S.B., Viergever, M.A., Leemans, A.: Multi-fiber tractography visualizations for diffusion MRI data. PLoS ONE **8**(11) (2013)
36. Goldau, M., Wiebel, A., Hlawitschka, M., Scheuermann, G., Tittgemeyer, M.: Visualizing DTI parameters on boundary surfaces of white matter fiber bundles. In: IASTED, pp. 53–61 (2011)
37. Cabeen, R.P., Immonen, R., Harris, N.G., Gröhn, O., Smith, G., Manninen, E., Garner, R., Duncan, D., Pitkänen, A., Toga, A.W.: A computational diffusion MRI framework for biomarker discovery in a rodent model of post-traumatic epileptogenesis. In: IEEE 17th International Symposium on Biomedical Imaging (ISBI). IEEE 2020, pp. 1013–1017 (2020)

Q-Space Quantitative Diffusion MRI Measures Using a Stretched-Exponential Representation

Tomasz Pieciak, Maryam Afzali, Fabian Bogusz, Santiago Aja-Fernández, and Derek K. Jones

Abstract Diffusion magnetic resonance imaging (dMRI) is a relatively modern technique used to study tissue microstructure in a non-invasive way. Non-Gaussian diffusion representation is related to the restricted diffusion and can provide information about the underlying tissue properties. In this paper, we analytically derive nth order statistics of the signal considering a stretched-exponential representation of the diffusion. Then, we retrieve the Q-space quantitative measures such as the Return-To-the-Origin Probability (RTOP), Q-space mean square displacement (QMSD), Q-space mean fourth-order displacement (QMFD). The stretched-exponential representation enables the handling of the diffusion contributions from a higher b-value regime under a non-Gaussian assumption, which can be useful in diagnosing or prognosis of neurodegenerative diseases in the early stages. Numerical implementation of the method is freely available at https://github.com/TPieciak/Stretched.

1 Introduction

Magnetic resonance imaging (MRI) is a powerful technique in clinical applications for diagnoses or prognoses of several diseases in the central nervous system [8, 19]. Diffusion-weighted MRI (dMRI) modality is sensitive to the random motion of water molecules in the tissue, and it is vastly used in both clinical and basic science to characterize tissue water behavior. In the category of dMRI studies, different imaging techniques can extract the microstructural features of the tissue, such as size, shape, and anisotropy. Many studies have shown a relationship between the changes in the

---

T. Pieciak (✉) · S. Aja-Fernández
LPI, ETSI Telecomunicación, Universidad de Valladolid, Valladolid, Spain
e-mail: pieciak@agh.edu.pl

T. Pieciak · F. Bogusz
AGH University of Science and Technology, Kraków, Poland

M. Afzali · D. K. Jones
School of Psychology, Cardiff University Brain Research Imaging Centre (CUBRIC), Cardiff University, Cardiff, UK

N. Gyori et al. (eds.), *Computational Diffusion MRI*, Mathematics and Visualization,
https://doi.org/10.1007/978-3-030-73018-5_10

diffusion properties of the tissue and the relevant alteration in the underlying tissue microstructure [28].

One of the most important features of dMRI is its sensitivity to the anisotropy in the tissue. Diffusion tensor imaging (DTI) [8] is the most common technique in clinical studies. The spins displacement in DTI is assumed to be Gaussian distributed, and some scalar anisotropy indices such as fractional anisotropy (FA), axial, and radial diffusivity (AD, RD), and mean diffusivity (MD) were defined directly from second-order tensor representation [7, 31]. The most common and simplest assumption in dMRI is the Gaussian assumption [18] on the spin displacement, which results in a mono-exponential decay of a diffusion signal versus the b-value parameter [10]. The Gaussian assumption in DTI is valid when the media is a simple fluid, and we have free diffusion. In more complicated structures such as tissue, there are restrictions in water diffusion, and therefore, the decay is much slower than a mono-exponential decay [9].

The presence of non-mono-exponential decay shows that the diffusion is restricted by the underlying microstructure of the tissue. If the structure's size is similar to the diffusion length-scale, then the diffusion deviates from the Gaussian towards a non-Gaussian displacement regime [15]. Therefore diffusion MRI can be used to probe the microstructural properties of the underlying tissue geometry by methods such as bi-exponential [11], stretched-exponential [9], composite hindered and restricted model of diffusion (CHARMED) [4], AxCaliber [5], ActiveAx [3] or neurite orientation dispersion and density imaging (NODDI) [34]. All these methods can be used to investigate tissue geometry, but they are not all equally applicable in all situations. Other methods such as the high angular resolution diffusion imaging (HARDI) [25, 30] and diffusion kurtosis imaging (DKI) [17] were also proposed. To obtain the non-Gaussian property of the signal higher b-values are required [20].

Alternative to model the tissue's underlying properties is the ensemble average propagator (EAP), which represents the probability that the water molecule moves in a specific direction under a certain diffusion time [12, 26]. From the EAP representation, one can retrieve different Q-space quantitative measures such as the Return-To-the-Origin Probability (RTOP), Q-space mean square displacement (QMSD), or Q-space mean fourth-order displacement (QMFD). For instance, the RTOP measure is shown to be a useful index for cellularity and diffusion restrictions [6], while QMSD and QMFD are sensitive to contributions from slow or restricted diffusion [22].

Different methods have been proposed so far to estimate the EAP and EAP-related features such as the multiple q-shell diffusion propagator Imaging (mq-DPI) [12], Hybrid Diffusion Imaging (HYDI) [33], Mean Apparent Propagator (MAP-MRI) [26], Radial Basis Functions (RBFs) [22], Laplacian-regularized MAP-MRI (MAPL) [13], and Generalized Diffusion Spectrum MRI (GDSI) [29]. These techniques are typically computationally intensive or require a huge amount of densely sampled Cartesian or multiple-shell data to estimate the EAP and its related features correctly. Recently, a single-shell technique that can estimate micro-structure diffusion scalar measures directly from the data has been proposed [2, 27]. This approach, although it enables one to estimate the measures rapidly and directly from the data, assumes a Gaussian profile of the signal, thus it might be problematic to recover higher b-value contributions to the signal.

In this paper, we analytically derive the nth order statistics of the signal considering a stretched-exponential decay to represent the Gaussian and non-Gaussian parts of the signal. In practice, when no information about the number of compartments is provided, the stretched exponential is a good choice [9]. Given the general formulation in the Q-space domain, we obtain closed-form formulas to retrieve basic indexes such as the RTOP, QMSD, or QMFD directly from the data in a manner analogous to direct techniques [2, 27, 33]. However, the proposal is no longer limited by a Gaussian assumption and can be used to retrieve the diffusion contributions under the higher b-values regime.

2 Theory

In this section, we start with the definition of the EAP and diffusion MR signal representation using a stretched-exponential function, and then we use this representation to extract the Q-space scalar measures such as the RTOP, QMSD, and QMFD.

2.1 Diffusion MR Signal Representation

The ensemble average propagator (EAP) is a three-dimensional probability density function that represents the average displacement of spins during the diffusion time. The EAP, $P(\mathbf{R})$, is related to the diffusion MR signal attenuation $E(\mathbf{q})$ via the Fourier transform [12, 26, 33]

$$P(\mathbf{R}) = \int_{\mathbb{R}^3} E(\mathbf{q}) \exp(-j2\pi \mathbf{q}^T \mathbf{R}) d^3\mathbf{q}, \quad j^2 = -1, \tag{1}$$

with $E(\mathbf{q}) = S(\mathbf{q})/S(0)$ being the normalized diffusion signal, $S(\mathbf{q})$ is the diffusion signal acquired at wave vector $\mathbf{q}$, $S(0)$ is the baseline measured without a diffusion sensitization.

The signal in Eq. (1) can be represented by a mono-exponential decay $E(\mathbf{g}) = \exp\left(-b\mathbf{g}^T \mathbf{D}\mathbf{g}\right)$ with $\mathbf{g}$ being a normalized vector $\mathbf{g} = \mathbf{q}/\|\mathbf{q}\|$ and $\mathbf{D}$ is a covariance matrix of a Gaussian EAP or a more general Kohlrausch–Williams–Watts function so-called a stretched-exponential representation given by [9, 21, 32]

$$E(\mathbf{g}) = \exp\left(-(bD(\mathbf{g}))^{\alpha(\mathbf{g})}\right), \quad \alpha(\mathbf{g}) \in (0, 1] \tag{2}$$

with the so-called the b-value $b = 4\pi^2\tau\|\mathbf{q}\|^2$ [s/mm^2] with $\tau = \Delta - \delta/3$ [s] being the effective diffusion time, $D(\mathbf{g})$ and $\alpha(\mathbf{g})$ being the apparent diffusion and stretching parameters at direction $\mathbf{g}$, respectively. Notice here once the stretching parameter tends to unitary, i.e., $\alpha(\mathbf{g}) \to 1$, the stretched-exponential representation (2) reduces to a mono-exponential signal decay.

2.2 Q-Space Domain Quantitative Measures

In what follows, we analytically derive nth order statistics of the stretched-exponential representation given by the Eq. (2). This enables to easily retrieve three quantitative Q-space measures namely the RTOP [mm^{-3}] being the probability in the origin, $P(\mathbf{0})$, QMSD [mm^{-5}] and QMFD [mm^{-7}] defined as the second- and fourth-order statistics of the signal $E(\mathbf{q})$ respectively [12, 22]

$$\text{RTOP} = \int_{\mathbb{R}^3} E(\mathbf{q}) d^3\mathbf{q}, \ \text{QMSD} = \int_{\mathbb{R}^3} \|\mathbf{q}\|^2 E(\mathbf{q}) d^3\mathbf{q}, \ \text{QMFD} = \int_{\mathbb{R}^3} \|\mathbf{q}\|^4 E(\mathbf{q}) d^3\mathbf{q}, \tag{3}$$

where $\|\cdot\|$ is the vector norm of the wave vector $\mathbf{q}$.

We specify now a more general equation in the Q-space domain related to the nth order statistics of the signal attenuation $E(\mathbf{q})$. Considering the stretched-exponential representation of the signal (2) and a spherical coordinate system (q, θ, φ) with polar θ and azimuthal φ angles, and a radial coordinate $q = \|\mathbf{q}\|$ [mm^{-1}] we define the nth order statistics of the signal attenuation $E(\mathbf{q})$

$$\begin{aligned} M_n &= \int_{\mathbb{R}^3} \|\mathbf{q}\|^n \exp\left(-(4\pi^2\tau\|\mathbf{q}\|^2 D(\mathbf{g}))^{\alpha(\mathbf{g})}\right) d^3\mathbf{q} \\ &= \int_0^{2\pi}\int_0^{\pi}\int_0^{\infty} \exp\left(-\left(4\pi^2\tau q^2 D(\theta,\varphi)\right)^{\alpha(\theta,\varphi)}\right) q^{n+2} \sin\theta \, dq \, d\theta \, d\varphi, \end{aligned} \tag{4}$$

where $D(\theta, \varphi)$ and $\alpha(\theta, \varphi)$ are the apparent diffusion coefficient and stretching parameter both defined in the spherical coordinate system. Next, we rewrite the integral (4) as follows (see Gradshteyn and Ryzhik [14, p. 370, Eq. 3.478(1)])

$$\begin{aligned} M_n &= C_n^\tau \int_0^{2\pi}\int_0^{\pi} \Gamma\left(\frac{n+3}{2\alpha(\theta,\varphi)}\right) \alpha^{-1}(\theta,\varphi)\, D^{-(n+3)/2}(\theta,\varphi) \sin\theta \, d\theta \, d\varphi \\ &= C_n^\tau \iint_\Sigma \Gamma\left(\frac{n+3}{2\alpha(\theta,\varphi)}\right) \alpha^{-1}(\theta,\varphi)\, D^{-(n+3)/2}(\theta,\varphi)\, d\Sigma, \end{aligned} \tag{5}$$

where $C_n^\tau = 2^{-n-4}\pi^{-n-3}\tau^{-(n+3)/2}$ is a diffusion time dependent constant and $\Gamma(\cdot)$ is the gamma function. Notice here that the last equation is a surface integral over the surface with a unitary radius, i.e., $q = 1$.

2.3 Numerical Implementation

To evaluate the surface integral (5), one can assume the surface area element, $\Delta\Sigma$, is inversely proportional to the number of sampled data points (e.g., the number of evenly distributed directions N_g, $\Delta\Sigma = 4\pi/N_g$). Transforming the Eq. (2) the diffusion becomes $D(\mathbf{g}) = 4^{-1}\pi^{-2}\tau^{-1}q^{-2}\left(-\log E(\mathbf{q})\right)^{1/\alpha(\mathbf{g})}$, and thus the Eq. (5) can be rewritten in the following form

$$M_n^{(1)} = \frac{1}{2}q^{n+3}\left\langle \Gamma\left(\frac{n+3}{2\alpha(\mathbf{g})}\right)\alpha^{-1}(\mathbf{g})\left(-\log E(\mathbf{q})\right)^{-\frac{n+3}{2\alpha(\mathbf{g})}}\right\rangle_{\mathbf{q}\in\mathbb{S}^2} \tag{6}$$

with $\langle\cdot\rangle_{\mathbf{q}\in\mathbb{S}^2}$ being a direction-averaged signal over a single acquisition shell. Notice that the Eq. (6) can be evaluated using the samples retrieved from the resampled data to uniformly cover the surface (e.g., the spherical harmonics [2]).

The numerical reciprocal of the negative log-diffusion function given in Eq. (6) might be prone to instabilities, i.e., the signal attenuation $E(\mathbf{q}) \to 1$ the function $(-\log(E(\mathbf{q})))^{-1} \to \infty$ [1, 27]. Alternatively, one can refine (6) to incorporate its second-order series expansion. To this end, we define a twice differentiable function $f : \mathbb{R} \to \mathbb{R}$ given by $f(X) = \Gamma\left(\frac{n+3}{2\alpha}\right)\alpha^{-1}X^{-(n+3)/(2\alpha)}$ with $n \geq 0$. The second-order series expansion of the expectation of the function $f(X)$ around the expectation $\mathbb{E}\{X\}$ is given then by $\mathbb{E}\{f(X)\} \approx f(\mathbb{E}\{X\}) + \frac{1}{2}\frac{d^2 f}{dX^2}\Big|_{X=\mathbb{E}\{X\}} \cdot \mathrm{Var}\{X\}$. After using some algebra we arrive at the following closed-form formula

$$\mathbb{E}\{f(X)\} \approx \frac{1}{8}\Gamma\left(\frac{n+3}{2\alpha}\right)\alpha^{-3}\,\mathbb{E}\{X\}^{-\frac{n+3}{2\alpha}}\Big((n+3)(n+3+2\alpha)\mathbb{E}\left\{X^2\right\}\mathbb{E}\{X\}^{-2} + 8\alpha^2 - (n+3)(n+3+2\alpha)\Big).$$

Given a stretched-exponential decay again (2) and a second-order series expansion of the expectation, we define an approximation to the measure (6)

$$M_n^{(2)} = \frac{1}{2}q^{n+3}\left\langle\frac{1}{8}\Gamma\left(\frac{n+3}{2\alpha(\mathbf{g})}\right)\alpha^{-3}(\mathbf{g})\right\rangle_{\mathbf{q}\in\mathbb{S}^2}\left\langle-\log E(\mathbf{q})\right\rangle_{\mathbf{q}\in\mathbb{S}^2}^{-\left\langle\frac{n+3}{2\alpha(\mathbf{q})}\right\rangle_{\mathbf{q}\in\mathbb{S}^2}} \times\left[\frac{(n+3)\left\langle(n+3)\mathbf{1}+2\alpha(\mathbf{g})\right\rangle_{\mathbf{q}\in\mathbb{S}^2}\left\langle(-\log E(\mathbf{q}))^2\right\rangle_{\mathbf{q}\in\mathbb{S}^2}}{\left\langle-\log E(\mathbf{q})\right\rangle_{\mathbf{q}\in\mathbb{S}^2}^2} + \left\langle 8\alpha^2(\mathbf{g}) - 2(n+3)\alpha(\mathbf{g}) - (n+3)^2\mathbf{1}\right\rangle_{\mathbf{q}\in\mathbb{S}^2}\right] \tag{7}$$

with $\mathbf{1}$ being the all-ones vector. Here, we have simplified our derivations in the series expansion procedure; thus, we direction average the stretching parameter to obtain the final formula. From Eq. (7) we can define basic Q-space domain measures such as the RTOP (M_0), QMSD (M_2) or QMFD (M_4; see Eq. (3)). The proposed stretched-exponential method requires a multiple-shell acquisition with at least two-shells at different b-values to fit the representation given by Eq. (2). Once the representation is fitted, a single-shell data at a fixed b-value is used to calculate the measures. In Sect. 2.4, we define a simple optimization cost function to retrieve the stretched representation of the diffusion signal.

2.4 *Optimization of Stretched-Exponential Representation*

To retrieve a stretched-exponential representation at direction $\mathbf{g}$ of the diffusion, we define an optimization cost function and solve it using a non-linear least squares procedure

$$(D(\mathbf{g}), \alpha(\mathbf{g})) = \underset{D'(\mathbf{g}),\, \alpha'(\mathbf{g})}{\operatorname{argmin}} \frac{1}{2} \sum_{\mathbf{q}:\, \mathbf{q} \| \mathbf{g}} \left[S(\mathbf{q}) - S(0) \exp\left(-(4\pi^2 \tau \|\mathbf{q}\|^2 D'(\mathbf{g}))^{\alpha'(\mathbf{g})} \right) \right]^2 .$$

We used a bound-constrained minimization via the trust-region reflective method with a linear loss function to find the optimal parameters. Notice here the procedure (2.4) applies for each direction $\mathbf{g}$ independently and might use the only a subset of $\mathbf{q}$-values employed to acquire the data.

3 Materials and Methods

In this study, we used ex vivo rat brain data as well as in vivo human brain data that was publicly available by Hansen and Jespersen [16].

3.1 *Ex Vivo rat brain data*

The ex vivo data were collected using a Bruker Biospec 9.4T (Bruker Biospin, Germany) with a 15 mm quadrature coil. Diffusion-weighted images were acquired in 15 b-value shells ranging from 0 to 5000 s/mm^2 with a step size of 200 s/mm^2 and 33 directions per each shell utilizing a spin echo sequence. Fifteen axial slices were collected at a resolution of $100 \times 100 \times 500\ \mu\text{m}^3$, matrix size 128×128, echo time of $\text{TE} = 23.3\,\text{ms}$, repetition time of $\text{TR} = 4\,\text{s}$, and diffusion timing of $\delta/\Delta = 4/14\,\text{ms}$. The data set was averaged twice to improve the signal-to-noise ratio being around 75 at the baseline.

3.2 In Vivo Human brain data

One healthy participant was scanned in an in vivo study using a Siemens Trio 3T equipped with a 32 channel head coil. The protocol comprised one $b = 0$ and 15 non-zero shells ranging from 200 s/mm^2 to 3000 s/mm^2 with the step size of 200 s/mm^2 and 33 directions per shell. Nineteen axial slices with a voxel size of 2.5 mm isotropic and a 96×96 matrix size, TE = 116 ms, TR = 7200 ms, TI = 2100 ms were obtained. The diffusion timings were estimated to be $\delta/\Delta = 29/58$ ms. The SNR of the baseline signal is around 39. In our experiments we used a five-shell acquisition with 200, 1000, 1800 2400, and 3000 s/mm^2.

3.3 Comparison to the Q-Space Measures from Different Methods

In Sect. 2.2 we introduced three measures that is the RTOP, QMSD and QMFD. In this work, we evaluate the proposed stretched-exponential Q-space measures and compare them to those obtained from the MAP-MRI technique [26] (positivity constraint), MAPL [13] (regularization parameter $\lambda = 0.2$), RBF [22] (l_1 regularization with $\lambda = 0.00055$), 3D-SHORE [24, 35] (scale factor $\zeta = 1/(8\pi^2 \tau D)$), and a single-shell approach [2, 27]. Except for the aforementioned frameworks, we calculate also the RTOP measure directly from diffusion tensor eigenvalues (a non-linear least squares fitting via the Levenberg-Marquardt method) as RTOP = $(4\pi\tau)^{-3/2} (\lambda_1\lambda_2\lambda_3)^{-1/2}$.

4 Results and Discussion

In the first experiment, we visually evaluate the measures using ex vivo rat brain data retrieved using different methodologies namely the DTI (at 1000 and 1400 s/mm^2), RBF, 3D-SHORE, MAP-MRI, MAPL (the EAPs in all four are fitted to a three-shells acquisition, i.e., $b = 1000$, 3000 and 5000 s/mm^2), single-shell technique, and proposed stretched-exponential one (see Fig. 1). For the proposed technique, we use three-shells for fitting the representation while a single-shell to calculate the measure. Visually inspecting the RTOP measure shows that DTI based ones (Fig. 1a, b) and single-shell technique (Fig. 1d–f) return smaller values compared to all other methods. The RBF and 3D-SHORE have the lowest contrast between the white matter and gray matter tissue in the measure (Fig. 1c, g). In the single-shell method by increasing the b-value from 1000 to 5000 s/mm^2, the RTOP value increases (Fig. 1d–f, RTOP). Comparing the EAP-based techniques and stretched-exponential representations (Fig. 1c, g–l, RTOP), 3D-SHORE and RBF provide the lowest while MAP-MRI gives the highest intensity in white matter areas. MAPL results of RTOP

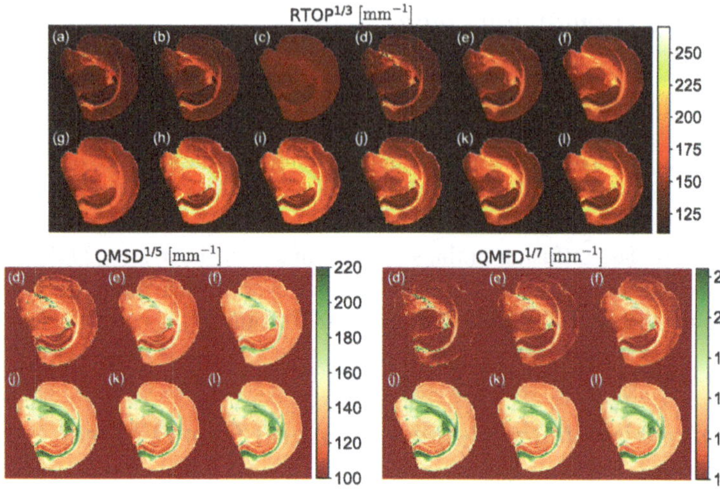

Fig. 1 Visual inspection of the RTOP, QMSD and QMFD measures on ex vivo rat brain data: **a** DTI ($b = 1000$ s/mm^2), **b** DTI ($b = 1400$ s/mm^2), **c** RBF, **d** single-shell ($b = 1000$ s/mm^2), **e** single-shell ($b = 3000$ s/mm^2), **f** single-shell ($b = 5000$ s/mm^2), **g** 3D-SHORE, **h** MAP-MRI, **i** MAPL, **j** stretched-exponential ($b = 1000$ s/mm^2), **k** stretched-exponential ($b = 3000$ s/mm^2) and **l** stretched-exponential ($b = 5000$ s/mm^2)

are similar to MAP-MRI while the RTOP values in MAPL are slightly lower than the MAP-MRI. Our proposed method of stretched-exponential provides similar RTOP maps for different b-values ($b = 1000$, 3000, and 5000 s/mm^2) and it preserves the consistency of the measures between different b-values which is not observed in single-shell technique (Fig. 1d–f, j–l, RTOP). Clearly, introducing the stretched-exponential representation enabled to improve contrast and kept the RTOP measure's uniformity across the b-values. In the QMSD/QMFD measures, single-shell method at $b = 1000$ s/mm^2 has the lowest value in the both gray matter and white matter while the $b = 5000$ s/mm^2 has the highest and $b = 3000$ s/mm^2 is the intermediate between the three alternatives of the single-shell method (Fig. 1d–f, QMSD/QMFD. Notice here that again the behavior of the QMSD/QMFD measures across the b-values are preserved, and the proposed stretched-exponential representation keeps the consistency of the quantities while changing the b-value used to calculate the measure.

In the second experiment, we extrapolate the previous one and evaluate the absolute changes in the measures due to the changes in the maximal b-value parameter. In this experiment we used six different acquisitions from three-shells (i.e., $b = 1000$, 2000 and $b = 3000$ s/mm^2) to eight-shells (up to $b = 5000$ s/mm^2 with a step of 400 s/mm^2). For the proposed methodology we fit the representation using k shells ($k = 3, \dots, 8$; similarly to the EAP-based methods) while calculate the measures using only one shell, i.e., $b = 3000$ s/mm^2 or $b_{\max}$. Figure 2 depicts the mean absolute changes of the RTOP, QMSD, and QMFD measures in terms of maximal b-value

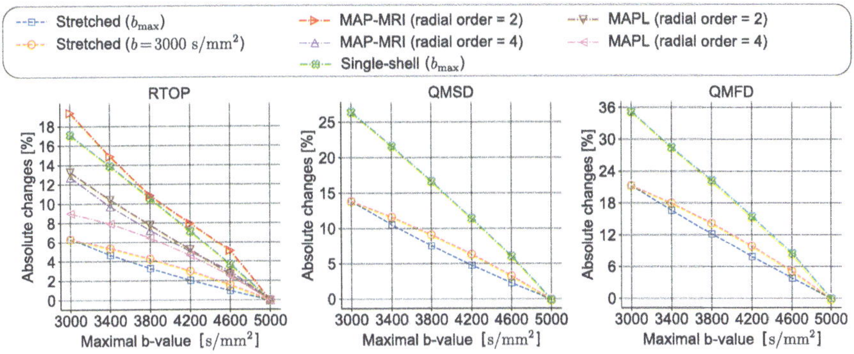

Fig. 2 The mean absolute changes of the RTOP, QMSD and QMFD measures in ex vivo rat brain data in terms of maximal b-value (b_{max}) used to estimate the EAP/calculate the measure under different techniques. For stretched-exponential representation two variants are used in measure calculation process namely b-value at b_{max} and $b = 3000 \mathrm{s/mm}^2$

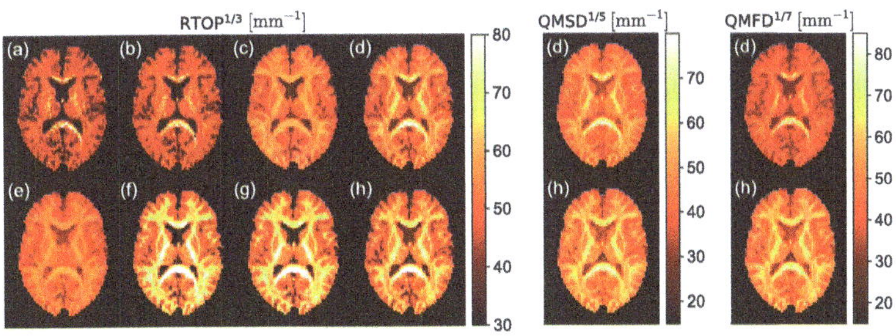

Fig. 3 Visual inspection of the measures on in vivo human brain data estimated using various approaches: **a** DTI ($b = 1000$ s/mm^2), **b** DTI ($b = 1400$ s/mm^2), **c** RBF, **d** single-shell ($b = 3000$ s/mm^2), **e** 3D-SHORE, **f** MAP-MRI, **g** MAPL and **h** stretched-exponential ($b = 3000$ s/mm^2)

(b_{max}) under different methodologies used to estimate the measures. For the estimation of RTOP, our proposed stretched-exponential methodology has the minimum mean absolute changes for both b_{max} and $b = 3000$ s/mm^2 used to retrieve the measure once fitted the Eq. (2.4). It is worth noticing here that the single-shell technique is heavy load due to the changes in the maximal b-value and our proposal improved the results though the measure is still calculated from a single-shell. As for the two other measures, our proposed method is superior to the single-shell method, while again, the one with b_{max} is slightly better than that with $b = 3000$ s/mm^2.

Figure 3 is devoted to visual inspection of the RTOP and QMSD/QMFD measures on in vivo human brain data estimated using various state-of-the-art approaches. The observed trend in the in vivo maps of RTOP is similar to the one observed in ex vivo data (Fig. 1; RTOP). Again, a comparable behavior of the QMSD/QMFD measures can be observed, i.e., the single-shell technique generally exhibits smaller values of the measure in white matter areas than stretched-exponential representation. Notice

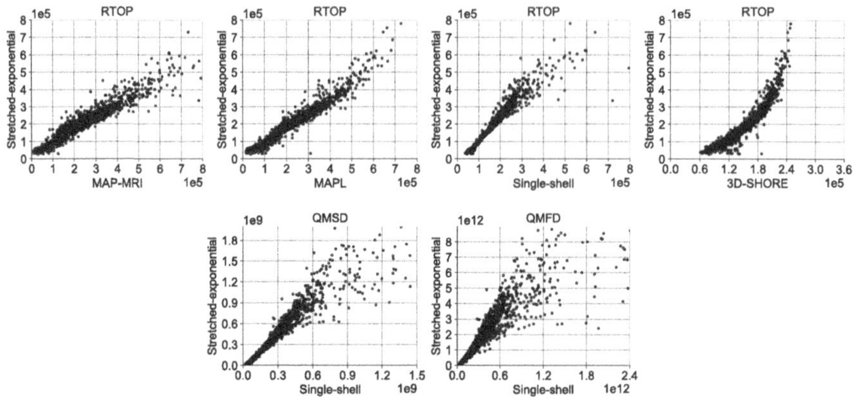

Fig. 4 Correlations between the RTOP retrieved from in vivo human brain data using the proposed stretched-exponential and state-of-the-art methods namely MAP-MRI, MAPL, single-shell and 3D-SHORE (top), and correlations between the proposal and single-shell approach in case of QMSD/QMFD (bottom)

Table 1 Pearson's correlation coefficient between different methodologies used to retrieve the RTOP measure from human brain data (see Fig. 3). Legend: Single (1), (2)—single-shell technique at $b = 2400$ and $b = 3000$ s/mm^2; SE (1), (2)—stretched-exponential at $b = 2400$ and $b = 3000$ s/mm^2, respectively

	DTI	MAP-MRI	MAPL	Single (1)	Single (2)	3D-SHORE	SE (1)	SE (2)
DTI	×							
MAP-MRI	0.681	×						
MAPL	0.674	0.981	×					
Single (1)	0.827	0.901	0.895	×				
Single (2)	0.797	0.936	0.932	0.978	×			
3D-SHORE	0.642	0.928	0.961	0.863	0.895	×		
SE (1)	0.739	0.951	0.965	0.944	0.954	0.931	×	
SE (2)	0.728	0.957	0.973	0.925	0.959	0.933	0.982	×

here that the QMFD measure obtained from the single-shell technique cannot catch heavy tails of the signal distribution properly from the data.

Lastly, we extrapolate the previous experiment using in vivo human data and evaluate the correlation between measures both visually and numerically. Here, we again use a five-shells acquisition to calculate the measures. Figure 4 (top row) illustrates correlations between the RTOP retrieved using the proposed approach and those obtained under different state-of-the-art methods, namely MAP-MRI, MAPL, the single-shell approach, and 3D-SHORE over the image mask (white matter, gray matter, and cerebrospinal fluid areas). Our method shows a good correlation with all the other four methods being characterized by Pearson's correlation coefficient equal to $\rho = 0.925$ is the worst case (see Table 1). Figure 4 also shows the correlation between our method and single-shell approach in the case of QMSD and QMFD measures. The correlograms for QMSD/QMFD measures clearly show the outliers generated by the single-shell technique, which are not present in the proposed one.

5 Conclusions

In this paper, we proposed a new approach based on stretched-exponentials to quantify EAP features such as RTOP, QMSD, and QMFD measures. From the results, it seems that the proposed method reduces the amount of data to be acquired, and therefore it can be clinically feasible. We have to mention that much more thorough validation and comparison are required to bring convincing evidence that comparable results can be obtained with fewer data in our method. In our experiments, we evaluated the method's stability, but the accuracy remains to be investigated in future work. Further, the convergence of the series expansion has to be verified as well. Finally, notice that in situations where the signal cannot be accurately described by the stretched-exponential [23], the scalar measures will be biased. Besides, the proposed method can be generalized to other Q-space factors such as return-to-the-axis or return-to-the-plane probability.

Acknowledgements Tomasz Pieciak acknowledges the Polish National Agency for Academic Exchange for grant PN/BEK/2019/1/00421 under the Bekker programme, the Ministry of Science and Higher Education (Poland) under the scholarship for outstanding young scientists (692/STYP/13/2018) and AGH Science and Technology, Kraków (16.16.120.773). Maryam Afzali and Derek K. Jones were supported by a Wellcome Trust Investigator Award (096646/Z/11/Z) and a Wellcome Trust Strategic Award (104943/Z/14/Z). Santiago Aja-Fernández was supported by Ministerio de Ciencia e Innovación of Spain (RTI2018-094569-B-I00).

References

1. Aja-Fernández, S., Pieciak, T., Tristán-Vega, A., Vegas-Sánchez-Ferrero, G., Molina, V., de Luis-García, R.: Scalar diffusion-MRI measures invariant to acquisition parameters: a first step towards imaging biomarkers. Magn. Reson. Imaging **54**, 194–213 (2018)
2. Aja-Fernández, S., de Luis-García, R., Afzali, M., Molendowska, M., Pieciak, T., Tristán-Vega, A.: Micro-structure diffusion scalar measures from reduced MRI acquisitions. PLOS ONE **15**(3) (2020)
3. Alexander, D.C.: A general framework for experiment design in diffusion MRI and its application in measuring direct tissue-microstructure features. Magn. Reson. Med. **60**(2), 439–448 (2008)
4. Assaf, Y., Basser, P.J.: Composite hindered and restricted model of diffusion (CHARMED) MR imaging of the human brain. Neuroimage **27**(1), 48–58 (2005)

5. Assaf, Y., Blumenfeld-Katzir, T., Yovel, Y., Basser, P.J.: AxCaliber: a method for measuring axon diameter distribution from diffusion MRI. Magn. Reson. Med. **59**(6), 1347–1354 (2008)
6. Avram, A.V., Sarlls, J.E., Barnett, A.S., Özarslan, E., Thomas, C., Irfanoglu, M.O., Hutchinson, E., Pierpaoli, C., Basser, P.J.: Clinical feasibility of using mean apparent propagator (MAP) MRI to characterize brain tissue microstructure. Neuroimage **127**, 422–434 (2016)
7. Basser, P.J.: Inferring microstructural features and the physiological state of tissues from diffusion-weighted images. NMR Biomed. **8**(7), 333–344 (1995)
8. Basser, P., Pierpaoli, C.: Microstructural features measured using diffusion tensor imaging. J. Magn. Reson. B **111**(3), 209–219 (1996)
9. Bennett, K.M., Schmainda, K.M., Bennett, R., Rowe, D.B., Lu, H., Hyde, J.S.: Characterization of continuously distributed cortical water diffusion rates with a stretched-exponential model. Magn. Reson. Med. **50**(4), 727–734 (2003)
10. Callaghan, P.T.: Translational Dynamics and Magnetic Resonance: Principles of Pulsed Gradient Spin Echo NMR. Oxford University Press, Oxford (2011)
11. Clark, C.A., Le Bihan, D.: Water diffusion compartmentation and anisotropy at high b values in the human brain. Magn. Reson. Med. **44**(6), 852–859 (2000)
12. Descoteaux, M., Deriche, R., Le Bihan, D., Mangin, J.F., Poupon, C.: Multiple q-shell diffusion propagator imaging. Med. Image Anal. **15**(4), 603–621 (2011)
13. Fick, R.H., Wassermann, D., Caruyer, E., Deriche, R.: MAPL: tissue microstructure estimation using Laplacian-regularized MAP-MRI and its application to HCP data. Neuroimage **134**, 365–385 (2016)
14. Gradshteyn, I.S., Ryzhik, I.M.: Table of Integrals, Series, and Products. Academic (2014)
15. Grebenkov, D.S.: Laplacian eigenfunctions in NMR. I. A numerical tool. Concepts Magn. Reson. Part A: Educ. J. **32**(4), 277–301 (2008)
16. Hansen, B., Jespersen, S.N.: Data for evaluation of fast kurtosis strategies, b-value optimization and exploration of diffusion MRI contrast. Sci. Data **3**(1), 1–5 (2016)
17. Jensen, J.H., Helpern, J.A., Ramani, A., Lu, H., Kaczynski, K.: Diffusional kurtosis imaging: the quantification of non-gaussian water diffusion by means of magnetic resonance imaging. Magn. Reson. Med. **53**(6), 1432–1440 (2005)
18. Jones, D.K., Knösche, T.R., Turner, R.: White matter integrity, fiber count, and other fallacies: the do's and don'ts of diffusion MRI. Neuroimage **73**, 239–254 (2013)
19. Kubicki, M., McCarley, R., Westin, C.F., Park, H.J., Maier, S., Kikinis, R., Jolesz, F.A., Shenton, M.E.: A review of diffusion tensor imaging studies in schizophrenia. J. Psychiatr. Res. **41**(1–2), 15–30 (2007)
20. Le Bihan, D.: Molecular diffusion nuclear magnetic resonance imaging. Magn. Reson. Q. **7**(1), 1 (1991)
21. Magin, R.L., Karani, H., Wang, S., Liang, Y.: Fractional order complexity model of the diffusion signal decay in MRI. Mathematics **7**(4), 348 (2019)
22. Ning, L., Westin, C.F., Rathi, Y.: Estimating diffusion propagator and its moments using directional radial basis functions. IEEE Trans. Med. Imaging **34**(10), 2058–2078 (2015)
23. Novikov, D.S., Fieremans, E., Jespersen, S.N., Kiselev, V.G.: Quantifying brain microstructure with diffusion MRI: Theory and parameter estimation. NMR Biomed. **32**(4) (2019)
24. Ozarslan, E., Koay, C., Shepherd, T., Blackb, S., Basser, P.: Simple harmonic oscillator based reconstruction and estimation for three-dimensional q-space MRI (2009)
25. Özarslan, E., Shepherd, T.M., Vemuri, B.C., Blackband, S.J., Mareci, T.H.: Resolution of complex tissue microarchitecture using the diffusion orientation transform (DOT). Neuroimage **31**(3), 1086–1103 (2006)
26. Özarslan, E., Koay, C.G., Shepherd, T.M., Komlosh, M.E., İrfanoğlu, M.O., Pierpaoli, C., Basser, P.J.: Mean apparent propagator (MAP) MRI: a novel diffusion imaging method for mapping tissue microstructure. Neuroimage **78**, 16–32 (2013)
27. Pieciak, T., Bogusz, F., Tristán-Vega, A., de Luis-García, R., Aja-Fernández, S.: Single-shell return-to-the-origin probability diffusion MRI measure under a non-stationary Rician distributed noise. In: IEEE 16th International Symposium on Biomedical Imaging, pp. 131–134. IEEE (2019)

28. Rovaris, M., Filippi, M.: Diffusion tensor MRI in multiple sclerosis. J. Neuroimaging **17**, 27S–30S (2007)
29. Tian, Q., Yang, G., Leuze, C., Rokem, A., Edlow, B.L., McNab, J.A.: Generalized diffusion spectrum magnetic resonance imaging (GDSI) for model-free reconstruction of the ensemble average propagator. Neuroimage **189**, 497–515 (2019)
30. Tuch, D.S., Reese, T.G., Wiegell, M.R., Wedeen, V.J.: Diffusion MRI of complex neural architecture. Neuron **40**(5), 885–895 (2003)
31. Westin, C.F., Maier, S.E., Mamata, H., Nabavi, A., Jolesz, F.A., Kikinis, R.: Processing and visualization for diffusion tensor MRI. Med. Imag. Anal. **6**(2), 93–108 (2002)
32. Williams, G., Watts, D.C.: Non-symmetrical dielectric relaxation behaviour arising from a simple empirical decay function. Trans. Faraday Soc. **66**, 80–85 (1970)
33. Wu, Y.C., Field, A.S., Alexander, A.L.: Computation of diffusion function measures in q-space using magnetic resonance hybrid diffusion imaging. IEEE Trans. Med. Imaging. **27**(6), 858–865 (2008)
34. Zhang, H., Schneider, T., Wheeler-Kingshott, C.A., Alexander, D.C.: NODDI: practical in vivo neurite orientation dispersion and density imaging of the human brain. Neuroimage **61**(4), 1000–1016 (2012)
35. Zucchelli, M., Brusini, L., Méndez, C.A., Daducci, A., Granziera, C., Menegaz, G.: What lies beneath? Diffusion EAP-based study of brain tissue microstructure. Med. Imag Anal. **32**, 145–156 (2016)

Repeatability of Soma and Neurite Metrics in Cortical and Subcortical Grey Matter

Sila Genc, Maxime Chamberland, Kristin Koller, Chantal M. W. Tax, Hui Zhang, Marco Palombo, and Derek K. Jones

Abstract Diffusion magnetic resonance imaging is a technique which has long been used to study white matter microstructure in vivo. Recent advancements in hardware and modelling techniques have opened up interest in disentangling tissue compartments in the grey matter. In this study, we evaluate the repeatability of soma and neurite density imaging in a sample of six healthy adults scanned five times on an ultra-strong gradient magnetic resonance scanner (300 mT/m). Repeatability was expressed as an intraclass correlation coefficient (ICC). Our findings reveal that measures of soma density (mean ICC = 0.976), neurite density (mean ICC = 0.959) and apparent soma size (mean ICC = 0.923) are highly reliable across multiple cortical and subcortical networks. Overall, we demonstrate the promise of moving advanced grey matter microstructural imaging towards applications of development, ageing, and disease.

1 Introduction

Conventional T1-weighted magnetic resonance imaging (MRI) is a useful tool in determining clinically relevant regional differences in grey matter volume, cortical thickness, surface area and gyrification. However, these crude macroscopic measures do not provide information on which distinct cellular features (e.g. cell bodies and neurites) and packing configurations drive differences in macroscopic measures.

Marco Palombo and Derek K. Jones—authors equally contributed to this work.

S. Genc (✉) · M. Chamberland · K. Koller · C. M. W. Tax · D. K. Jones
Cardiff University Brain Research Imaging Centre (CUBRIC), Cardiff University, Cardiff, UK
e-mail: GencS@cardiff.ac.uk

H. Zhang · M. Palombo
Department of Computer Science, Centre for Medical Image Computing, University College London (UCL), London, UK

D. K. Jones
Mary MacKillop Institute for Health Research, Australian Catholic University, Melbourne, Australia

N. Gyori et al. (eds.), *Computational Diffusion MRI*, Mathematics and Visualization,
https://doi.org/10.1007/978-3-030-73018-5_11

Diffusion MRI (dMRI) can enhance sensitivity to much smaller structures by probing water diffusion that is modulated by the presence of micrometer-scale compartments. Previous studies have applied the commonly used diffusion tensor imaging (DTI) technique to profile microstructure in the grey matter (e.g. [1, 2]), however biological interpretations are limited as DTI metrics are non-specific to the aforementioned microstructural compartments.

Progress in acquisition and modelling methods using ultra-strong gradient and ultra-high b-value dMRI [3, 4] hold promise for disentangling and quantifying biologically meaningful cellular components in vivo [5, 6]. One recent model-based method to study grey matter microstructure is Soma and Neurite Density Imaging (SANDI) [5], which aims to disentangle microstructural contributions from cellular projections (neurites: including axons, dendrites and glial processes), soma (neuronal cell bodies and glia) density and their apparent size, and extracellular space.

The original SANDI paper demonstrated results in humans using ultra-high b-value data (up to 10,000 s/mm^2) [5]. In this study, we utilise a rich repeatability database of scan-re-scan dMRI data acquired from 6 healthy participants, each across 5 sessions [7] on an ultra-strong gradient MR scanner [3, 4]. Our primary aim is to establish whether SANDI metrics are repeatable at lower b-values (up to 6,000 s/mm^2) to establish the translatability and utility of advanced microstructural imaging in cortical and subcortical grey matter.

2 Methods

2.1 Image Acquisition and Pre-processing

The data used for this study were previously reported by Koller et al. [7], comprising a sample of 6 healthy adults (3 female) aged 24–30 years. This study was approved by a local ethics board. Each participant was scanned five times in the span of two weeks on a 3.0T Siemens Connectom system with ultra-strong (300 mT/m) gradients.

Structural data were acquired using a magnetization-prepared rapid acquisition with gradient echo (MPRAGE, voxel-size $= 1 \times 1 \times 1$ mm) and multi-shell dMRI data were collected (TE/TR = 59/3000 ms; voxel size $= 2 \times 2 \times 2$ mm; b-values = 0 (14 vols), 200;500 (20 dirs), 1200 (30 dirs), and 2400; 4000; 6000 (60 dirs) s/mm^2). dMRI data were acquired in an anterior-posterior (AP) phase-encoding direction, with additional $b = 0$ s/mm^2 images acquired in the PA direction. Pre-processing involved: noise estimation using Marchenko–Pastur Principles Component Analysis (MP-PCA) [8] and subsequent denoising in *MRtrix3* [9], correction for signal drift [10], motion, eddy, and susceptibility-induced distortions [11, 12], gradient non-linearities [13, 14], Gibbs ringing artefacts [15], and bias field [9, 16].

2.2 Image Processing and Analysis

The SANDI compartment model was fitted to the pre-processed dMRI dataset for each subject using the machine learning approach described in [5], based on random forest regression. Four parameters of interest were investigated:

(i) the intraneurite signal fraction, $f_{intraneurite}$
(ii) the intrasoma signal fraction, $f_{intrasoma}$
(iii) the soma radius, R_{soma} (μm)
(iv) the extracellular signal fraction, $f_{extracellular}$

Additionally, an estimate of model uncertainty was obtained using the quartile deviation of predictions (QD) from the ensemble of regression trees. To complement the SANDI estimates, diffusion tensor estimation was performed on the $b = 1200\,s/mm^2$ shell using an iteratively reweighted linear least squares estimator. The tensor-derived parameters fractional anisotropy (FA) and mean diffusivity (MD) were computed.

T1 data were co-registered to an upsampled $b = 0\,s/mm^2$ image (1mm isotropic) and processed through Freesurfer [17] to obtain cortical and subcortical parcellations using the Destrieux atlas [18]. This resulted in 74 cortical regions per hemisphere, alongside subcortical regions. We studied seven different functionally-defined networks from the Yeo functional network atlas [19] (Fig. 1). The subcortical parcellation was treated as a single sub network, resulting in eight total subnetworks for each participant for further statistical analysis. Follow-up analyses of individual subcortical regions were restricted to the amygdala, caudate, hippocampus, pallidum and thalamus. Network labels (L) were resampled to each individual subject's diffusion space, and we computed the intersection between the cortical ribbon (R) and resampled network labels (L∩R).

Statistical analyses were performed within R (v3.4.3) and RStudio (v1.2.1335). The intra-class correlation coefficient (ICC; two-way random effects, absolute agreement) was computed for assessment of test-re-test repeatability of SANDI and DTI metrics (Table 1). Summary statistics were computed using an analysis of variance

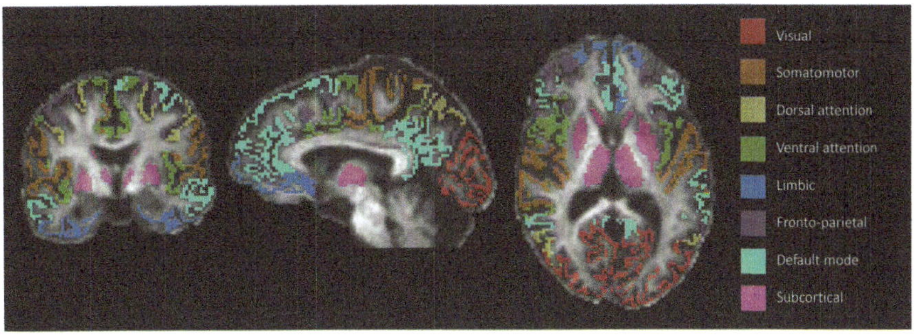

Fig. 1 A representation of the eight cortical and subcortical sub-networks [19] on a representative participant

Table 1 Statistics on test-re-test repeatability of DTI and SANDI metrics. Statistics summarise the mean, median absolute deviation (MAD), intra-class correlation coefficient (ICC) and p-value across all repeated measurements

Network	Metric							
	FA				MD (10^{-3} mm^2/s)			
	Mean	MAD	ICC	p-value	Mean	MAD	ICC	p-value
Visual	0.12	0.009	0.964	<0.001	0.90	0.015	0.991	<0.001
Somatomotor	0.12	0.006	0.906	<0.001	1.00	0.072	0.997	<0.001
Dorsal attention	0.12	0.010	0.975	<0.001	0.99	0.069	0.999	<0.001
Ventral attention	0.14	0.005	0.935	<0.001	0.90	0.042	0.996	<0.001
Limbic	0.16	0.006	0.520	0.11	0.83	0.020	0.949	<0.001
Fronto-parietal	0.14	0.009	0.911	<0.001	0.95	0.078	0.998	<0.001
Default	0.14	0.005	0.928	<0.001	0.94	0.065	0.996	<0.001
Subcortical	0.23	0.014	0.879	<0.001	0.73	0.010	0.850	<0.001
	$f_{extracellular}$				$f_{intraneurite}$			
	Mean	MAD	ICC	p-value	Mean	MAD	ICC	p-value
Visual	0.41	0.014	0.984	<0.001	0.14	0.008	0.950	<0.001
Somatomotor	0.45	0.031	0.995	<0.001	0.12	0.007	0.954	<0.001
Dorsal attention	0.44	0.029	0.997	<0.001	0.11	0.008	0.962	<0.001
Ventral attention	0.43	0.023	0.987	<0.001	0.11	0.010	0.965	<0.001
Limbic	0.41	0.015	0.935	<0.001	0.18	0.012	0.962	<0.001
Fronto-parietal	0.45	0.033	0.995	<0.001	0.11	0.004	0.934	<0.001
Default	0.44	0.023	0.992	<0.001	0.11	0.005	0.965	<0.001
Subcortical	0.34	0.009	0.627	0.04	0.29	0.034	0.977	<0.001
	$f_{intrasoma}$				R_{soma} (μm)			
	Mean	MAD	ICC	p-value	Mean	MAD	ICC	p-value
Visual	0.46	0.017	0.985	<0.001	8.81	0.110	0.934	<0.001
Somatomotor	0.43	0.033	0.997	<0.001	8.75	0.130	0.983	<0.001
Dorsal attention	0.45	0.033	0.996	<0.001	8.81	0.130	0.985	<0.001
Ventral attention	0.46	0.018	0.991	<0.001	8.90	0.130	0.939	<0.001
Limbic	0.42	0.015	0.891	<0.001	8.68	0.040	0.656	0.04
Fronto-parietal	0.45	0.024	0.996	<0.001	8.86	0.130	0.974	<0.001
Default	0.45	0.020	0.996	<0.001	8.84	0.130	0.962	<0.001
Subcortical	0.37	0.040	0.957	<0.001	8.37	0.250	0.953	<0.001

Note Bonferroni adjusted level of significance was set to $p < 0.001$

(ANOVA), and lower and upper estimates of each ICC represent the bounds of the 95% confidence interval (CI). Based on the number of comparisons (8 networks × 6 metrics = 48 comparisons) we adjusted our p-value threshold of significance using a Bonferroni correction to $p < 0.001$.

3 Results

The results of the repeatability analysis and estimated values for $f_{intraneurite}$, $f_{intrasoma}$, R_{soma}, and $f_{extracellular}$ are reported in Table 1. These values were comparable to previously reported values estimated using ultra-high b-value data [5]. Intra-subject variability was generally very low for all metrics across all grey matter networks (Figs. 2 and 3), reflected by high ICC values for SANDI metrics (mean ICC = 0.95) and DTI metrics (mean ICC = 0.93). Regions and metrics with lower repeatability and greater intra-subject variability included FA in the limbic network (ICC = 0.52, p = 0.11), $f_{extracellular}$ in the subcortical grey matter (ICC = 0.63, p = 0.04) and R_{soma} in the limbic network (ICC = 0.66, p = 0.04).

Despite high repeatability across both DTI and SANDI metrics in the grey matter, DTI metrics exhibited larger uncertainty around ICC estimates, indicated by larger error bars (Fig. 4a). In terms of SANDI model uncertainty, ICC values in the limbic network for all QD estimates had a wide variation indicated by larger error bars for the bounds of each ICC estimate (Fig. 4b), and similar patterns were observed for $f_{intrasoma}$ in the subcortical grey matter.

The results of the regional subcortical analysis are presented in Fig. 5. We observed low repeatability of $f_{extracellular}$ in all regions apart from the left amygdala and left caudate (Fig. 5a, b). For R_{soma}, only the left amygdala, left pallidum, and right hippocampus exhibited low repeatability (Fig. 5a, b). The distribution of QD estimates for $f_{intrasoma}$ suggest potential variation in model fit between regions (Fig. 5c).

4 Discussion

Our findings reveal that estimates of grey matter microstructure using soma and neurite density imaging are highly stable across repeated imaging sessions. We demonstrate high repeatability in a number of functional networks, known to share structural covariance [19]. In addition, the soma signal fraction variation across limbic and visual networks follows the estimated anterior to posterior gradient of cell density in the cortex of human and other primates [20]. Overall, our findings of high repeatability of dMRI metrics in the grey matter suggest the increased power to detect group differences in applications of this technique.

The limbic network showed consistently lower repeatability and model uncertainty amongst both DTI and SANDI metrics. Given the anatomical location of these fronto-temporal structures, it is likely that susceptibility-induced distortions may detrimentally influence the repeatability of certain diffusion MRI metrics. The effect of gradient non-linearities and spatiotemporally varying b-values could impact repeatability, if the subject is placed in a slightly different position in the scanner. Despite this general observation of the SANDI metrics studied here, only soma radius exhibited low repeatability in this network. Therefore, R_{soma} estimates in fronto-

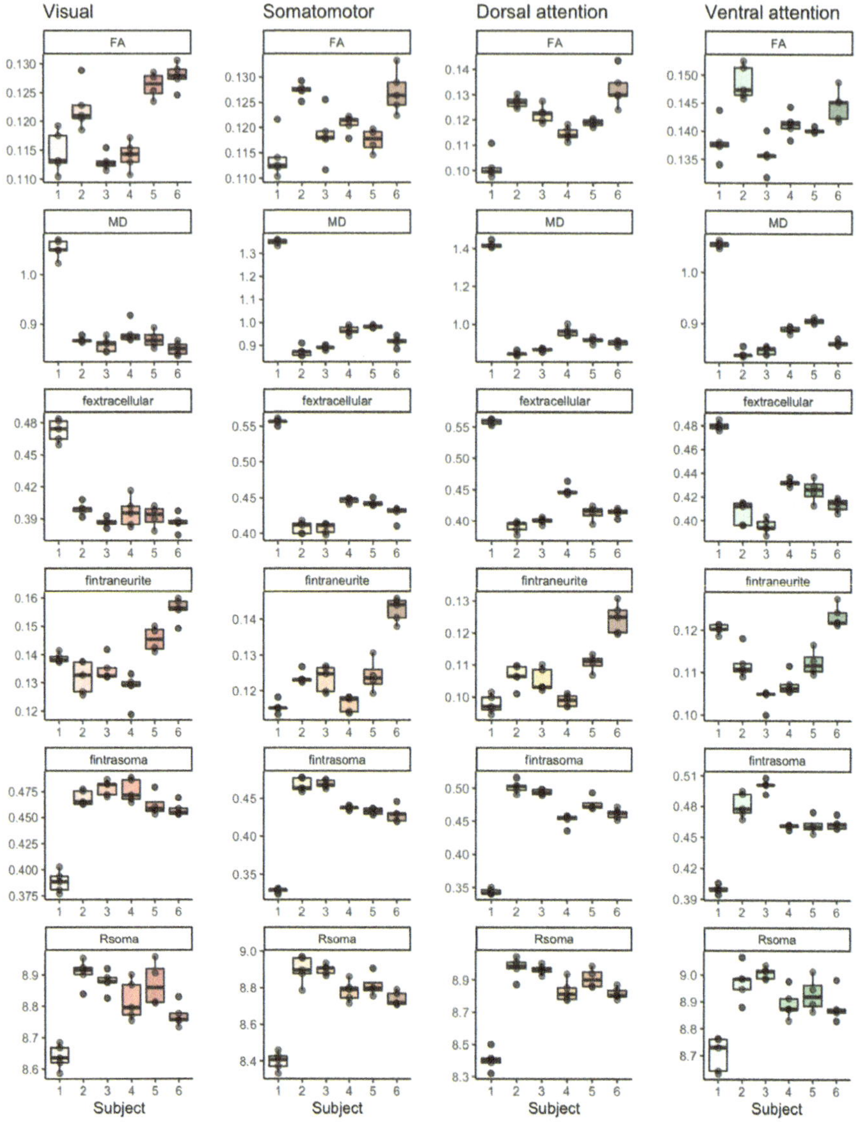

Fig. 2 Spread of values for DTI and SANDI metrics in networks 1–4. Each subject (on the x-axis) has 5 data points representing each scan. The y-axis represents the point estimate of each microstructural metric: FA, MD (10^{-3} mm^2/s), $f_{extracellular}$, $f_{intraneurite}$, $f_{intrasoma}$, and R_{soma} (μm)

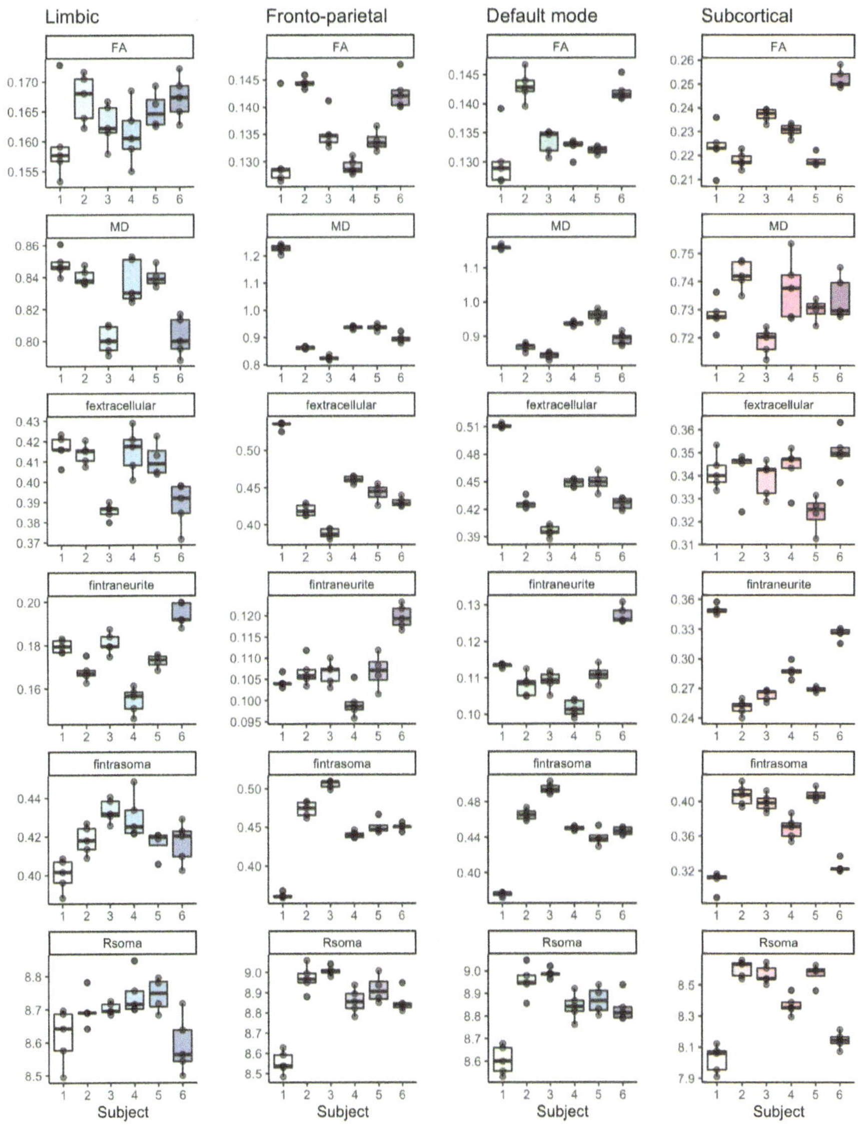

Fig. 3 Spread of values for DTI and SANDI metrics in networks 5–8. Each subject (on the x-axis) has 5 data points representing each scan. The y-axis represents the point estimate of each microstructural metric: FA, MD (10^{-3} mm^2/s), $f_{extracellular}$, $f_{intraneurite}$, $f_{intrasoma}$, and R_{soma} (μm)

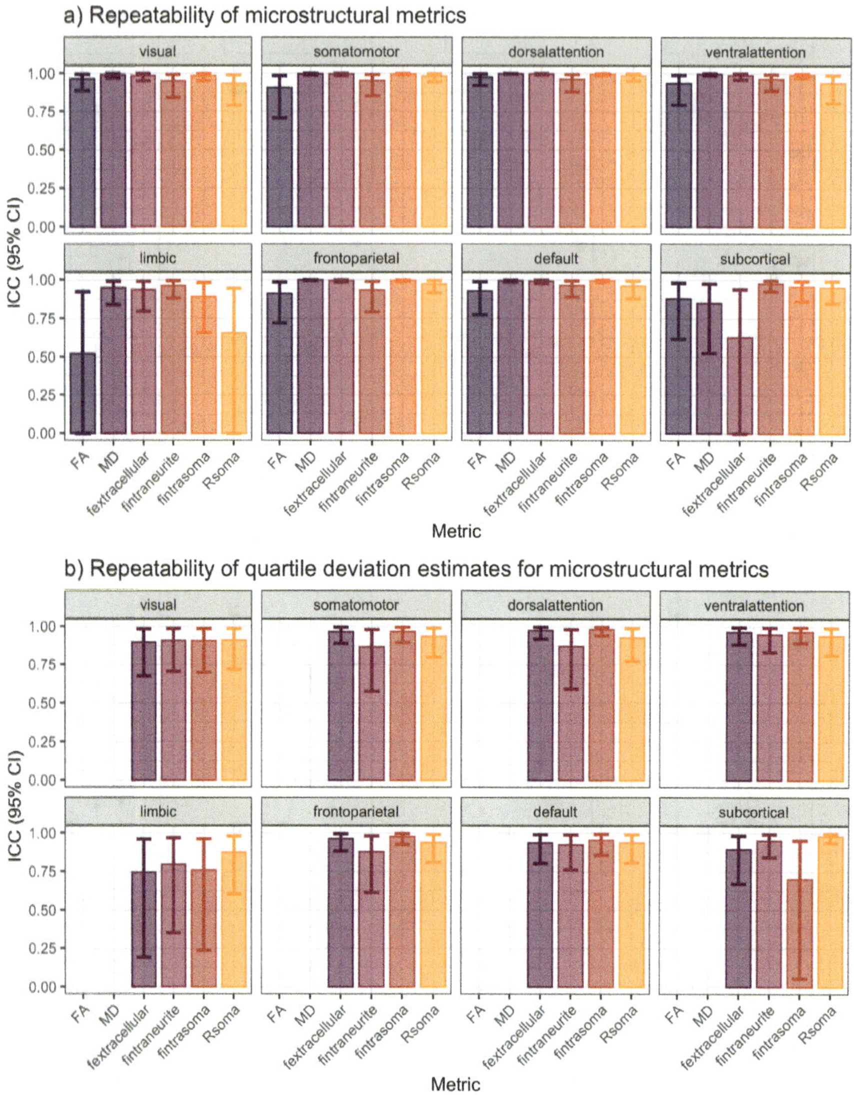

Fig. 4 Intra-class correlation coefficients (two-way random effects, absolute agreement) for test-re-test repeatability of **a** DTI and SANDI metrics, and **b** the quartile deviation of SANDI metrics, in cortical and subcortical grey matter networks. Error bars represent the bounds of the 95% confidence interval (CI) for each ICC estimate

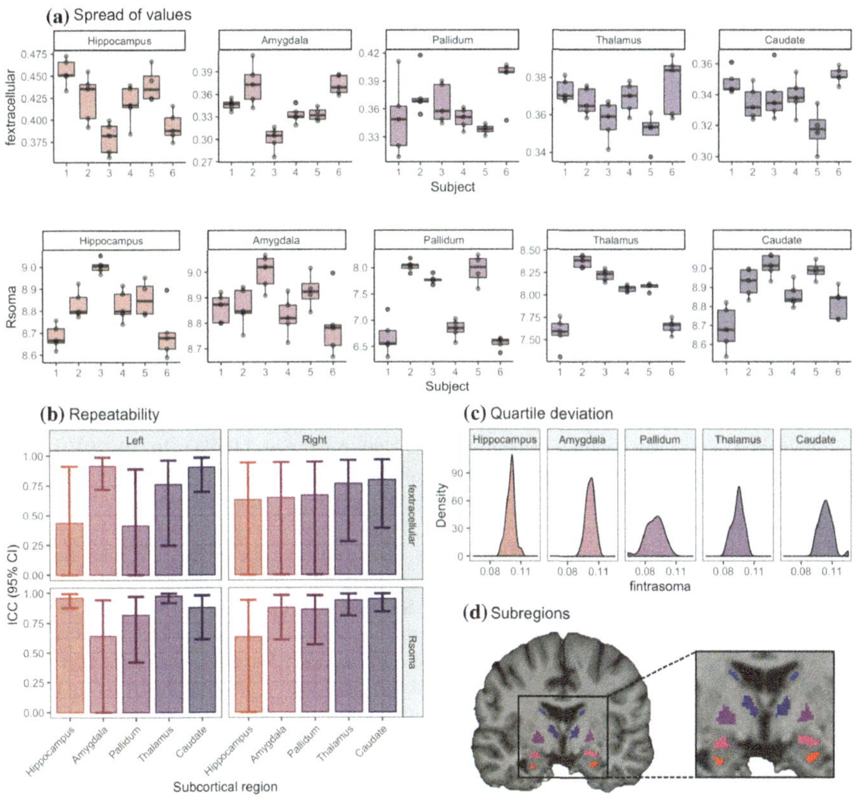

Fig. 5 Regional analysis in subcortical grey matter for the hippocampus, amygdala, pallidum, thalamus and caudate. **a** Spread of $f_{extracellular}$ and R_{soma} values in the left hemisphere; **b** intra-class correlation coefficients (two-way random effects, absolute agreement) for test-re-test repeatability of $f_{extracellular}$ and R_{soma}; **c** distribution of quartile deviation (QD) estimates for $f_{intrasoma}$; **d** regions of interest used in the analysis, obtained using Freesurfer [17] and eroded by 1 mm

temporal structures should be interpreted with caution, particularly in populations where these artefacts may be exaggerated (e.g. fronto-temporal dementia).

Upon further analysis of individual subcortical regions, the repeatability of $f_{extracellular}$ was generally low. Tissue properties within subcortical regions are heterogeneous, as sub-segments can differ in their neurite and soma composition [21, 22]. These anatomical variations may influence the estimates reported here, and as such, even finer parcellation of individual subcomponents would be an important avenue of future research.

Finally, we have demonstrated that SANDI estimates obtained from moderate-to-high b-values (up to b = 6000 s/mm^2) are comparable in terms of magnitude to previous estimates derived from ultra-high b-values (e.g. up to b = 10,000 s/mm^2). Whilst a direct comparison between multiple sampling schemes tested on the same participant across repeated scans would be required to confirm similar magnitudes of repeatability, based on our findings we are confident that the repeatability is high enough to be acceptable for research applications using high b-values. Now that we have established that these novel markers of grey matter microstructure are stable across repeated sessions, the next step is to pinpoint the underlying tissue properties driving rapidly changing grey matter macrostructure, such as that observed in neurodevelopment and neurodegeneration.

References

1. Ball, G., Srinivasan, L., Aljabar, P., et al.: Development of cortical microstructure in the preterm human brain. Proc. Natl. Acad. Sci. **110**(23), 9541 (2013)
2. Natu, V.S., Gomez, J., Barnett, M., et al.: Apparent thinning of human visual cortex during childhood is associated with myelination. Proc. Natl. Acad. Sci. **116**(41), 20750 (2019)
3. Jones, D.K., Alexander, D.C., Bowtell, R., et al.: Microstructural imaging of the human brain with a "super-scanner": 10 key advantages of ultra-strong gradients for diffusion MRI. Neuroimage **182**, 8–38 (2018)
4. Setsompop, K., Kimmlingen, R., Eberlein, E., et al.: Pushing the limits of in vivo diffusion MRI for the Human Connectome Project. Neuroimage **80**, 220–233 (2013)
5. Palombo, M., Ianus, A., Guerreri, M., et al.: SANDI: a compartment-based model for non-invasive apparent soma and neurite imaging by diffusion MRI. Neuroimage **215**, 116835 (2020)
6. Tax, C.M.W., Szczepankiewicz, F., Nilsson, M., Jones, D.K.: The dot-compartment revealed? Diffusion MRI with ultra-strong gradients and spherical tensor encoding in the living human brain. Neuroimage **210**, 116534 (2020)
7. Koller, K., Rudrapatna, S.U., Chamberland, M., et al.: Powering Up microstructural imaging: assessing cross-metric and cross-tract statistical power on an ultra-strong gradient MRI system. In: Paper Presented at: Proceedings of the International Society for Magnetic Resonance in Medicine (ISMRM), vol. 3577, Montreal, Canada (2019)
8. Veraart, J., Fieremans, E., Novikov, D.S.: Diffusion MRI noise mapping using random matrix theory. Magn. Reson. Med. **76**(5), 1582–1593 (2016)
9. Vos, S.B., Tax, C.M., Luijten, P.R., Ourselin, S., Leemans, A., Froeling, M.: The importance of correcting for signal drift in diffusion MRI. Magn. Reson. Med. **77**(1), 285–299 (2017)
10. Andersson, J.L.R., Sotiropoulos, S.N.: An integrated approach to correction for off-resonance effects and subject movement in diffusion MR imaging. Neuroimage **125**, 1063–1078 (2016)
11. Andersson, J.L.R., Skare, S., Ashburner, J.: How to correct susceptibility distortions in spin-echo echo-planar images: application to diffusion tensor imaging. Neuroimage **20**(2), 870–888 (2003)
12. Rudrapatna, S., Parker, G., Roberts, J., Jones, D.: Can we correct for interactions between subject motion and gradient-nonlinearity in diffusion MRI. In: Paper Presented at Proceedings of the International Society for Magnetic Resonance in Medicine (2018)
13. Glasser, M.F., Sotiropoulos, S.N., Wilson, J.A., et al.: The minimal preprocessing pipelines for the Human Connectome Project. Neuroimage **80**, 105–124 (2013)
14. Kellner, E., Dhital, B., Kiselev, V.G., Reisert, M.: Gibbs-ringing artifact removal based on local subvoxel-shifts. Magn. Reson. Med. **76**(5), 1574–1581 (2016)
15. Tustison, N.J., Avants, B.B., Cook, P.A., et al.: N4ITK: improved N3 bias correction. IEEE Trans. Med. Imaging **29**(6), 1310–1320 (2010)
16. Tournier, J.D., Smith, R., Raffelt, D., et al.: MRtrix3: a fast, flexible and open software framework for medical image processing and visualisation. Neuroimage **202**, 116137 (2019)
17. Fischl, B., van der Kouwe, A., Destrieux, C., et al.: Automatically Parcellating the Human Cerebral Cortex. Cereb. Cortex **14**(1), 11–22 (2004)

18. Destrieux, C., Fischl, B., Dale, A., Halgren, E.: Automatic parcellation of human cortical gyri and sulci using standard anatomical nomenclature. Neuroimage **53**(1), 1–15 (2010)
19. Yeo, B.T., Krienen, F.M., Sepulcre, J., et al.: The organization of the human cerebral cortex estimated by intrinsic functional connectivity. J. Neurophysiol. **106**(3), 1125–1165 (2011)
20. Collins, C.E., Turner, E.C., Sawyer, E.K., et al.: Cortical cell and neuron density estimates in one chimpanzee hemisphere. Proc. Natl. Acad. Sci. **113**(3), 740 (2016)
21. McDonald, A.J.: Cytoarchitecture of the central amygdaloid nucleus of the rat. J. Comp. Neurol. **208**(4), 401–418 (1982)
22. Vogt, B.A., Pandya, D.N., Rosene, D.L.: Cingulate cortex of the rhesus monkey: I. Cytoarchitecture and thalamic afferents. J. Comp. Neurol. **262**(2), 256–270 (1987)

DW-MRI Microstructure Model of Models Captured Via Single-Shell Bottleneck Deep Learning

Vishwesh Nath, Karthik Ramadass, Kurt G. Schilling, Colin B. Hansen, Rutger Fick, Sudhir K. Pathak, Adam W. Anderson, and Bennett A. Landman

Abstract In the last two decades, numerous methods have been proposed to capture biological meaning from observed diffusion-weighted magnetic resonance imaging (DW-MRI) signals each addressing recovery of specific tissue properties (e.g. pathology based, volume fraction of tissues). Generically, specific tissue properties are recovered via a category of methods termed as tissue compartment modeling. Many recent compartmental approaches require two or more diffusivity shells. We hypothesize existence of a low dimension common representation for a wide variety of commonly used microstructural measures in common use. To test this hypothesis, we constructed 13 voxel-wise measurements from 5 distinct model-based approaches and used a multi-task deep convolutional network with a variable width bottleneck to infer these metrics from empirical DW-MRI. This is the first study to use data-driven exploration to map a common basis among DW-MRI modeling approaches. We propose to capture a compact feature space in the form of a bottleneck that preserves common features to all methods and retrieve information from single shell DW-MRI. We train on 3D patches of 40 Human Connectome Project (HCP) subjects ($\sim$27 million patches) where input is based on single shell DW-MRI and ground truth is estimated from all three shells of HCP data. We validate on 24 subjects and test

V. Nath (✉) · K. Ramadass · C. B. Hansen · B. A. Landman
Computer Science, Vanderbilt University, Nashville, TN, USA
e-mail: vishwesh.nath@vanderbilt.edu

K. G. Schilling
Radiology, Vanderbilt University Medical Center, TN, USA

A. W. Anderson · B. A. Landman
Biomedical Engineering, Vanderbilt University, Nashville, TN, USA

R. Fick
Therapanacea, Paris, France

S. K. Pathak
Learning Research and Development Center, University of Pittsburgh, Pittsburgh, PA, USA

N. Gyori et al. (eds.), *Computational Diffusion MRI*, Mathematics and Visualization,
https://doi.org/10.1007/978-3-030-73018-5_12

on 25 subjects. The error reported for 5 microstructure methods on test set: 4.0%, 2.7%, 6.3%, 4.5% and 3.6%. We find that 6 features in the bottleneck layer efficiently capture an intrinsic feature space for the range of DW-MRI metrics explored.

1 Introduction

Currently, diffusion-weighted magnetic resonance imaging (DW-MRI) is the only imaging modality that indirectly probes the 3-D microstructural geometries of the tissue in a non-invasive fashion [1]. Applications in clinical research for DW-MRI are increasing at an exponential rate [2]. An upcoming application domain for DW-MRI is estimation of compartmental tissue properties at an intra-voxel milli-metric scale [3]. Voxel-based compartmental models of the DW-MRI signal are in common use to assess tissue properties and reveal associations between tissue microstructure in health and disease. Common measures include the ball and stick [4], intra-voxel incoherent motion (IVIM) [5], neurite orientation density distribution imaging (NODDI) [6], spherical mean technique (SMT), [7] and diffusion tensor imaging (DTI) [8] approaches. The compartment methods estimate volume fractions, diffusivities and dispersions. These measures are useful for estimating the complexity of dendrites and axons and provide more specific markers of brain tissue when compared to relatively simple anisotropic measures such as fractional anisotropy (FA) from diffusion tensor imaging (DTI) [8]. NODDI [6] offers different kinds of volume fractions such as intra-cellular, extra-cellular and cerebrospinal fluid compartment measures along with a orientation dispersion index, as reviewed by [9]. A typical caveat for a majority of these methods is that they require multiple diffusion shells in terms of acquisitions in [3, 6, 7].

There have been multiple attempts at the recovery of tissue microstructural measures with observations at two non-zero shells of DW-MRI data [10–13]. We hypothesize the existence of a compact low dimensional intrinsic feature space which could be used to understand commonalities and differences among DW-MRI modeling approaches (Fig. 1). To test this hypothesis, we use a data-driven approach for estimation of the unified feature space via a convolutional neural network. As a part of this hypothesis, we also chose to model utilizing only a single shell DW-MRI for the estimation of compartmental microstructural measures.

We choose to focus on data-driven single shell learning targeting metrics that are typically estimated from multi-shell data to address the following challenges. (1) A plethora of new DW-MRI methods requires at least a minimum of 2 diffusion shells of information [3]. Moreover, if a method can be estimated by one shell, then often multiple shells preferable for a more robust fit. (2) The compartmental methods are computationally expensive on CPUs. For example, NODDI takes approximately 30 hours of computation time on a single subject with high-quality acquisitions from the Human Connectome Project (HCP) data. Parallel computing or GPU-based processing have been proposed for computational speed up. (3) We currently lack a

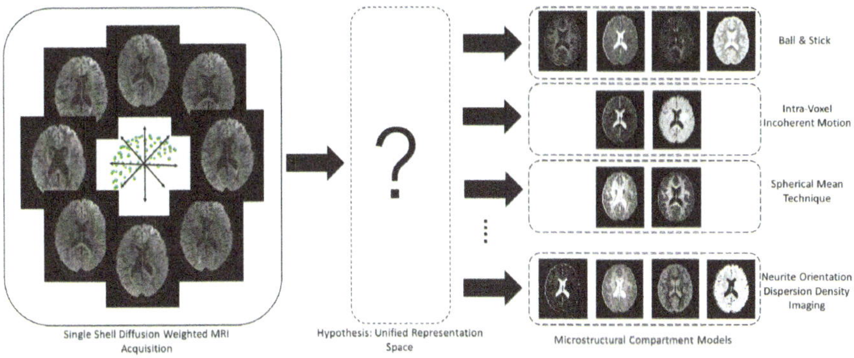

Fig. 1 Multi-compartment models rely or prefer multi-shell DW-MRI. Here, we depict the hypothesis of a compact low dimensional unified feature space that generalize to multiple compartment microstructural measures of dispersion, diffusivities and volume fractions for different compartments

unified framework where multiple methods can be computed simultaneously or in a joint fashion.

To overcome these limitations, we propose a proof of concept for a unified framework that can recover 13 different intra-voxelmicrostructural measures from 5 different DW-MRI modeling methods. We show that these methods rely on a common feature space that is low dimensional and this manifold exhibits the generalizability for multiple microstructural methods that are incorporated in this joint framework.

2 Related Work

There are multiple prior methods that have investigated the problem for a deep learned estimation of microstructural measures. The work of [10] operates on DW-MRI volumes directly without any signal representation and shows that NODDI metrics can be recovered with 12 DWI volumes with promising accuracy. The work was limited in the capacity that a single microstructure method NODDI was being utilized to train a data-driven version of NODDI implying that for multiple microstructure methods multiple data-driven models would be needed. In [11], the authors propose to use a two-stage deep learning method (MEDN) to learn the microstructural measures of NODDI. They model three measures of intra-cellular, isotropic diffusivity and orientation dispersion, but the method is limited as it utilizes two shells of information and is specific to only a single microstructural method. The authors use training data for five subjects and testing for five subjects. Error reported is 0.04 for v_{ic} (intra-cellular volume fraction), 0.02 for v_{iso} (cerebrospinal fluid volume fraction) and 0.06 for v_{od} (orientation dispersion). In [12], the authors follow a similar proposition as [11] with improved results, but were still limited as two shells of information were required for input. In [13], LSTM's were incorporated for a better deep learning network architecture (MESC). The authors utilize two shells of information 1000 and 2000 s/mm^2 with 30 gradient directions per shell. The method is applied independently for NODDI, SMT and SHORE reconstructions. However, the work was still limited by the use of two shells of information and it does not represent a joint framework. The network still needs to be independently trained for each microstructure method using 25 HCP subjects, 5 for training and validation both and 20 for testing.

3 Data Acquisition

Dataset 1: A random selection of 89 subjects from the (HCP [14] was used. The data were split into 40 subjects for training, 24 for validation, and 25 for testing. The data has three diffusivity shells of 1000, 2000 and 3000 s/mm^2 with 90 gradient directions per shell with an additional 18 non diffusion-weighted volumes. All diffusivity shells with all gradient directions were used for fitting of microstructural compartment models for ground truth. However, for the training of the unified deep model, we only used the single shell at b-value of 1000 s/mm^2 for input.

Considering in-brain voxels, there are over $\approx$680,000 voxels in a single brain volume. Assuming that a cubic patch of $3 \times 3 \times 3$ mm^3 can be selected per voxel the total number of training patches were $\sim$27 million across the 40 subjects in the training dataset. We randomly sample a consistent number of patches from each subject and similarly for validation. More detail is covered in Sect. 4.

4 Proposed Method

DW-MRI Signal Representation: We are tackling the problem of the non-linear mapping between the DW-MRI volumes and the microstructural scalar measures. First, for a unified representation of the DW-MRI volumes, we need to make it independent of scanning acquisition parameters. For single shell DW-MRI, spherical harmonics (SH) are well known to represent the DW-MRI signal with minimal representation error [15]. The SH basis is formed by gradient direction table through the utilization of Legendre polynomials. Once the basis is formed we can use regularized linear least squares fitting to solve for C_{lm} in $E = C_{lm}Y_l^m(g)$. Here, E represents the normalized DW-MRI signal, $Y_l^m(g)$ represent the SH basis and C_{lm} are the SH coefficients that represent the DW-MRI signal. l and m represents the SH order and the phase respectively, we used even 8th order SH (45 coefficients) as HCP dataset has 90 gradient directions per single shell of DW-MRI data. The SH coefficients were fitted independently per acquisition shell.

Microstructural Measures: The 13 microstructural measures we used are from 5 different methods: Ball and Stick [4], IVIM [5], NODDI [6], SMT [7] and diffusion tensor imaging (DTI) [8]. All these methods estimate different microstructural scalar measures and for the sake of generalizability, we will refer to these as $\mathcal{M}_b^a$ here $\mathcal{M}^a$

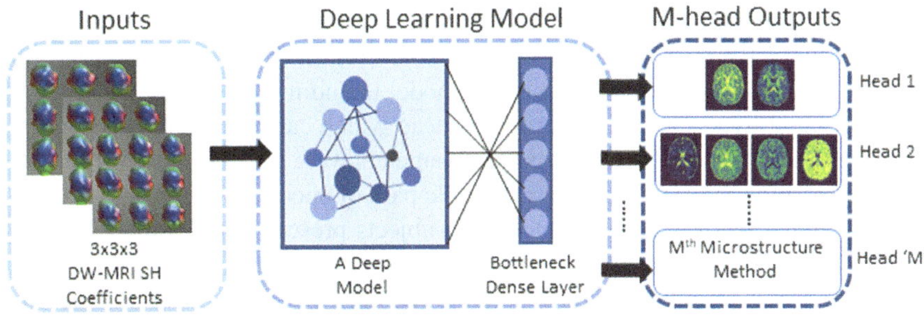

Fig. 2 A schematic overview of the proposed method from left to right showing SH coefficient of a 3D cubic patch as input to the deep learning model which has a bottleneck dense layer with M-heads at output for different microstructural methods

represents a unique microstructure method a representing its index and b representing the number of microstructural measures estimated by $\mathcal{M}$. As microstructural measures for a method could have different units and scales, they were rescaled specifically for in-brain voxels to [0, 1]. To mitigate the energy of the loss function everything was normalized. The diffusivity measures were linearly scaled as rescaling has shown to be effective as in prior work [13]. All microstructural measures estimated have been utilized for experiments except for orientational estimates.

Bottleneck Deep Learning: We use a deep learning network architecture which is inspired by prior work from [16]. The SHResNet was proposed as a harmonization model and has been shown to be capable of also predicting fiber orientation distributions. We adapt the network with two modifications: First, we introduce a dense layer as a 'true' bottleneck (Fig. 2). The number of neurons in this layer were varied for training different models. We would also like to express clarity with the information bottleneck line of work by Tishby et al. in [18, 19] as those methods were developed to study informational content of deep learning layers. In this work, we restrict the feature space as low as possible and the terms of information bottleneck and the bottleneck here should not be confused.Second, inspired by the M-heads network architecture [17], we attach the number of heads based on the number of microstructural methods. Each head has a dense layer of 50 neurons, and the output layer neurons are dependent upon the number of microstructural measures b of the method $\mathcal{M}^a$. Both layers are linearly activated.

For the input of the deep learning framework, we use a patch size of $3 \times 3 \times 3 \times 45$ of SH coefficients. A similar patch technique has been used before in [12, 13]. We define a joint loss function $\mathcal{L}$ covering the terms of microstructural measures from different heads as below:

$$\mathcal{L} = \sum_{a=1}^{M} \sum_{b=1}^{N} \gamma^a |\hat{\mathcal{M}}_b^a - \mathcal{M}_b^a| \quad (1)$$

Since each set of microstructural measures is calculated per head, the gradients are dependent upon the microstructural method. To overcome this problem, all gradients are constrained to pass through the entire model including the bottleneck layer. γ^a are loss weight hyper-parameter terms per head which can be tuned if needed. For this experiment, all weight tuning hyper-parameters were set to 1.

Deep learning hyper-parameters: A single pass of epoch constitutes 50,000 randomly sampled patches per subject for all subjects present in the training dataset. Similarly, 50,000 patches were randomly sampled per subject for estimating validation errors across all subjects in the validation dataset. Other deep learning parameters: batch size of 1000, learning rate was kept consistent at 0.0001, RMSProp [21] was used as the optimizer and epochs were set at 30. Convergence criteria were based on the least amount of validation error.

Inference Testing: The deep learning model was trained with an input of 8th order SH coefficients and same was utilized for all testing subjects with multiple bottleneck neurons with 90 gradient directions. For additional validation, gradient directions were uniformly randomly sampled in sets of 15, 30, 45 and 60 per subject from the testing set. The corresponding SH orders that were applied are: 4th order SH for 15 directions, 6th order SH for 30, 45 directions and 8th order SH for 60 directions. For the specified set of gradient directions the bottleneck model was tested at 50 and 6 neurons. It should be noted that since the input was tied to 45 coefficients as for training, the reduced order SH were padded with zeros to compensate as additional input.

For implementation purposes, Dmipy was used for microstructural model fitting [3]. The deep learning models were implemented with Keras module of Tensorflow 2.0 [20].

5 Results

Quantitatively, the number of neurons are increased in the bottleneck layer the error decreases for all methods in a similar trend for training (Left) and validation (Right) plots (Fig. 3). Although, the error decreases as more neurons are added however, the differences are minimal after 6 neurons in the bottleneck layer.

In Table 1, the mean values of all metrics per method across all 25 subjects. The mean error with 6 neurons versus 50 neurons for 90 gradient directions the change in error is on the order of the third decimal. The highest change in error can be seen for NODDI which is 0.006. The standard deviations for both variants are consistent. The error from prior work where multi-shell DW-MRI were used as input [13]. These results are comparable as they were computed on 20 HCP subjects. The individual metric errors were reported across all 20 subjects are 0.05, 0.04 and 0.03 for v_{od},v_{ic} and v_{iso}. Relatively, we achieve competitive errors of 0.053, 0.037 and 0.033 for v_{od},v_{ic} and v_{iso} respectively with a bottleneck width of 50.

Comparing across different sets of gradient directions of 15, 30, 45 and 60, when intra-compared for a set of gradient directions. The mean error between 6 versus

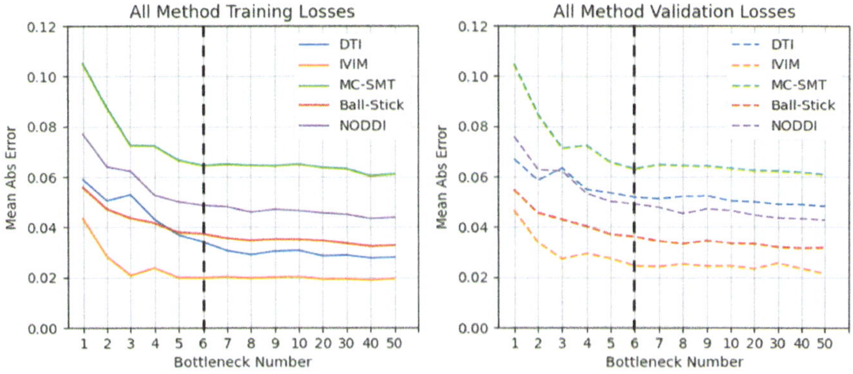

Fig. 3 Training and validation errors across all microstructural methods per bottleneck number individually for each method. At each bottleneck number the model with the lowest validation error was selected. The black line represents a bottleneck number of 6 after which increasing the bottleneck width the changes in error are minimal

50 neurons is similar as was observed for 90 gradient directions in the first 2 rows. Mean error as the gradient directions are being decreased, the error steadily increases. However, the incremental difference is in the 3rd order of the decimal till 30 gradient directions for all methods. SMT shows a little larger increase in error relatively as compared to other methods. However, at 15 gradient directions the errors go up by a multiple of 2 indicating a huge drawback and feasibility is only indicated till 30 gradient directions.

The ground truth of one metric per method using all DW-MRI shells in the first row (Fig. 4). The second row and fourth row depict the predictions with 50 and 6 neurons in the bottleneck layer, qualitatively the predictions are quite similar for both variants. The spatial absolute error for the variants of 50 and 6 neurons can be observed in the third and the fifth row respectively. Comparing between the variants for DTI (FA), Ball and Stick (intra-axonal fraction) and IVIM (diffusivity), the change in error is negligible. However, for SMT (intra-neurite fraction) and NODDI (orientation dispersion), there is a slight increase in error towards the frontal lobe (SMT) and the ventricles (NODDI).

Qualitatively, Fig. 5 visualizes the intrinsic features captured by the bottleneck layer (in this case, a layer with 6 features). These features are visually representative of contrasts that micro-structural models aim to represent. For example, parallels can be drawn between the contrast seen in Feature #4 and traditional anisotropy maps. Feature #1 shows contrast that may be typical of a mean or overall diffusivity, while #5 shows trends expected in neurite dispersion maps. Importantly, this space of 6 features captures the trends and variations of the diffusion signal (i.e., enough to decode to the individual metrics captured in the studied models) and shows the ability to distinguish contrast both between and within tissue types. Thus, we posit that this compact representation of the multi-orientation multi-diffusion weighted

Table 1 Mean absolute error for all metrics between ground truth and predictions per method for bottleneck variants of 6 and 50 neurons (N). The different bottleneck variants are tested with reduced set of gradient directions (G) of 60, 45, 30 and 15. The mean and standard deviation reported here are across all 25 subjects from the test split of the data

Mean Abs error across 25 test subjects					
Bottleneck Variant	Ball & Stick	IVIM	SMT	NODDI	DTI
90G and 6N	0.040 ± 0.001	0.024 ± 0.001	0.068 ± 0.003	0.051 ± 0.001	0.040 ± 0.001
90G and 50N	0.035 ± 0.001	0.027 ± 0.001	0.063 ± 0.002	0.045 ± 0.001	0.036 ± 0.001
60G and 6N	0.041 ± 0.001	0.029 ± 0.002	0.072 ± 0.004	0.055 ± 0.002	0.043 ± 0.001
60G and 50N	0.037 ± 0.001	0.030 ± 0.001	0.068 ± 0.002	0.049 ± 0.002	0.038 ± 0.001
45G and 6N	0.044 ± 0.002	0.032 ± 0.003	0.082 ± 0.006	0.059 ± 0.003	0.044 ± 0.001
45G and 50N	0.040 ± 0.002	0.031 ± 0.002	0.074 ± 0.004	0.053 ± 0.003	0.040 ± 0.002
30G and 6N	0.049 ± 0.007	0.037 ± 0.008	0.085 ± 0.010	0.063 ± 0.008	0.047 ± 0.005
30G and 50N	0.043 ± 0.004	0.035 ± 0.005	0.082 ± 0.007	0.056 ± 0.006	0.043 ± 0.003
15G and 6N	0.092 ± 0.007	0.052 ± 0.005	0.187 ± 0.016	0.115 ± 0.009	0.064 ± 0.003
15G & 50N	0.103 ± 0.008	0.056 ± 0.003	0.213 ± 0.014	0.127 ± 0.008	0.064 ± 0.003

signal contains the principal features (or some combination of features) which are shared by standard modeling approaches.

6 Discussion

We showed a unified framework that can recover microstructural measures with relatively low error. The errors are comparable to prior deep learning approaches where multiple shells of information have been used [12, 13]. The results lend support to our hypothesis that shows these methods can be computed from a common (latent) representation. A deeper consideration is required on the intrinsic representation as they resemble the computed microstructural measures closely in terms of diffusivity, anisotropic and fractional components. This consideration leads us to the question of the existence of a unified biophysical model that can capture the essence of various microstructure models. On a parallel, data-driven strategies can show utility of compressed SH coefficients and this has been mathematically shown in [22]. The microstructure methods can be recovered from relatively sparse or low number of gradient directions, e.g as low as 30 (Table 1).

There are quite a few limitations to be addressed in future work. The model used in our approach for scalar microstructure metrics applied a straight forward mean absolute error loss function. This loss function would not be appropriate for orientation estimates because it does not consider the symmetric nature of diffusion and ambiguity of the sign of orientations estimated with DW-MRI data. A full consideration of these metrics requires incorporation of symmetry invariant loss functions. To

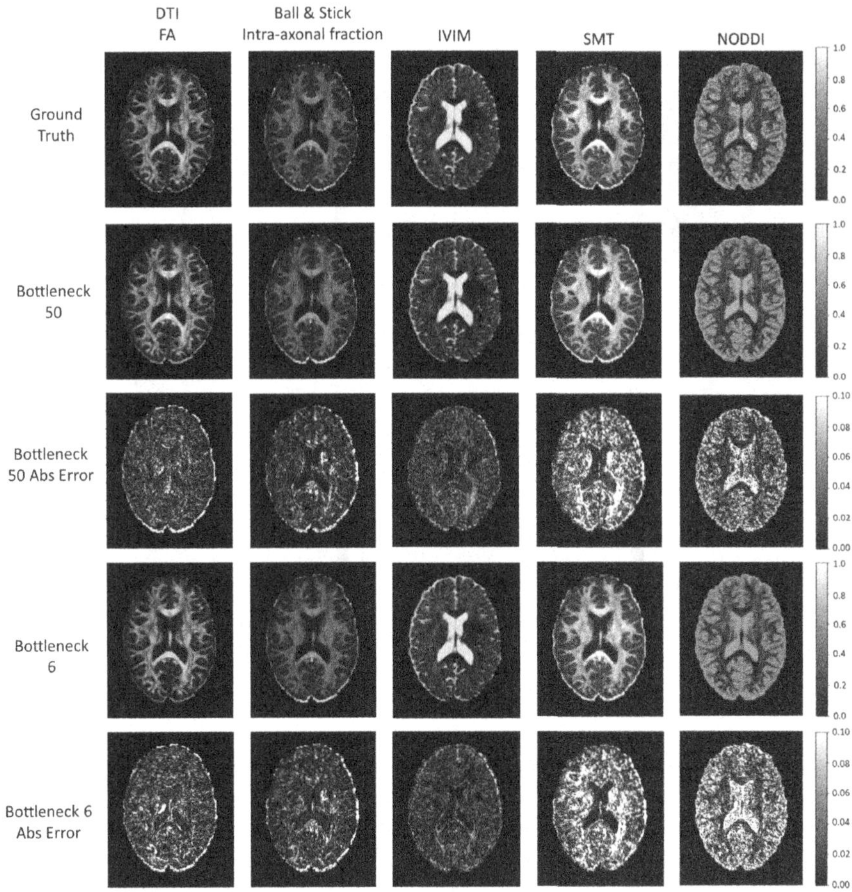

Fig. 4 Middle axial slice from a randomly selected subject. The ground truth estimates of a microstructural measure per method from all diffusion shells of data are in first row. The second row shows the predictions of the corresponding metric with a bottleneck width of 50. Third row depicts absolute error spatial maps for predictions with bottleneck width of 50. Fourth row shows predictions of corresponding metric with a bottleneck width of 6. The last row shows spatial absolute error for predictions with bottleneck width of 6

our best knowledge, the compartment based orientation estimates have not yet been recovered with low error from deep learning based approaches to date and would be a promising area of exploration. While we show the utility of the spatial neighborhood information for recovery of multi-shell information, additional validation is necessary to determine optimal patch sizes. An alternative perspective would be to observe that errors are relatively higher for microstructural measures when compared to their corresponding ground truths. There are multiple future directions that could be pursued such as improving network architectures, utilization of a better signal representation technique to mitigate such errors.

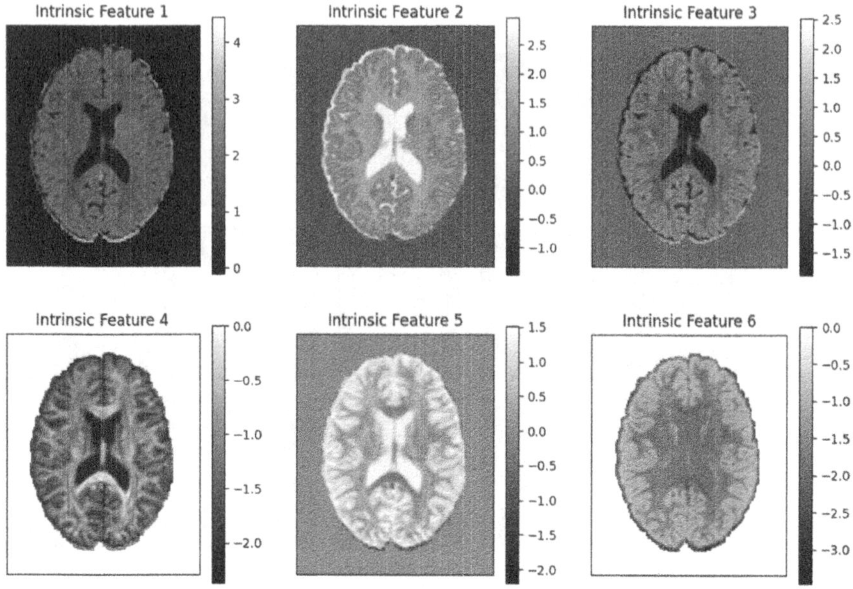

Fig. 5 Middle axial slice from a randomly selected subject. The 6 components of the bottleneck layer depicting latent representations/features that are common for the different learned microstructural measures

Conclusion: We show the evidence of a data-driven latent biophysical model that connects existing DW-MRI micro-structural metrics. Our approach captures intrinsic features specific to compartmental microstructural measures and reveals contrasts that are similar to those in parametric models. Additionally, we showed the recovery of multi-shell based microstructural measures from a single DW-MRI shell with comparable error to prior approaches that have used more data and been task-specific. This work is the first unified deep learning architecture for multiple microstructural measures from different methods and should not be considered the last.

Acknowledgements This work was supported by R01EB017230 (Landman). This work was conducted in part using the resources of the Advanced Computing Center for Research and Education at Vanderbilt University, Nashville, TN. This project was supported in part by the National Center for Research Resources, Grant UL1 RR024975-01, and is now at the National Center for Advancing Translational Sciences, Grant 2 UL1 TR000445-06. This research was supported in part by the Intramural Research Program, National Institute on Aging, National Institutes of Health. The content is solely the responsibility of the authors and does not necessarily represent the official views of the NIH. This work has been supported by Nvidia with supplement of hardware resources (GPUs) in the form of a Titan Xp. This work was supported by the U.S. Army Medical Research and Material Command and from the U.S. Department of Veterans Affairs Chronic Effects of Neurotrauma Consortium under Award No. W81XWH-13-2-0095. The U.S. Army Medical Research Acquisition Activity, and the Chronic Effects of Neurotrauma Consortium/Veterans Affairs Rehabilitation Research & Development project F1880, US Army 12342013 (W81XWH-12-2-0139), Office of Naval Research and Naval Health Research Center (W911QY-15-C-0043).

References

1. Le Bihan, D.: Looking into the functional architecture of the brain with diffusion MRI. Nat. Rev. Neurosci. **4**(6), 469–480 (2003)
2. Novikov, D.S., Kiselev, V.G., Jespersen, S.N.: On modeling. Mag. Resonance Med. **79**(6), 3172–3193 (2018)
3. Fick, R.H.J., Wassermann, D., Deriche, R.: The dmipy toolbox: diffusion mri multi-compartment modeling and microstructure recovery made easy. Frontiers Neuroinf. **13**, 64 (2019)
4. Behrens, T.E.J., et al.: Characterization and propagation of uncertainty in diffusion-weighted MR imaging. Mag. Resonance Med. Official J. Int. Soc. Mag. Resonance Med. **50**(5), 1077–1088 (2003)
5. Gurney-Champion, O.J., et al.: Comparison of six fit algorithms for the intra-voxel incoherent motion model of diffusion-weighted magnetic resonance imaging data of pancreatic cancer patients. PloS one **13**(4) (2018)
6. Zhang, H., et al.: NODDI: practical in vivo neurite orientation dispersion and density imaging of the human brain. Neuroimage **61**(4), 1000–1016 (2012)
7. Kaden, E., et al.: Multi-compartment microscopic diffusion imaging. NeuroImage **139**, 346–359 (2016)
8. Basser, P.J., Mattiello, J., LeBihan, D.: MR diffusion tensor spectroscopy and imaging. Biophys. J. **66**(1), 259–267 (1994)
9. Panagiotaki, E., et al.: Compartment models of the diffusion MR signal in brain white matter: a taxonomy and comparison. Neuroimage **59**(3), 2241–2254 (2012)
10. Golkov, V., et al.: q-Space deep learning: twelve-fold shorter and model-free diffusion MRI scans. IEEE Trans. Med. Imaging **35**(5), 1344–1351 (2016)
11. Ye, C.: Estimation of tissue microstructure using a deep network inspired by a sparse reconstruction framework. In: International Conference on Information Processing in Medical Imaging. Springer, Cham (2017)
12. Ye, C.: Tissue microstructure estimation using a deep network inspired by a dictionary-based framework. Med. Image Anal. **42**, 288–299 (2017)
13. Ye, C., Li, X., Chen, J.: A deep network for tissue microstructure estimation using modified LSTM units. Med. Image Anal. **55**, 49–64 (2019)
14. Van Essen, D.C., et al.: The WU-Minn human connectome project: an overview. Neuroimage **80**, 62–79 (2013)
15. Descoteaux, M., et al.: Apparent diffusion coefficients from high angular resolution diffusion imaging: Estimation and applications. Mag. Resonance Med. Official J. Int. Soc. Mag. Resonance Med. **56**(2), 395–410 (2006)
16. Koppers, S., et al.: Spherical harmonic residual network for diffusion signal harmonization. In: International Conference on Medical Image Computing and Computer-Assisted Intervention. Springer, Cham (2018)
17. Lee, S., et al.: Why M heads are better than one: Training a diverse ensemble of deep networks (2015). arXiv:1511.06314
18. Tishby, N., Zaslavsky, N.: Deep learning and the information bottleneck principle. IEEE Inf. Theory Workshop (ITW), IEEE (2015)
19. Tishby, N., Pereira, F.C., Bialek, W.: The information bottleneck method (2000). ArXiv: physics/0004057
20. Abadi, M., et al.: Tensorflow: A system for large-scale machine learning. In: 12th USENIX Symposium on Operating Systems Design and Implementation (OSDI 16) (2016)
21. Tieleman, T., Hinton, G.: Lecture 6.5-rmsprop: Divide the gradient by a running average of its recent magnitude. COURSERA: Neural Netw. Mach. Learn. **4**(2), 26–31 (2012)
22. Zucchelli, M., Deslauriers-Gauthier, S., Deriche, R.: A computational Framework for generating rotation invariant features and its application in diffusion MRI. Med. Image Anal. **60** (2020)

Above, $p_i^{\min}$ and $p_i^{\max}$ are again the lower/upper bounds of p_i, while $h(v_i)$ is

$$h(v_i) = \frac{2}{1 + e^{-\alpha_i \left(\log\left(\mathrm{softplus}(v_i)\right) - \log\left(\log(2)\right)\right)}} - 1, \tag{6}$$

where $\mathrm{softplus}(x) = \log(1 + e^x)$ and where α_i is an actual network parameter that is learnt during training, similarly to all linear layer weights and biases. The mapping in Eq. 5 prevents infinite or not-a-number values to occur when computing Eq. 4. Training is performed on synthetic or actual qMRI measurements and no knowledge of ground truth tissue parameters is required.

2.2 *In Silico Study*

Signal synthesis We performed numerical simulations to compare DNN fitting implemented according to strategies 1 and 2. To this end, we consider one among several potential qMRI signal models and focus on increasingly popular DRI, which exploits the complementary information from different qMRI contrasts in a unified acquisition [11, 12]. Specifically, we consider the case of joint diffusion and T1 mapping in the brain, as achievable in inversion recovery (IR) [13] or saturation IR (SIR) [14] DWI. We borrow a previously proposed tensor approach [13] and adapt it to directionally-averaged (i.e. spherical mean) DW signals, as this removes the dependence on the underlying fibre orientation distribution [15]. The forward model $s(\mathbf{p}, \mathbf{u})$ is

$$s = \frac{\sqrt{\pi}}{2} s_0 \left| 1 - e^{-\frac{\mathrm{TI}}{T_1}} - \left(1 - e^{-\frac{\mathrm{TS}}{T_1}}\right) e^{-\frac{\mathrm{TI}}{T_1}} \right| e^{-b\,k\,d_{||}} \frac{\mathrm{erf}\left(\sqrt{b\,(d_{||} - k\,d_{||})}\right)}{\sqrt{b\,(d_{||} - k\,d_{||})}}, \tag{7}$$

with tissue parameters $\mathbf{p} = \left[d_{||}\ k\ T_1\ s_0\right]^{\mathrm{T}}$ (parallel diffusivity $d_{||}$, defined in [0.01; 3.2] $\mu\mathrm{m}^2\mathrm{ms}^{-1}$; anisotropy parameter k, defined in [0; 1] and such that the perpendicular diffusivity $d_\perp$ is $d_\perp = k\,d_{||}$ (anisotropy $\propto 1/k$); longitudinal relaxation time T_1, defined [100; 4000] ms; apparent proton density s_0, defined in [0.5; 5]) and sequence parameters $\mathbf{u} = \left[\mathrm{TS}\ \mathrm{TI}\ b\right]^{\mathrm{T}}$ (saturation-inversion delay TS; inversion-excitation delay, i.e. inversion time TI; b-value b). Tissue parameters are defined within biologically plausible ranges at 3T.

We generated synthetic SIR DWI signals with Eq. 7 (160,000 as training set; 40,000 as validation set; 16,000 as test set). For this, we used tissue parameters drawn from uniform distributions, using the same parameter bounds reported above. A protocol made of $N = 32$ measurements was used, replicating acquisitions performed in vivo and described in Sect. 2.3. These were: $(b, \mathrm{TI}) = \{0, 1000, 2000, 3000\}$ $\mathrm{s\,mm}^{-2} \times \{70, 320, 570, 820, 1070, 1320, 1570, 1820\}$ ms, with TS fixed to TS = 300 ms. Signals were corrupted with both Gaussian and Rician noise, varying the signal-to-noise ratio (SNR) across voxels uniformly within the range SNR $= \frac{s_0}{\sigma} \in [40; 100]$.

DNN fitting qMRI measurements $\mathbf{m} = [\, m_1(b_1, \mathrm{TI}_1) \; ... \; m_{32}(b_{32}, \mathrm{TI}_{32}) \,]^{\mathrm{T}}$ from each synthetic voxel were normalised as $\frac{\mathbf{m}}{\underset{n}{max}(m_n)}$, and then used to train DNNs according to both strategy 1 and 2. DNNs consisted of 7 hidden layers with input/output sizes of (32×28), (28×24), (24×20), (20×16), (16×12), (12×8), (8×4), and were trained with ADAM [16] for 250 epochs (learning rate: 0.0005) grouping voxels in mini-batches (one DNN update per mini-batch of 100 voxels). We repeated training 8 times, each time with a different random DNN initialisation, to minimise the risk of incurring in local minimina. The DNN providing the lowest validation loss was used to predict tissue parameters and signals on the test data. Root mean squared error (RMSE) with respect to synthetic noisy signals and ground truth tissue parameters were computed.

2.3 *In Vivo study*

MRI acquisition We performed brain DRI on a 3T Philips Ingenia CX system. Scans were performed after obtaining informed written consent, and were approved by a local Research Ethics Board. We recruited 3 healthy volunteers (2 females) and used a multi-slice SIR [14] DW EPI sequence, with a 32-channel head coil for signal detection. Sequence parameters were: 48 axial slices, 2.4mm-thick; field-of-view: $230 \times 230\,\mathrm{mm}^2$; in plane resolution: $2.4 \times 2.4\,\mathrm{mm}^2$; TR $= 2563$ ms; TE $=$ 90 ms; TS $= 300$ ms; SENSE $= 2$; MB $= 3$; BW $= 2.51 \frac{\mathrm{KHz}}{\mathrm{pixel}}$. 528 images were acquired with 32 unique (b, TI) values ($(b, \mathrm{TI}) = \{0, 1000, 2000, 3000\}\,\mathrm{s\,mm}^{-2} \times \{70, 320, 570, 820, 1070, 1320, 1570, 1820\}$ ms, 21 directions for non-zero b, 3 images for each $b = 0$) in 46 minutes, which included an image with reversed phase encoding for EPI distortion mitigation.

MRI post-processing We denoised scans with the MP-PCA technique [17], subsequently mitigating noise floor [18] and Gibbs ringing [19]. We corrected for motion and eddy current via affine co-registration based on NiftyReg [20] and EPI distortions with FSL [21], and obtained a brain mask [22]. Finally, we averaged images corresponding to different gradient directions at fixed (b, TI), obtaining a data set made of $N = 32$ unique (b, TI) measurements.

DNN fitting The DNNs trained for in silico experiments with Rician noise were also used for fitting Eq. 7 on in vivo data. Moreover, for training strategy 2, we also trained a DNN on actual in vivo measurements extracted from within the brain mask (which included cerebrospinal fluid (CSF)), using the same learning settings described in Sect. 2.2. We followed a leave-one-out strategy, training on 2 subjects and using the DNN for model fitting on the other (number of training voxels comparable to in silico training). Fitting provided voxel-wise tissue parameter maps and MRI signal estimates, from which RMSEs with respect to actual in vivo MRI measurements were computed. Differences between parametric maps obtained with strategy 2 with respect to those obtained with strategy 1 were evaluated. Moreover, we also calculated the difference between DNN signal prediction and actual MRI measurements (i.e. signal residuals).

3 Results

Figure 2 shows predictions of MRI signals and of tissue parameters for simulations conducted with Gaussian noise. Inspection of the plots suggests that strategy 1 and 2 perform similarly. However, when Rician noise is considered (Fig. 3), strategy 1 enables more accurate tissue parameter estimation. Nonetheless, strategy 2 still provides a better estimation of the MRI signal as compared to strategy 1, which underestimates slightly the signal level.

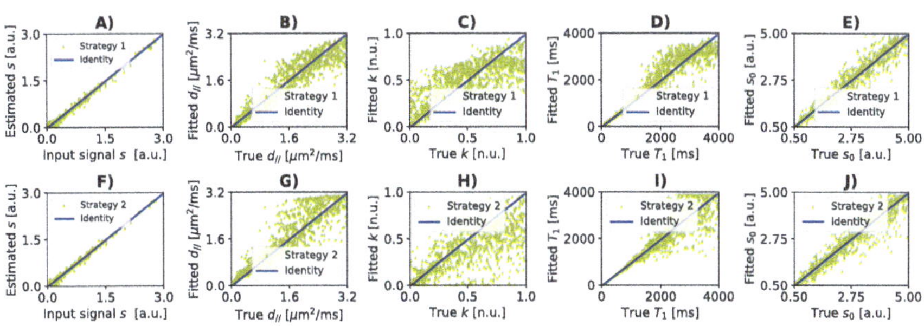

Fig. 2 Scatter plots of simulation results obtained on the test set when Gaussian noise is considered. Panels A and F show predicted signal (y-axis) against input measurements (x-axis). Panels B–E and G–J show fitted parameters (y-axis) against ground truth tissue parameters (x-axis) ($d_{||}$ in B, G; k in C, H; T_1 in D, I; s_0 in E, J). Top (A–E): training strategy 1. Bottom (F–J): training strategy 2

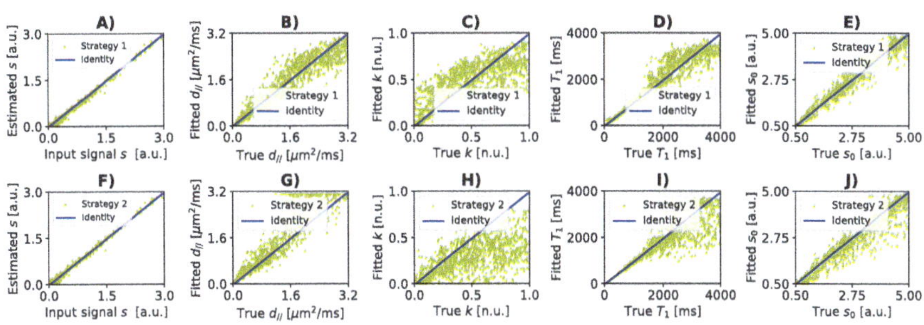

Fig. 3 Scatter plots of simulation results obtained on the test set when Rician noise is considered. As in Fig. 2, panels A and F show predicted signal against input measurements, while panels B–E and G–J fitted against ground truth tissue parameters

Table 1 Root mean squared errors (RMSEs) from simulations. RMSE between MRI measurements and signal predictions as well as between ground truth and fitted parameters. Median and 2.5–97.5 percentiles (in brackets) over the test set are reported (lowest median RMSE highlighted). Both Gaussian/Rician noise cases are reported

	Signal (a.u.)	$d_{\|\|}$ ($\mu m^2 ms^{-1}$)	k (n.u.)	T_1 (ms)	s_0 (a.u.)
Strategy 1 (Gauss. noise)	0.043 (0.009; 0.129)	**0.192** (0.008; 0.862)	0.125 (0.005; 0.445)	155.7 (4.1; 958.0)	0.126 (0.004; 1.05)
Strategy 2 (Gauss. noise)	**0.038** (0.008; 0.104)	0.205 (0.009; 1.109)	**0.120** (0.003; 0.647)	**134.0** (1.5; 1329)	**0.125** (0.004; 1.25)
Strategy 1 (Rician noise)	0.048 (0.010; 0.144)	**0.194** (0.008; 0.859)	**0.131** (0.004; 0.441)	155.6 (4.3; 976.7)	**0.124** (0.004; 1.05)
Strategy 2 (Rician noise)	**0.039** (0.008; 0.107)	0.266 (0.009; 1.543)	0.176 (0.004; 0.704)	**150.5** (1.7; 1696)	0.135 (0.003; 1.38)

Table 1 reports statistics of signal and tissue parameter RMSE distributions obtained on the test set with both Gaussian and Rician noise. Strategy 2 provides lower median RMSE for almost all tissue parameters as well as for the signal estimation than strategy 1 when Gaussian noise is considered. However, it also provides wider RMSE distributions. With Rician noise instead, strategy 1 provides overall lower RMSE figures for tissue parameter estimates, while strategy 2 still provides lower RMSE for MRI signal predictions. For both strategies, Rician noise leads to worse performances compared to Gaussian noise.

Figure 4 shows examples of voxel-wise tissue parameter maps obtained in vivo. The two training strategies provide metrics that are qualitatively similar and that exhibit similar between-tissue contrasts. However, some differences between the approaches are seen (e.g. slightly higher $d_{||}$ and lower k in strategy 2 compared to strategy 1), especially in areas with strong CSF partial volume. For training strategy 2, training the DNN on synthetic signals or actual in vivo measurements essentially provides the same results.

Figure 5 shows examples of voxel-wise differences between parametric maps obtained with strategy 2 with respect to those obtained from strategy 1. The strongest differences are observed in areas with strong CSF contamination (e.g. cortical grey matter and above all ventricles), and are the most intense for parameters $d_{||}$ and k.

Figure 6 shows examples of images acquired in vivo as well as DNN predictions (leave-one-out experiment). All training strategies enable learning of a variety of diffusion and T1 contrasts, as predictions are qualitatively similar to the acquired images. On visual inspection, predictions from strategy 2 more closely resemble in vivo images as compared to strategy 1. This latter observation is confirmed by Fig. 7, which shows examples of voxel-wise differences between DNN signal prediction and acquired MR images (i.e. signal residuals). The figure confirms that strategy 2 provides signal predictions that are closer to the acquired images than strategy 1. Moreover, the figure also highlights that both signal underestimation and overesti-

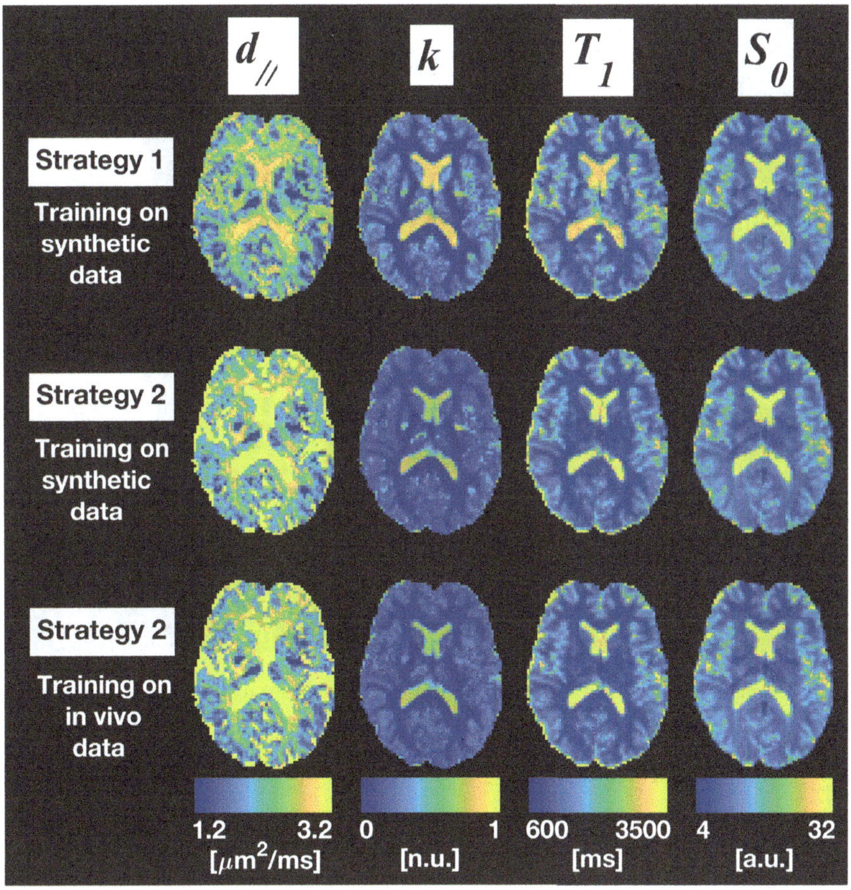

Fig. 4 Examples of in vivo parametric maps. From left to right: $d_{||}$, k, T_1 and s_0. Top row: strategy 1. Central row: strategy 2, with DNN trained on synthetic MRI signals. Bottom row: strategy 2, with DNN trained on actual in vivo MRI measurements

mation are possible. These occur mainly in areas with strong CSF partial volume, especially in the ventricles.

Finally, Table 2 reports signal prediction RMSE as measured on in vivo data. Results are in line with RMSE figures from simulations. In all cases, training strategy 2 has slightly lower RMSE compared to training strategy 1. For training strategy 2, the lowest RMSE figures are obtained when training on synthetic data. However, training on actual in vivo measurements yields almost identical performances (difference of median RMSE less than 1%).

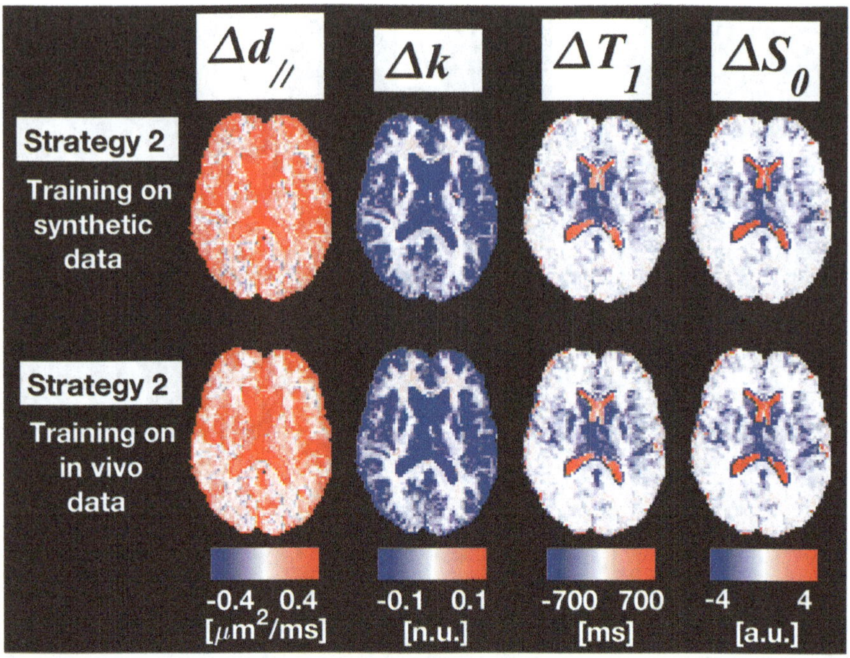

Fig. 5 Examples of differences of in vivo parametric maps as obtained from strategy 2 and strategy 1. From left to right: $\Delta d_{||}$, Δk, ΔT_1 and Δs_0 (i.e. difference of a metric from strategy 2 minus the corresponding metric from strategy 1). Top row: strategy 2, with DNN trained on synthetic MRI signals. Bottom row: strategy 2, with DNN trained on actual in vivo MRI measurements

4 Discussion

In this work we studied DNN-based qMRI signal model fitting. We used DNNs to implement an estimator of tissue parameters from input MRI measurements, and ran experiments on synthetic and in vivo DRI data. Specifically, we considered a model describing directionally-averaged SIR DWI data for joint diffusion and T1 mapping. We performed computational experiments on synthetic and in vivo scans, which were acquired at 3T on 3 healthy volunteers. Both in silico and in vivo data were used to explore two alternative training strategies: in strategy 1, the training loss is defined as the l^2-norm of tissue parameter estimation errors; in strategy 2, as the l^2-norm of MRI signal reconstruction errors.

Our results suggest that both strategies are viable options for DNN-based fitting. When Gaussian noise is considered, the performances of the two strategies are comparable: strategy 2 provides higher accuracy than strategy 1 in parameter estimation, at the expenses of lower precision. However, when more realistic noise distributions (i.e. Rician) are considered, the accuracy of strategy 1 is superior to strategy 2. We speculate that such a reduction in accuracy for strategy 2 results from the fact that DNNs end up learning features that originate from the noise floor, and adjust tissue

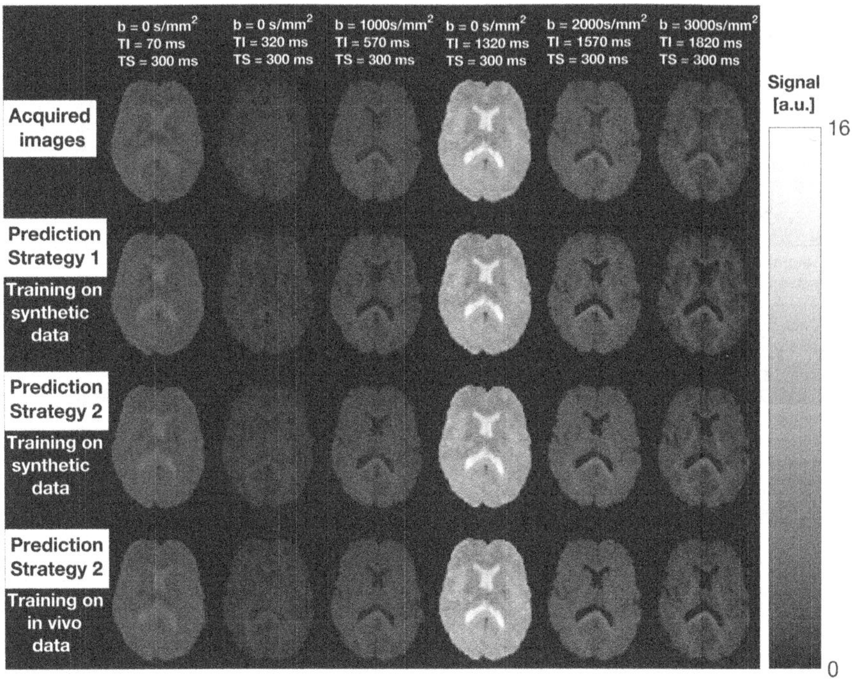

Fig. 6 Examples of images acquired in vivo and DNN predictions. Left to right: different contrasts (spherical mean images; sequence parameters reported on top). Top to bottom: acquired images; images predicted by DNN trained with strategy 1; images predicted by DNN trained with strategy 2 on synthetic signals; images predicted by DNN trained with strategy 2 on in vivo signals

parameter accordingly to accommodate for this. Moreover, noise floor bias could also explain the fact that the MRI signal is slightly underestimated in strategy 1 when Rician noise is considered.

Importantly, it should be noted that unlike in simulations, one does not have access to ground truth tissue parameters in vivo. Therefore, care is needed when extrapolating the better parameter estimation obtained in strategy 1 from simulations to the in vivo case.

Another key observation relates to the fact that strategy 1 requires paired examples of MRI measurements and corresponding tissue parameters. Such examples could be gathered from previous NLLS fitting, or synthesised in silico, as done here. The former is anything but convenient, while the latter would lead to assumptions on the underlying noise level and distribution. Such assumptions can instead be avoided in strategy 2, as demonstrated here when training is performed on actual in vivo measurements, obtaining robust fitting results.

Finally, we acknowledge a number of potential limitations. Firstly, we considered only one qMRI signal model. In future, we will assess the generalisability of our findings by considering additional models. Secondly, we did not compare our fit-

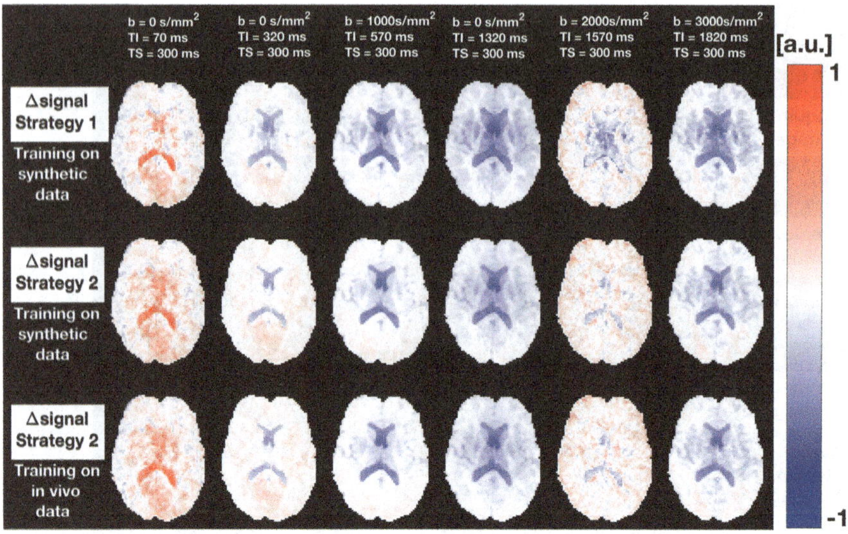

Fig. 7 Examples of differences between DNN signal predictions and images acquired in vivo (i.e. signal residuals of spherical mean images). Left to right: different contrasts (sequence parameters reported on top). Top to bottom: difference between images predicted by DNN trained with strategy 1 minus acquired images; images predicted by DNN trained with strategy 2 on synthetic signals minus acquired images; images predicted by DNN trained with strategy 2 on in vivo signals minus acquired images

Table 2 In vivo root mean squared error (RMSEs) between MRI measurements and signal predictions. Median and 2.5–97.5 percentiles (in brackets) obtained within the brain are reported, highlighting the approach providing the lowest median RMSE

Signal RMSE (a.u.)	Strategy 1 (synthetic training)	Strategy 2 (synthetic training)	Strategy 2 (in vivo training)
Subject 1	0.208 (0.078; 0.674)	**0.142** (0.061; 0.626)	0.143 (0.060; 0.640)
Subject 2	0.191 (0.070; 0.589)	**0.125** (0.056; 0.519)	0.126 (0.054; 0.524)
Subject 3	0.233 (0.086; 0.742)	**0.164** (0.068; 0.665)	0.165 (0.067; 0.675)

ting to non-DNN-based alternatives, e.g. standard NLLS. While this would certainly be interesting to put DNN performance in context, previous literature has focussed extensively on comparisons between NLLS and DNN fitting [7, 9], demonstrating that DNN fitting is a valid alternative to NLLS. Here we focus on specific design choices of DNN-based fitting for DRI, since this has not been investigated so far. Thirdly, we acknowledge that residual noise floor bias may still affect in vivo MRI measurements even after Rician bias mitigation. In future, we will explore more effective noise floor correction techniques. Furthermore, in our comparative study some tissue parameters (e.g. s_0) were estimated better than others (e.g. k). This may be a result of the intrinsic poor sensitivity of the MRI protocol used here with respect to certain parameters. In future we will overcome this limitation by optimising the (b, TI) sampling scheme. Finally, we acknowledge that it cannot be concluded whether our results hold in pathology, e.g. in small focal lesions affecting only a small percentage of the voxels in the image, since here we only consider healthy subjects. We speculate that in those cases strategy 1 may provide additional advantages as compared to strategy 2 in the under-represented lesion, especially if training is performed on actual in vivo measurements. Future work will address this question, and also consider the performance of these strategies with more complex models or larger in vivo data sets, and clarify whether an hybrid of the two strategies can be advantageous.

5 Conclusion

DNN-based fitting is a viable approach for brain DRI modelling. For the specific DRI model considered in this study, training by minimising parameter estimation errors or signal reconstruction errors are both valid choices. The latter is more practical as it does not require pre-calculation or synthesis of tissue parameters, but may provide worse parameter estimation.

Acknowledgements This project was funded by the Engineering and Physical Sciences Research Council (EPSRC EP/R006032/1, M020533/1, EP/N018702/1 G007748, I027084) and by the UK Research and Innovation (MR/T020296/1, funding M.P.). This project has received funding under the European Union's Horizon 2020 research and innovation programme under grant agreement No. 634541, and from: Rosetrees Trust (UK, funding F.G.); Spinal Research (UK), Wings for Life (Austria), Craig H. Neilsen Foundation (USA) for INSPIRED; Wings for Life (169111); UK Multiple Sclerosis Society (grants 892/08 and 77/2017); the Department of Health's National Institute for Health Research (NIHR) Biomedical Research Centres and UCLH NIHR Biomedical Research Centre.

References

1. Cercignani, M., Bouyagoub, S.: Brain microstructure by multi-modal MRI: Is the whole greater than the sum of its parts? NeuroImage **182**, 117–127 (2018)
2. Jelescu, I.O., Veraart, J., Fieremans, E., Novikov, D.S.: Degeneracy in model parameter estimation for multi-compartmental diffusion in neuronal tissue. NMR Biomed. **29**(1), 33–47 (2016)
3. Sjölund, J., Eklund, A., Özarslan, E., Herberthson, M., Bånkestad, M., Knutsson, H.: Bayesian uncertainty quantification in linear models for diffusion MRI. NeuroImage **175**, 272–285 (2018)
4. Liu, H., Xiang, Q.S., Tam, R., Dvorak, A.V., MacKay, A.L., Kolind, S.H., Traboulsee, A., Vavasour, I.M., Li, D.K.B., Kramer, J.K., Laule, C.: Myelin water imaging data analysis in less than one minute. NeuroImage **210**, 116551 (2020)

5. Wright, S., Nocedal, J.: Numerical Optimization. Springer, New York (1999). ISBN 0-387-98793-2
6. Nedjati-Gilani, G.L., Schneider, T., Hall, M.G., Cawley, N., Hill, I., Ciccarelli, O., Drobnjak, I., Gandini Wheeler-Kingshott, C.A.M., Alexander, D.C.: Machine learning based compartment models with permeability for white matter microstructure imaging. NeuroImage **150**, 119–135 (2017)
7. Barbieri, S., Gurney-Champion, O.J., Klaassen, R., Thoeny, H.C.: Deep learning how to fit an intravoxel incoherent motion model to diffusion-weighted MRI. Mag. Resonance Med. **83**(1), 312–321 (2020)
8. Palombo, M., Ianus, A., Guerreri, M., Nunes, D., Alexander, D.C., Shemesh, N., Zhang, H.: SANDI: a compartment-based model for non-invasive apparent soma and neurite imaging by diffusion MRI. NeuroImage **215**, 116835 (2020)
9. Golkov, V., Dosovitskiy, A., Sperl, J.I., Menzel, M.I., Czisch, M., Sämann, P., Brox, T., Cremers, D.: Q-space deep learning: twelve-fold shorter and model-free diffusion MRI scans. IEEE Trans. Med. Imaging **35**(5), 1344–1351 (2016)
10. Cybenko, G.: Approximation by superpositions of a sigmoidal function. Math. Control, Signals Syst. **2**(4), 303–314 (1989)
11. Hutter, J., Slator, P.J., Christiaens, D., Teixeira, R.P.A.G., Roberts, T., Jackson, L., Price, A.N., Malik, S., Hajnal, J.V.: Integrated and efficient diffusion-relaxometry using ZEBRA. Sci. Rep. **8**(1), 1–13 (2018)
12. Grussu, F., Battiston, M., Veraart, J., Schneider, T., Cohen-Adad, J., Shepherd, T.M., Alexander, D.C., Fieremans, E., Novikov, D.S., Gandini Wheeler-Kingshott, C.A.M.: Multi-parametric quantitative in vivo spinal cord MRI with unified signal readout and image denoising. NeuroImage **217**, 116884 (2020)
13. De Santis, S., Assaf, Y., Jeurissen, B., Jones, D.K., Roebroeck, A.: T1 relaxometry of crossing fibres in the human brain. NeuroImage **141**, 133–142 (2016)
14. Wang, H., Zhao, M., Ackerman, J.L., Song, Y.: Saturation-inversion-recovery: a method for T1 measurement. J. Mag. Resonance **274**, 137–143 (2017)
15. Kaden, E., Kruggel, F., Alexander, D.C.: Quantitative mapping of the per-axon diffusion coefficients in brain white matter. Mag. Resonance Med. **75**(4), 1752–1763 (2016)
16. Kingma, D.P., Ba J.: ADAM: a method for stochastic optimization. In: Proceedings of the 3rd International Conference on Learning Representations (2015)
17. Veraart, J., Novikov, D.S., Christiaens, D., Ades-Aron, B., Sijbers, J., Fieremans, E.: Denoising of diffusion MRI using random matrix theory. NeuroImage **142**, 394–406 (2016)
18. Koay, C.G., Basser, P.J.: Analytically exact correction scheme for signal extraction from noisy magnitude MR signals. J. Mag. Resonance **179**(2), 317–322 (2006)
19. Kellner, E., Dhital, B., Kiselev, V.G., Reisert, M.: Gibbs-ringing artifact removal based on local subvoxel-shifts. Mag. Resonance Med. **76**(5), 1574–1581 (2016)
20. Ourselin, S., Roche, A., Subsol, G., Pennec, X., Ayache, N.: Reconstructing a 3D structure from serial histological sections. Image Vis. Comput. **19**, 25–31 (2001)
21. Andersson, J.L.R., Skare, S., Ashburner, J.: How to correct susceptibility distortions in spin-echo echo-planar images: application to diffusion tensor imaging. NeuroImage **20**(2), 870–888 (2003)
22. Smith, S.M.: Fast robust automated brain extraction. Human Brain Map. **17**(3), 143–155 (2002)

Pretraining Improves Deep Learning Based Tissue Microstructure Estimation

Yuxing Li, Yu Qin, Zhiwen Liu, and Chuyang Ye

Abstract *Diffusion magnetic resonance imaging* (dMRI) is commonly used to non-invasively estimate brain tissue microstructure, which provides important biomarkers for studying the structural changes of the brain. Due to the constraint of imaging time, the quality of dMRI scans can be limited by the number of diffusion gradients and the spatial resolution, and deep learning based approaches have been developed to provide high-quality estimation of tissue microstructure from the low-quality diffusion signals. In existing deep learning based methods, the estimation models are trained from scratch. However, it has been shown in various tasks that pretraining can improve the performance of deep networks, and it may also be used to improve deep learning based tissue microstructure estimation because there are abundant publicly available high-quality dMRI datasets. Moreover, for many datasets where acquisitions of a large number of high-quality training dMRI scans are not convenient, pretraining may also allow deep learning based methods to be applied with only a small number of training samples. Motivated by these potential benefits of pretraining, in this work, we explore whether pretraining improves deep learning based tissue microstructure estimation and how to achieve such improvement. Suppose we are given an auxiliary dataset with high-quality dMRI scans for pretraining and a target dMRI dataset of interest with a certain amount of high-quality training data. To generate inputs for pretraining, the diffusion signals of the auxiliary dataset are first downsampled in the spatial domain. Then, since the acquisition scheme is usually different between the two datasets, we interpolate the downsampled signals in the q-space using a dictionary-based signal representation. Finally, the downsampled and interpolated diffusion signals are used for pretraining the estimation network and the pretrained model is fine-tuned with the training data of the target dataset. Experiments were performed on brain dMRI scans, where we show that pretrain-

Y. Li · Y. Qin · Z. Liu · C. Ye (✉)
School of Information and Electronics, Beijing Institute of Technology, Beijing, China
e-mail: chuyang.ye@bit.edu.cn

N. Gyori et al. (eds.), *Computational Diffusion MRI*, Mathematics and Visualization,
https://doi.org/10.1007/978-3-030-73018-5_14

ing leads to improved accuracy of tissue microstructure estimation under different settings and may reduce the burden of training data acquisition.

1 Introduction

Diffusion magnetic resonance imaging (dMRI) is an image modality that is widely used to noninvasively probe brain tissue microstructure, which can reveal structural changes of brain tissue [10]. However, due to the constraint of imaging time in clinical settings, typically low-quality dMRI scans are acquired with a limited number of diffusion gradients and a relatively low spatial resolution. High-quality tissue microstructure estimation, for example, for the *neurite orientation dispersion and density imaging* (NODDI) model [25], from low-quality dMRI scans can be challenging using conventional model-based approaches [3, 25].

To improve the quality of tissue microstructure estimation from low-quality dMRI scans, deep learning based approaches have been developed, where low-quality diffusion signals are mapped to high-quality tissue microstructure. For example, deep networks have been developed to map diffusion signals undersampled in the q-space to high-quality tissue microstructure [7, 21–23]. These works are then extended to estimate *high resolution* (HR) tissue microstructure from *low resolution* (LR) diffusion signals acquired with a reduced number of diffusion gradients, where the network is improved to allow resolution enhancement [24].

In existing deep learning based methods of tissue microstructure estimation, the network is trained from scratch using high-quality—HR, densely sampled in the q-space—training dMRI data acquired for the dataset of interest [24]. It has been observed in various tasks that pretraining using a different dataset can lead to improved performance of deep networks [9], yet this has not been explored for tissue microstructure estimation.

The benefit of such pretraining may be two-fold. First, pretraining allows better initialization of deep networks so that the network parameters are better learned, which may improve the accuracy of tissue microstructure estimation. Second, pretraining may also alleviate the need to acquire a large number of training scans for the dataset of interest, which can be inconvenient in clinical settings. Such reduction of training data acquisitions can widen the applicability of deep learning based tissue microstructure estimation to different datasets and increase its impact. Since there are abundant publicly available high-quality dMRI datasets [20], it is possible to explore pretraining for deep learning based tissue microstructure estimation.

Motivated by the potential benefits of pretraining, in this work, we explore whether pretraining improves deep learning based tissue microstructure estimation and how to achieve such improvement. In particular, we focus on HR tissue microstructure estimation from LR diffusion signals undersampled in the q-space. Suppose we have an auxiliary dataset, for example, the publicly available *Human Connectome Project* (HCP) dataset [20], where high-quality dMRI data is acquired, and this dataset is used to pretrain a model for the target dataset of interest. To generate inputs for pretraining, the diffusion signals of the auxiliary dataset are first downsampled in the spatial domain. Since the set of diffusion gradients of the target dataset is usually not a subset of that of the auxiliary dataset, the downsampled diffusion signals of the auxiliary dataset cannot be directly used for network pretraining. Therefore, it is important to interpolate the LR diffusion signals in the q-space to produce LR undersampled diffusion signals that match the acquisition scheme of the target dataset. Specifically, a dictionary-based representation of diffusion signals with the SHORE basis [13] is used to perform the interpolation [17]. Training inputs are then extracted from the interpolated LR diffusion signals, and they are used for pretraining together with the high-quality tissue microstructure computed from the original auxiliary dataset. Finally, the pretrained model is fine-tuned with the high-quality training samples acquired for the target dataset. For demonstration, we applied the pretraining strategy to the network developed in [24], which is later described with more details in Sect. 2.3, and estimated tissue microstructure for the widely used NODDI model [1, 6, 14, 15]. The results show that pretraining leads to improved accuracy of tissue microstructure estimation under different settings and allows accurate tissue microstructure estimation given a reduced number of training samples.

2 Methods

2.1 Problem Formulation

In this work, we aim to investigate whether and how pretraining with an auxiliary dataset can improve deep learning based tissue microstructure estimation for a target dataset of interest. In particular, we focus on the estimation of HR tissue microstructure from low-quality diffusion signals acquired with a reduced number of diffusion gradients and a low spatial resolution [24]. For convenience, we denote the target dataset by $\mathcal{D}$ and the auxiliary dataset by $\mathcal{D}_{\mathrm{a}}$.

In the auxiliary dataset $\mathcal{D}_{\mathrm{a}}$, diffusion signals are acquired with a set $\mathcal{G}_{\mathrm{a}}$ of diffusion gradients densely sampling the q-space and a high spatial resolution. In the target dataset $\mathcal{D}$ the diffusion signals are acquired with a set $\mathcal{G}$ of diffusion gradients that undersample the q-space and a relatively low spatial resolution. In addition, high-quality dMRI scans are also acquired for the training subjects in $\mathcal{D}$ with a set $\mathcal{G}_{\mathrm{t}}$ ($\mathcal{G}_{\mathrm{t}} \supset \mathcal{G}$) of diffusion gradients densely sampling the q-space and a high spatial resolution. Generally, $\mathcal{G}_{\mathrm{t}}$ and $\mathcal{G}_{\mathrm{a}}$ are not the same. If the auxiliary dataset is not used, like in [24], the network is trained with the low-quality diffusion signals of the training subjects and the corresponding high-quality tissue microstructure computed from the high-quality dMRI scans, which are the network inputs and outputs, respectively. With the auxiliary dataset $\mathcal{D}_{\mathrm{a}}$, our goal is to pretrain an estimation network, so that better initialization can be achieved when the network is trained with the training data for $\mathcal{D}$.

2.2 Signal Generation for Pretraining

Since we estimate HR tissue microstructure for LR undersampled diffusion signals, to pretrain the estimation network with the auxiliary dataset, we need to downsample the diffusion signals of $\mathcal{D}_a$ in the spatial domain. Here, the block-wise mean [18] is used to downsample the HR diffusion signals to generate LR diffusion signals. The size of the block is set according to the desired upsampling rate γ (an integer). Note that generally $\mathcal{G}$ is not a subset of $\mathcal{G}_t$. Thus, the downsampled diffusion signals in $\mathcal{D}_a$ cannot be directly used to pretrain a network that is compatible with the diffusion signals of the target dataset.

To allow network pretraining using the auxiliary dataset, we interpolate the downsampled diffusion signals of $\mathcal{D}_a$ so that they match the acquisition scheme of the undersampled diffusion signals in the target dataset $\mathcal{D}$. Specifically, we follow the interpolation strategy proposed in [17], and its detailed description is given below.

The interpolation is based on the representation of diffusion signals using the SHORE basis [13], which has been shown to effectively reconstruct diffusion signals. Specifically, a set of SHORE basis functions can be used to represent the diffusion signal vector $\boldsymbol{y}_a$ at each voxel for $\mathcal{D}_a$. Mathematically the representation can be expressed as follows

$$\boldsymbol{y}_a = \Phi_a \boldsymbol{c}_a, \tag{1}$$

where Φ_a represents the dictionary computed from the SHORE basis functions according to $\mathcal{G}_a$, and the corresponding vector of coefficients is $\boldsymbol{c}_a$. If $\boldsymbol{c}_a$ is known, we can then use the dictionary Φ computed from the SHORE basis according to $\mathcal{G}$ to interpolate the diffusion signals in the q-space, so that the interpolated signals correspond to the undersampled diffusion signals in the target dataset. Like in Eq. (1), the interpolated diffusion signal vector $\boldsymbol{y}$ is computed as

$$\boldsymbol{y} = \Phi \boldsymbol{c}_a. \tag{2}$$

$\boldsymbol{c}_a$ can be estimated from $\boldsymbol{y}_a$ via a regularized least squares problem [13]. Here, we use the implementation given by Dipy [5] with the default parameters.

2.3 Backbone Deep Network

Before we introduce the specific pretraining and fine-tuning strategy using the auxiliary dataset, we first describe the backbone network for tissue microstructure estimation and how it is conventionally trained.

Here, we select the network proposed in [24], which estimates HR tissue microstructure from LR dMRI scans acquired with a reduced number of diffusion gradients and outperforms competing networks. The network in [24] is patch-based and comprises two functional components. The first component computes the sparse representation of diffusion signals for each voxel in the LR input patch, and it is constructed by unfolding the iterative hard thresholding strategy [2]. The second component computes HR tissue microstructure from the LR sparse representation obtained by the first component with the 3D-ESPCN architecture [18]. The HR output patches are finally concatenated to obtain the complete HR tissue microstructure map.

In [24], the deep network is trained from scratch for the target dataset $\mathcal{D}$ using the training scans acquired with diffusion gradients $\mathcal{G}_\mathrm{t}$ and a high spatial resolution. Specifically, high-quality tissue microstructure maps are computed from the training scans using conventional model-based approaches [3]. Then, nonoverlapping patches of the high-quality tissue microstructure are used together with the corresponding patches of low-quality diffusion signals for training.

2.4 Pretraining with the Auxiliary Dataset and Fine-Tuning

To improve network training with the auxiliary dataset, we extract training samples from the interpolated and spatially downsampled diffusion signals and high-quality tissue microstructure maps of $\mathcal{D}_\mathrm{a}$ similarly to the way that training samples are extracted from the target dataset. First, a set of training patch samples centering on the voxels in the brain mask is extracted from $\mathcal{D}_\mathrm{a}$, and this set is denoted by $\mathcal{A} = \{\mathbf{Y}_\mathrm{a}^{(i)}, \mathbf{X}_\mathrm{a}^{(i)}\}_{i=1}^{N_\mathrm{a}}$, where N_a is the number of training samples, $\mathbf{Y}_\mathrm{a}^{(i)}$ is the LR input patch for the ith training sample, and $\mathbf{X}_\mathrm{a}^{(i)}$ is the output HR tissue microstructure patch for the ith training sample. $\mathbf{Y}_\mathrm{a}^{(i)}$ has a larger spatial extent than $\mathbf{X}_\mathrm{a}^{(i)}$, and $\mathbf{X}_\mathrm{a}^{(i)}$ corresponds to the center region of $\mathbf{Y}_\mathrm{a}^{(i)}$. In this way, redundant information can be exploited for resolution enhancement [24]. Then, the set $\mathcal{A}$ is used for pretraining the deep network.

The pretrained network weights are used to initialize the deep network that estimates tissue microstructure for the target dataset. For fine-tuning, we extract training samples from $\mathcal{D}$ and the set of training samples is denoted by $\mathcal{T} = \{\mathbf{Y}^{(i)}, \mathbf{X}^{(i)}\}_{i=1}^{N}$, where N is the number of training samples in $\mathcal{T}$, $\mathbf{Y}^{(i)}$ stands for the input patches of low-quality diffusion signals for the ith training sample, and $\mathbf{X}^{(i)}$ represents the output patch of high-quality tissue microstructure for the ith training sample. The network initialized by the pretrained weights is then fine-tuned with the training set $\mathcal{T}$, and the resulting network estimates high-quality tissue microstructure for $\mathcal{D}$. Note that the number N can be small if a large number of training scans for $\mathcal{D}$ cannot be achieved. In that case, if the network is trained from scratch, the estimation performance can be poor.

2.5 Implementation Details

Following the settings in [24], we consider the case where the upsampling rate is two ($\gamma = 2$) and set the sizes of the input patch ($\mathbf{Y}_\mathrm{a}^{(i)}$ and $\mathbf{Y}^{(i)}$) and output patch ($\mathbf{X}_\mathrm{a}^{(i)}$ and

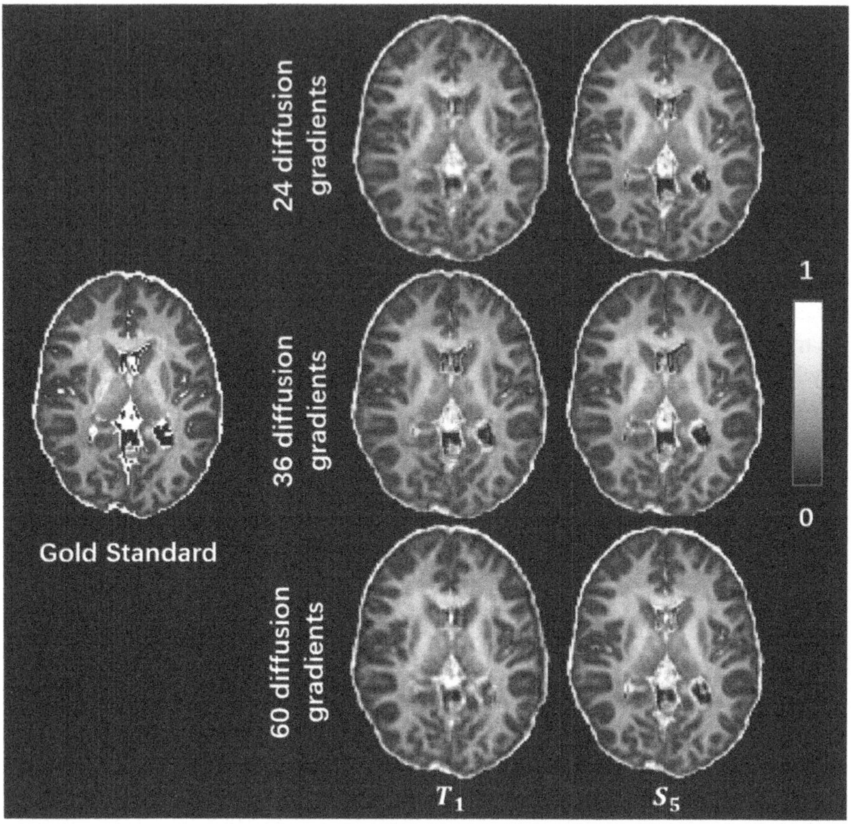

Fig. 1 A representative axial view of the v_{ic} map estimated under different settings. The gold standard is shown for reference

A more comprehensive evaluation of the benefit of pretraining for tissue microstructure estimation can be performed in future works. For example, statistical significance and effect sizes can be computed for the comparison of results, and a more detailed analysis of the areas with improved estimation quality can be performed. Also, in addition to estimation accuracy, the reproducibility of the estimation results can be investigated using test-retest scans. Finally, the original NODDI model uses a fixed parallel diffusivity for all voxels, which could be a limitation of the method, and it would be interesting to explore the benefit of pretraining for revised NODDI models [8] with optimized parallel diffusivity values.

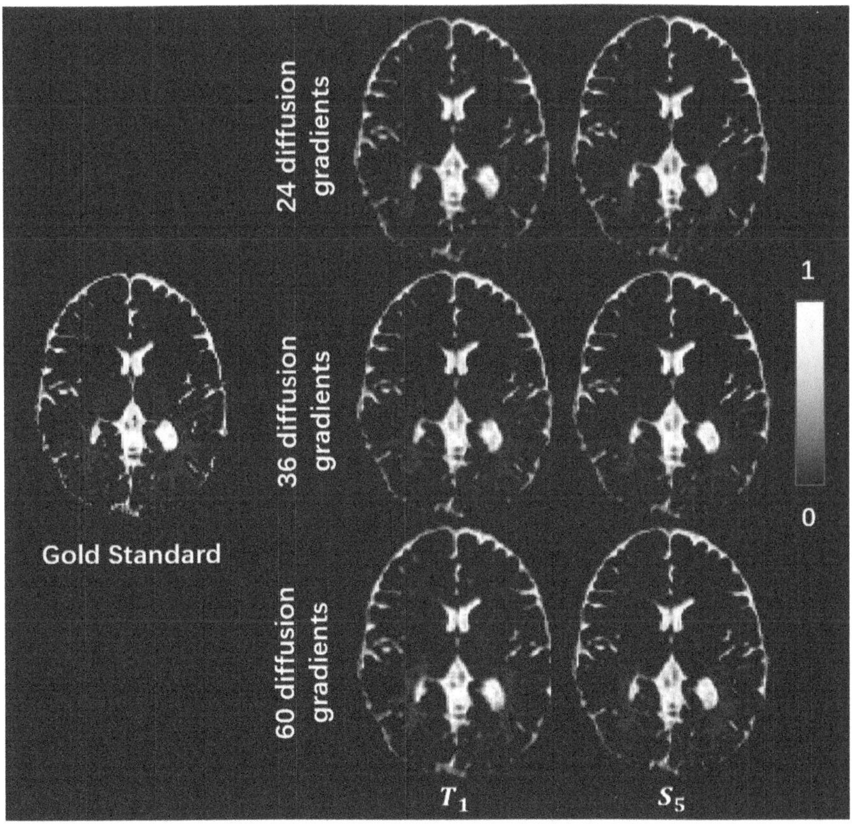

Fig. 2 A representative axial view of the v_{iso} map estimated under different settings. The gold standard is shown for reference

5 Conclusion

We have explored whether and how pretraining with an auxiliary dataset can improve deep learning based tissue microstructure estimation for a target dataset of interest. The high-quality diffusion signals of the auxiliary dataset are downsampled in the spatial domain and interpolated in the q-space to pretrain the deep network. Then, the pretrained deep network is fine-tuned with the training scans acquired for the target dataset, and the fine-tuned network estimates high-quality tissue microstructure for the target dataset. Although this approach is simple, results on brain dMRI scans show that it can achieve more accurate HR tissue microstructure estimation under different settings. In addition, the strategy may also be used to reduce the burden of training data acquisition, which could widen the applicability of deep learning based tissue microstructure estimation.

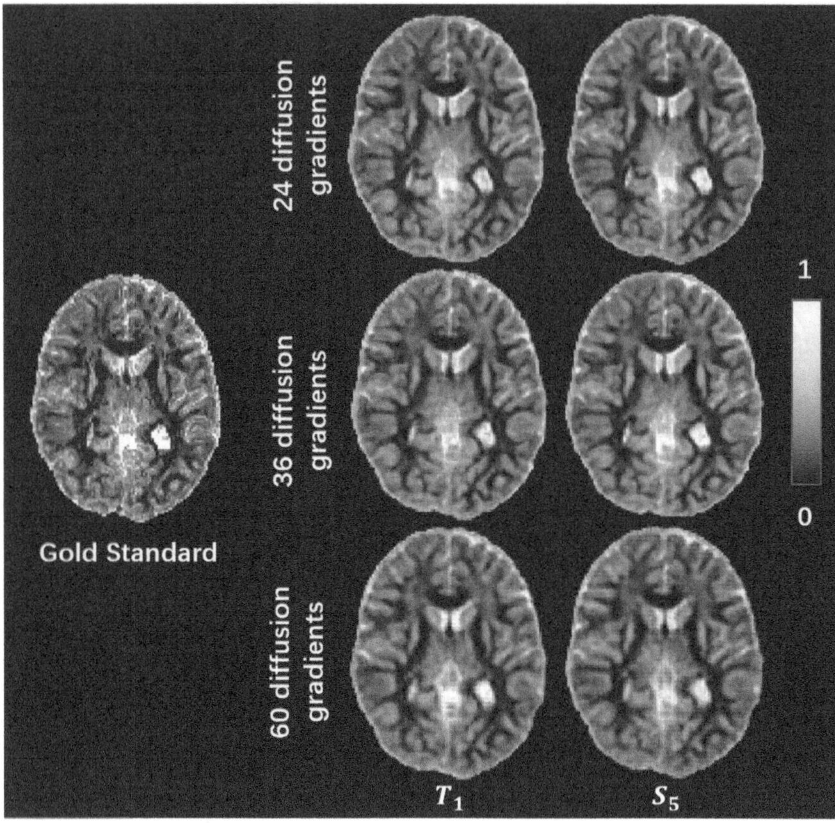

Fig. 3 A representative axial view of the OD map estimated under different settings. The gold standard is shown for reference

Acknowledgements This work is supported by Beijing Natural Science Foundation (7192108 & L192058) and Beijing Institute of Technology Research Fund Program for Young Scholars. The HCP-Minn dataset was provided by the Human Connectome Project, WU-Minn Consortium and the McDonnell Center for Systems Neuroscience at Washington University. The HCP-MGH dataset was provided by the MGH-USC Human Connectome Project and the Laboratory of Neuro Imaging at the University of Southern California.

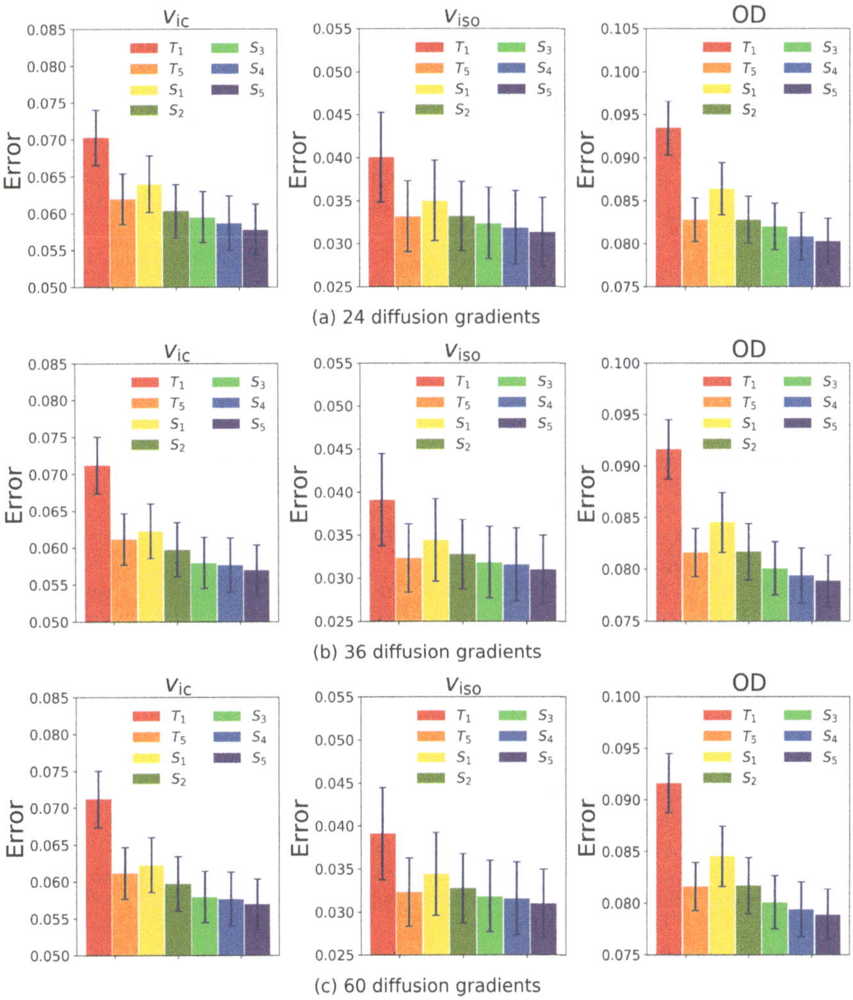

Fig. 4 Means and standard deviations of the average estimation errors in the brains of test subjects under different settings

References

1. Batalle, D., O'Muircheartaigh, J., Makropoulos, A., Kelly, C.J., Dimitrova, R., Hughes, E.J., Hajnal, J.V., Zhang, H., Alexander, D.C., Edwards, A.D., Counsell, S.J.: Different patterns of cortical maturation before and after 38 weeks gestational age demonstrated by diffusion MRI in vivo. NeuroImage **185**, 764–775 (2019)
2. Blumensath, T., Davies, M.E.: Iterative thresholding for sparse approximations. J. Fourier Anal. Appl. **14**(5–6), 629–654 (2008)
3. Daducci, A., Canales-Rodríguez, E.J., Zhang, H., Dyrby, T.B., Alexander, D.C., Thiran, J.P.: Accelerated microstructure imaging via convex optimization (AMICO) from diffusion MRI

data. NeuroImage **105**, 32–44 (2015)
4. Fan, Q., Witzel, T., Nummenmaa, A., Van Dijk, K.R., Van Horn, J.D., Drews, M.K., Somerville, L.H., Sheridan, M.A., Santillana, R.M., Snyder, J., Hedden, T., Shaw, E.E., Hollinshead, M.O., Renvall, V., Zanzonico, R., Keil, B., Cauley, S., Polimenia, J.R., Rosen, B.R.: MGH-USC human connectome project datasets with ultra-high b-value diffusion MRI. NeuroImage **124**, 1108–1114 (2016)
5. Garyfallidis, E., Brett, M., Amirbekian, B., Rokem, A., Van Der Walt, S., Descoteaux, M., Nimmo-Smith, I.: Dipy, a library for the analysis of diffusion MRI data. Frontiers Neuroinformatics **8**, 8 (2014)
6. Genç, E., Fraenz, C., Schlüter, C., Friedrich, P., Hossiep, R., Voelkle, M.C., Ling, J.M., Güntürkün, O., Jung, R.E.: Diffusion markers of dendritic density and arborization in gray matter predict differences in intelligence. Nat. Commun. **9**(1), 1–11 (2018)
7. Golkov, V., Dosovitskiy, A., Sperl, J.I., Menzel, M.I., Czisch, M., Sämann, P., Brox, T., Cremers, D.: Q-space deep learning: twelve-fold shorter and model-free diffusion MRI scans. IEEE Trans. Med. Imaging **35**(5), 1344–1351 (2016)
8. Guerrero, J.M., Adluru, N., Bendlin, B.B., Goldsmith, H.H., Schaefer, S.M., Davidson, R.J., Kecskemeti, S.R., Zhang, H., Alexander, A.L.: Optimizing the intrinsic parallel diffusivity in NODDI: An extensive empirical evaluation. PloS one **14**(9), e0217118 (2019)
9. Hendrycks, D., Lee, K., Mazeika, M.: Using pre-training can improve model robustness and uncertainty (2019). arXiv:1901.09960
10. Johansen-Berg, H., Behrens, T.E.: Diffusion MRI: From Quantitative Measurement to in Vivo Neuroanatomy. Academic (2013)
11. Karayumak, S.C., Kubicki, M., Rathi, Y.: Harmonizing diffusion MRI data across magnetic field strengths. In: International Conference on Medical Image Computing and Computer-Assisted Intervention, pp. 116–124. Springer, Berlin (2018)
12. Kingma, D.P., Ba, J.: Adam: A method for stochastic optimization (2014). arXiv:1412.6980
13. Merlet, S.L., Deriche, R.: Continuous diffusion signal, EAP and ODF estimation via compressive sensing in diffusion MRI. Med. Image Anal. **17**(5), 556–572 (2013)
14. Ocklenburg, S., Friedrich, P., Fraenz, C., Schlüter, C., Beste, C., Güntürkün, O., Genç, E.: Neurite architecture of the planum temporale predicts neurophysiological processing of auditory speech. Sci. Adv. **4**(7), eaar6830 (2018)
15. Parker, T.D., Slattery, C.F., Zhang, J., Nicholas, J.M., Paterson, R.W., Foulkes, A.J., Malone, I.B., Thomas, D.L., Modat, M., Cash, D.M., Crutch, S.J., Alexander, D.C., Ourselin, S., Fox, N.C., Zhang, H., Schott, J.M.: Cortical microstructure in young onset Alzheimer's disease using neurite orientation dispersion and density imaging. Hum. Brain Map. **39**(7), 3005–3017 (2018)
16. Poulin, P., Cote, M.A., Houde, J.C., Petit, L., Neher, P.F., Maier-Hein, K.H., Larochelle, H., Descoteaux, M.: Learn to track: deep learning for tractography. In: International Conference on Medical Image Computing and Computer-Assisted Intervention, pp. 540–547. Springer, Berlin (2017)
17. Qin, Y., Li, Y., Liu, Z., Ye, C.: Knowledge transfer between datasets for learning-based tissue microstructure estimation. In: International Symposium on Biomedical Imaging, pp. 1530–1533 (2020)
18. Tanno, R., Worrall, D.E., Ghosh, A., Kaden, E., Sotiropoulos, S.N., Criminisi, A., Alexander, D.C.: Bayesian image quality transfer with CNNs: exploring uncertainty in dMRI super-resolution. In: International Conference on Medical Image Computing and Computer-Assisted Intervention, pp. 611–619. Springer, Berlin (2017)
19. Van Essen, D.C., Smith, S.M., Barch, D.M., Behrens, T.E.J., Yacoub, E., Ugurbil, K.: The WU-Minn human connectome project: an overview. NeuroImage **80**, 62–79 (2013)
20. Van Essen, D.C., Ugurbil, K., Auerbach, E., Barch, D., Behrens, T., Bucholz, R., Chang, A., Chen, L., Corbetta, M., Curtiss, S.W., Penna, S.D., Feinberg, D., Glasser, M.F., Harel, N., Heath, A.C., Larson-Prior, L., Marcus, D., Michalareas, G., Moeller, G., Oostenveld, R., Petersen, S.E., Prior, F., Schlaggar, B.L., Smith, S.M., Synder, A.Z., Xu, J., Yacoub, E.: WU-Minn HCP consortium: the human connectome project: a data acquisition perspective. NeuroImage **62**(4), 2222–2231 (2012)

21. Ye, C.: Estimation of tissue microstructure using a deep network inspired by a sparse reconstruction framework. In: International Conference on Information Processing in Medical Imaging, pp. 466–477. Springer, Berlin (2017)
22. Ye, C.: Tissue microstructure estimation using a deep network inspired by a dictionary-based framework. Med. Image Anal. **42**, 288–299 (2017)
23. Ye, C., Li, X., Chen, J.: A deep network for tissue microstructure estimation using modified LSTM units. Med. Image Anal. **55**, 49–64 (2019)
24. Ye, C., Qin, Y., Liu, C., Li, Y., Zeng, X., Liu, Z.: Super-resolved q-space deep learning. In: International Conference on Medical Image Computing and Computer-Assisted Intervention, pp. 582–589. Springer, Berlin (2019)
25. Zhang, H., Schneider, T., Wheeler-Kingshott, C.A., Alexander, D.C.: NODDI: practical in vivo neurite orientation dispersion and density imaging of the human brain. NeuroImage **61**(4), 1000–1016 (2012)
26. Zhu, G., Jiang, B., Tong, L., Xie, Y., Zaharchuk, G., Wintermark, M.: Applications of deep learning to neuro-imaging techniques. Frontiers Neurol. **10**, 869 (2019)

Signal Augmentation and Super Resolution

pre-processing, individualized loss functions and individual layers are among the most influential factors [13].

In the field of diffusion imaging, most deep learning methods are still based on regular convolutional neural networks and/or ordinary neural networks, if no spatial neighborhood information shall be included. At the same time, however, a 3D diffusion signal within a voxel consists of a large number of individual signals that have been sampled using different gradient directions and different gradient field strengths (b-values). If every individual signal is acquired using a single b-value, this is referred to as single-shell acquisition, while a measurement based on different b-values is called multi-shell acquisition. Due to the fact that a spherical diffusion signal on a single shell is assumed to be continous, gradient direction signals based on adjacent gradient directions show a higher correlation than those which are further away from each other.

If a conventional neural network is employed to process such diffusion signals, gradient specific neighborhood information is usually remains unused. Therefore, novel deep learning methods are required to exploit this additional spherical information, in order to maintain the spherical character of the input 3D diffusion signal while only slightly altering it. This is highly relevant for applications such as signal augmentation, signal denoising or signal harmonization.

In this work, three different approaches to realize spherical convolutions are investigated, exemplified using signal augmentation: The most straightforward approach projects the sphere onto a 2D plane, while afterwards regular 2D convolutions can be utilized to augment the spherical diffusion signal. The second approach is called local spherical convolution [16], which applies a convolution kernel directly to the sphere's surface. Finaly, the third approach transforms the diffusion signal into the Fourier space, which is why a convolution can be performed using a simple multiplication.

2 Signal Augmentation

Signal Augmentation attempts to accelerate acquisition sequences, by reducing the number of acquired gradient directions, or is employed to replace noisy signals by interpolating outliers. For this purpose the diffusion signal is usally modeled using either a tissue model or a data driven approach [7]. In this work we focus on the data driven approaches.

Currently, there are two strategies to augment a diffusion signal: The first employs a predefined model [5, 22, 23] to cope with the reduced number of gradient directions. In addition, a strong regularization [6] or useful constraints, such as the positivity of diffusion [25], are usually utilized. However, due to intravoxel incoherent motion [18] and kurtosis effects [14], the relationship between different gradient directions and different b-values shows a highly non-linear behavior.

Alternatively, data-driven methods based on deep learning can be used. These methods are ideally suited to detect highly non-linear relationships in a diffusion signal and to interpolate non-measured gradient direction signals based on training data [17]. Furthermore, deep learning is not constrained to rigid models, which is why they are able to reconstruct even very complex diffusion signals.

To train a diffusion signal augmentation model that is able to predict a different shell or refine an existing ones, a high-resolution diffusion dataset is required. While this high-resolution dataset forms the target that has to be predicted during training, a subsampled version of the same dataset is used as input.

To represent an individual diffusion signal, each shell is substituted using Spherical Harmonic (SH) coefficients, where the SH order was selected based on the number of gradient directions to ensured that there are less SH coefficients than number of sampled gradient directions. In addition, a Laplace-Beltrami regularization of $\lambda = 0.006$ is employed to be robust against strong noise. This is very important as the trained neural network would be bound to the input gradient directions it was trained on, which would limit its usabillity.

2.1 Deep Learning Models

All deep learning networks investigated in this work are based on the neural network presented in [9], which is also utilized as baseline model for comparison. This utilizes four fully connected (FC) layers and rectified linear units (ReLU) as activation function. An overview over the utilized base network is given in Table 1. Every fully connected layer is replaced with the respective spherical convolution layer, presented in the following subsections, resulting in three different networks. The activation function of these networks is chosen individually such that the spherical character is maintained within the network. If the diffusion signal is in the Fourier domain after applying a spherical convolution, the tangent hyperbolic function (TanH) is used as activation function (resulting in values between -1 and 1), while a diffusion signal within the signal domain is activated utilizing a ReLU (resulting in values between 0 and Inf).

Training: Every network is trained using a high-resolution dataset as target data, while the same, but subsampled, dataset is employed as input data. Training is performed by randomly looping over each subsampled voxel of all training subjects, attempting to predict its original high-resolution data. Here, the RAdam [20] optimizer (learning rate 0.001, batch size: 128 single voxels) is utilized to reduce the

Table 1 Topology of the neural network presented in [9]. It contains 27845 trainable parameters

#	Type	Parameters	Activation
0	FC	100 nodes	ReLU
1	FC	100 nodes	ReLU
2	FC	100 nodes	ReLU
3	FC	#SHCoefficients×#OutputShells	–

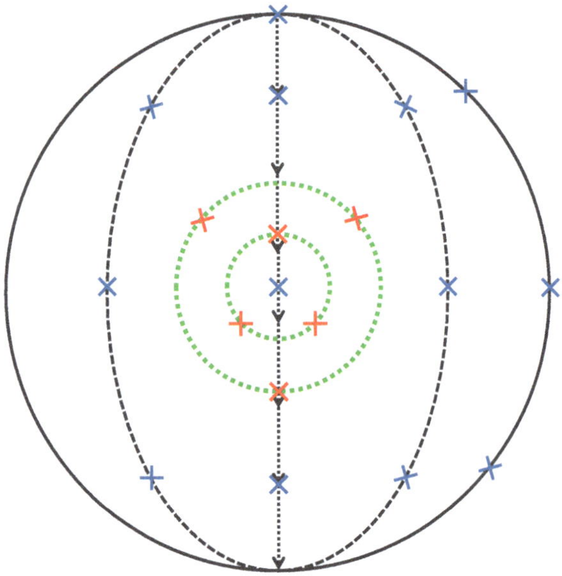

Fig. 2 Exemplary visualization of a local spherical convolution. Real aquired gradient directions are marked in blue, while interpolated gradient directions are marked in red. Furthermore, convolution trajectories are marked in black dotted lines that go from top to bottom

Table 4 Adapted topology of the neural network utilizing Local Spherical Convolutions. It contains 23923 trainable parameters

#	Type	Parameters	Activation
0	LSC	40 kernels, SH order: 4, kernel size: 3, 3	TanH
1	LSC	40 kernels, SH order: 4, kernel size: 3, 3	TanH
2	LSC	40 kernels, SH order: 4, kernel size: 3, 3	TanH
3	LSC	#OutputShells kernels, SH order: 4, kernel size: 3, 3	–

between adjacent circles is chosen to be the average distance between neighboring gradient directions. Since not all support points have been acquired during the scan, missing gradient directions are interpolated using spherical harmonics.

However, this kind of convolution is also not flawless, as it exhibits two singularities at the poles of the sphere. The behavior at these poles is defined by its implementation, which is why it is important to employ the same implementation for inference that was used during training.

A full overview over the utilized network is given in Table 4.

2.3 Material

Within this work, the publicly available Human Connectome Project database is utilized, which consists of very high resolution datasets, including a very high number of gradient directions, to investigate the human connectome [26]. Since evaluating all available HCP subjects would be too time-consuming, a subgroup of 20 female subjects was randomly selected for the evaluation of all methods. Male subjects were excluded to remove possible gender bias.

Every subject was scanned using three different b-values ($b = 1000\frac{s}{mm^2}$, $b = 2000\frac{s}{mm^2}$, $b = 3000\frac{s}{mm^2}$) with 90 gradients directions each and 18 non-weighted measurements ($b = 0\ \frac{s}{mm^2}$). All voxels have an isotropic resolution of 1.25 mm^3. All acquisitions were corrected for eddy current and susceptibility distortions [1]. To train only on relevant diffusion signals, only white matter voxels were extracted for training and testing using FAST [27].

3 Evaluation

In order to evaluate the augmentation performance of each network, we reduced the number of sampled gradients direction artifically to $n = 15$ for each shell. This subset is chosen such that the 15 gradient directions are well distributed over the hemisphere [15]; each gradient direction itself remains unchanged. In addition, the third shell ($b = 3000\frac{s}{mm^2}$) is removed from the input diffusion signal to investigate if a neural network is able to identify and enhance the remaining kurtosis part of the diffusion signal, which increases for higher b-values. With this evaluation setup, a clinical scenario is imitated, which is usually limited in terms of acquisition time and higher b-values. Furthermore, clincal trials often show a high level of motion noise within their acquisitions, which is why many gradient directions have to be removed from the diffusion signal after acquisition.

All computations are carried out on a PC with 3.4 GHz Intel i5-4670 processor, 64GB RAM and NVIDIA GeForce RTX 2080 Ti. Furthermore, all models are evaluated using a five-fold cross-validation, always using twelve subjects to train and four subjects to validate the network. After training, the remaining four unseen subjects are evaluated for testing to asses the quality of the diffusion signal augmentation for this fold. This is repeated for each fold, e.g. all subjects have been in testing at least once.

For the purpose of comparison, we investigate the difference in fractional anisotropy (FA) and mean diffusivity (MD) based on regular diffusion tensor imaging [2], but also the mean kurtosis (MK) of the resulting diffusion signals. For completness, we also evaluate the difference of the diffusion signal itself. Here, however, only the firth two shells ($b = 1000\frac{s}{mm^2}$ and $b = 2000\frac{s}{mm^2}$) and compared, since there is no third shell inside the input signal to compare against.

While standard diffusion kurtosis imaging is often affected by microstructural properties (e.g. crossing fibers within a voxel), it also depends on mesoscopic properties (e.g. fiber dispersion or crossing angles) [11]. We therefore utilize the mean signal diffusion kurtosis reconstruction [11] implemented in dipy [8], which shows a high degree of robustness when reconstructing the MK of a diffusion signal, even in situations of crossing fibers.

For evaluation, the mean absolute error (MAE), defined by

$$\mathrm{MAE} = \frac{1}{N} \cdot \sum_{i} \left| \mathrm{X}_i - \hat{\mathrm{X}}_i \right|, \tag{2}$$

is utilized. Here, X_i defines the reconstruction for the high-resolution target and $\hat{\mathrm{X}}_i$ reconstruction for the low-resolution dataset with or without employing an augmentation for every voxel i, divided by the number of all voxels N.

3.1 Results

Figure 3 shows the MAE for FA (see Fig. 3a), for MD (see Fig. 3b) and MSK (see Fig. 3c), but also for the diffusion signal (see Fig. 3d) itself, comparing all models as well as the MAE if no augmentation is applied. The resulting MAE if no augmentation is utilized is marked in red and the basic neural network without spherical convolutional layers is marked in blue, respectively. In addition, the FSC based model is marked in green, the LSC based model marked in purple, while the 2D projection based model (2DP) is marked in orange. Evaluating all three markers diffusion markers shows that any of the presented models achieves a very good result in comparison to no augmentation. At the same time, it can be seen that the difference in diffusion signal is reduced by a smaller fraction in comparison to the diffusion markers. On average, the MAE for FA is reduced by $\approx$30%, for MD by $\approx$50%, for MSK by $\approx$70% and for the diffusion signal by $\approx$5%. In addition, the baseline neural network shows a higher variance in comparison to the FSC, the LSC and the 2DP model. For FA and the MSK, the lowest values are achieved using the LSC model (average MAE for FA: 0.049 and the average MAE for MSK: 3×10^{-5}), while for MD and the diffusion signal, the LSC and 2DP model achieve the best results (average MAE for MD with LSC: 3.05×10^{-5} and average MAE for the diffusion signal: 0.042, while the 2DP achieves an average MAE for MD of 3.04×10^{-5} and 0.042 for the diffusion signal.).

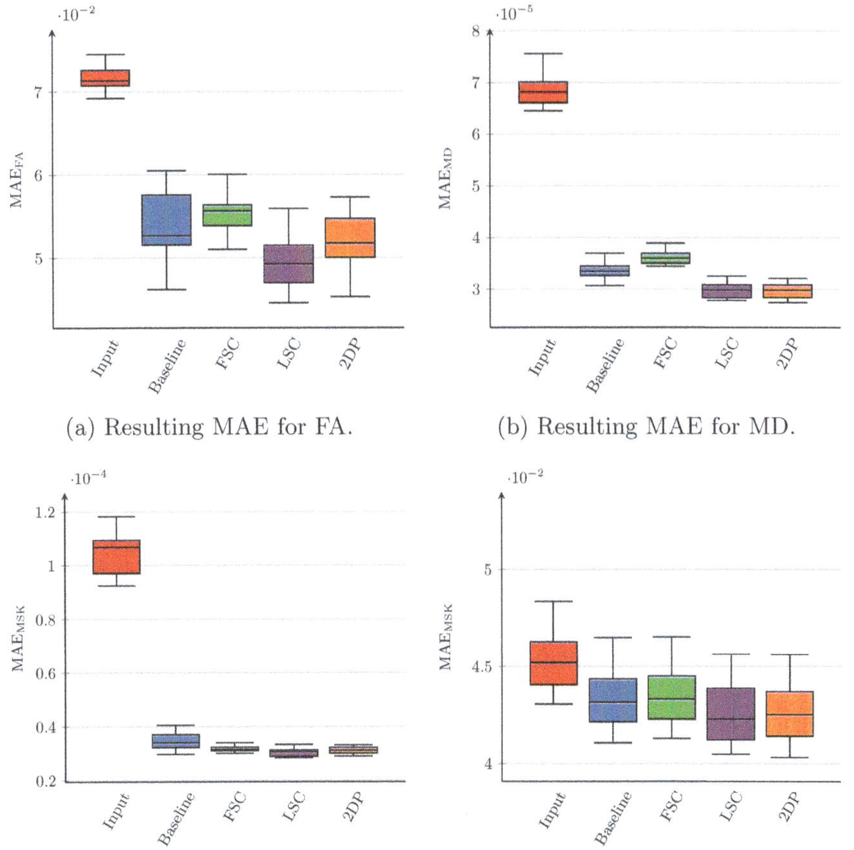

(a) Resulting MAE for FA.

(b) Resulting MAE for MD.

(c) Resulting MAE for MSK.

(d) Resulting MAE for the diffusion signal.

Fig. 3 Resulting MAE for FA, MD, MSK and the diffusion signal before and after augmentation using four different deep learning models (baseline model without spherical convolution, as well as three different models that implement spherical convolutions, namely FSC, LSC and 2DP). For the diffusion signal, only the first two shells have been evaluated, since the third shell was removed from the input signal

4 Discussion

Comparing the results reveals that all methods are very effective in predicting diffusion signals and clearly improve a subsequent reconstruction. Here, it is also visible that all models are able to identify and enhance the remaining kurtosis part of the diffusion signal as we see a huge improvement comparing the initial MSK in comparison to all models. Additionally, taking a closer look at the baseline model in comparion to the LSC or projective approach, including spherical information results in a positive effect on the reconstruction. However, no distinctly superior spherical convolution can be defined on the basis of these results. Interestingly, the difference in the first two shells of the diffusion signal is reduced for all models. However, the MAE reduces only by ≈6% on average, leading to the assumption that predicting the third shell might be the major component for improving the subsequent markers.

It can be seen that the FSC performed slightly worse than other spherical convolutions. This is probably caused by strong limitation based on the utilized zonal kernels. In order to improve this convolution, higher SH Orders could be employed. To ensure that this also applies to diffusion signals based on few gradient directions, additional constraints could be employed (similar to constraints introduced in [25]).

On the other hand, it should be noted that only very small neural networks (≈25 000 trainable parameters) are evaluated to be comparable against the baseline network. Small models are beneficial if only a small dataset is available for training. In addition, small models are optimal to show the effect of addition constrains. If, however, more training data is available the effect of constrains might be lost, as this would allow bigger networks, which don't have to focus on specified information.

In general, it is also important to note that information that is not available in the input signal, will not be available after augmentation. Due to this, signal augmentation across multiple b-values is only applicable, if enough b-values have been acquired initially. If, for example, only small b-values are acquired, no Kurtosis effect would be introduced into the diffusion signal, which is why neural networks will not be able to predict this part of the signal. The same effect might occur for a clinical study, where the clinical subjects show alterations of the diffusion signal that have not been in the training dataset. In this case, further inverstigations have to show if these neural networks are able to generalize and predict all occuring diffusion signals.

In order to investigate the remaining questions, a more in-depth evaluation of different scenarios (e.g. the effect of signal augmentation on the resulting diffusion orientations) is required. In this context, the Grouped Convolution (presented in [4]) should also be evaluated, as it appears to be particularly useful for classification or prediction of single markers. Subsequently, every spherical convolution can also be combined with conventional convolutional neural networks to include additional spatial neighborhood information.

5 Conclusion

Overall, this work demonstrates that simple neural networks based on spherical convolutions can effectively augment a diffusion signal. Thus, the difference between a low-resolution and a high-resolution diffusion signal in terms of number of aquired gradient directions can be drastically reduced. According to our, currently still very limited, experiments on signal augmentation, the benefit of including spherical information is clearly visible. In this context, we see that the LSC model provides the most favorable results.

Acknowledgements Data were provided by the Human Connectome Project, WU-Minn Consortium (Principal Investigators: David Van Essen and Kamil Ugurbil; 1U54MH091657) funded by the 16 NIH Institutes and Centers that support the NIH Blueprint for Neuroscience Research; and by the McDonnell Center for Systems Neuroscience at Washington University. This work was funded by the Deutsche Forschungsgemeinschaft (DFG, German Research Foundation) - Project ID 417063796.

References

1. Andersson, J.L., et al.: How to correct susceptibility distortions in spin-echo echo-planar images: application to diffusion tensor imaging. Neuroimage **20**(2), 870–888 (2003)
2. Basser, P.J., et al.: MR diffusion tensor spectroscopy and imaging. Biophys. J. **66**(1), 259 (1994)
3. Bronstein, M.M., Bruna, J., LeCun, Y., Szlam, A., Vandergheynst, P.: Geometric deep learning: going beyond euclidean data. IEEE Signal Process. Mag. **34**(4), 18–42 (2017)
4. Cohen, T.S., Geiger, M., Köhler, J., Welling, M.: Spherical CNNs (2018). arXiv:1801.10130
5. Descoteaux, M., Angelino, E., Fitzgibbons, S., Deriche, R.: Apparent diffusion coefficients from high angular resolution diffusion imaging: estimation and application. Magn. Reson. Med. **56**(2), 395–410 (2006)
6. Descoteaux, M., Angelino, E., Fitzgibbons, S., Deriche, R.: Regularized, fast, and robust analytical Q-ball imaging. Mag. Reson. Med. Official J. Int. Soc. Mag. Reson. Med. **58**(3), 497–510 (2007)
7. Ferizi, U., Scherrer, B., Schneider, T., Alipoor, M., Eufracio, O., Fick, R.H., Deriche, R., Nilsson, M., Loya-Olivas, A.K., Rivera, M., et al.: Diffusion MRI microstructure models with in vivo human brain Connectome data: results from a multi-group comparison. NMR Biomed. **30**(9), e3734 (2017)
8. Garyfallidis, E., Brett, M., Amirbekian, B., Rokem, A., Van Der Walt, S., Descoteaux, M., Nimmo-Smith, I.: Dipy, a library for the analysis of diffusion MRI data. Frontiers Neuroinf. **8**, 8 (2014)
9. Golkov, V., Dosovitskiy, A., Sperl, J.I., Menzel, M.I., Czisch, M., Sämann, P., Brox, T., Cremers, D.: Q-space deep learning: twelve-fold shorter and model-free diffusion MRI scans. IEEE Trans. Med. Imaging **35**(5), 1344–1351 (2016)
10. He, K., Zhang, X., Ren, S., Sun, J.: Deep residual learning for image recognition. In: Proceedings of the IEEE Conference on Computer Vision and Pattern Recognition, pp. 770–778 (2016)
11. Henriques, R.N., Jespersen, S.N., Shemesh, N.: Microscopic anisotropy misestimation in spherical-mean single diffusion encoding MRI. Mag. Reson. Med. **81**(5), 3245–3261 (2019)
12. Iandola, F., Moskewicz, M., Karayev, S., Girshick, R., Darrell, T., Keutzer, K.: DenseNet: Implementing Efficient ConvNet Descriptor Pyramids (2014). arXiv:1404.1869
13. Isensee, F., Kickingereder, P., Wick, W., Bendszus, M., Maier-Hein, K.H.: No New-Net. In: International MICCAI Brainlesion Workshop, pp. 234–244. Springer, Berlin (2018)
14. Jensen, J.H., Helpern, J.A., Ramani, A., Lu, H., Kaczynski, K.: Diffusional Kurtosis imaging: the quantification of non-gaussian water diffusion by means of magnetic resonance imaging. Mag. Reson. Med. Official J. Int. Soc. Mag. Reson. Med. **53**(6), 1432–1440 (2005)
15. Jones, D.K., Horsfield, M.A., Simmons, A.: Optimal strategies for measuring diffusion in anisotropic systems by magnetic resonance imaging. Mag. Reson. Med. Official J. Int. Soc. Mag. Reson. Med. **42**(3), 515–525 (1999)
16. Koppers, S., Merhof, D.: Delimit pytorch—an extension for deep learning in diffusion imaging (2018)
17. Koppers, S., et al.: Diffusion MRI signal augmentation: from single shell to multi shell with deep learning. In: CDMRI, pp. 61–70. Springer, Berlin (2016)

18. Le Bihan, D., Breton, E., Lallemand, D., Grenier, P., Cabanis, E., Laval-Jeantet, M.: MR imaging of intravoxel incoherent motions: application to diffusion and perfusion in neurologic disorders. Radiology **161**(2), 401–407 (1986)
19. Lin, T.Y., Goyal, P., Girshick, R., He, K., Dollár, P.: Focal loss for dense object detection. In: Proceedings of the IEEE International Conference on Computer Vision, pp. 2980–2988 (2017)
20. Liu, L., Jiang, H., He, P., Chen, W., Liu, X., Gao, J., Han, J.: On the variance of the adaptive learning rate and beyond (2019). arXiv:1908.03265
21. Menegatto, V.: Approximation by spherical convolution. Numer. Funct. Anal. Optim. **18**(9–10), 995–1012 (1997)
22. Michailovich, O., Rathi, Y.: Fast and accurate reconstruction of HARDI data using complessed sensing. In: International Conference on Medical Image Computing and Computer-assisted Intervention, pp. 607–614. Springer, Berlin (2010)
23. Özarslan, E., Koay, C.G., Shepherd, T.M., Komlosh, M.E., İrfanoğlu, M.O., Pierpaoli, C., Basser, P.J.: Mean Apparent Propagator (MAP) MRI: a novel diffusion imaging method for mapping tissue microstructure. NeuroImage **78**, 16–32 (2013)
24. Ronneberger, O., Fischer, P., Brox, T.: U-net: Convolutional networks for biomedical image segmentation. In: International Conference on Medical Image Computing and Computer-assisted Intervention, pp. 234–241. Springer, Berlin (2015)
25. Tournier, J.D., Calamante, F., Connelly, A.: Robust determination of the fibre orientation distribution in diffusion MRI: non-negativity constrained super-resolved spherical deconvolution. Neuroimage **35**(4), 1459–1472 (2007)
26. Van Essen, D.C., Smith, S.M., Barch, D.M., Behrens, T.E., Yacoub, E., Ugurbil, K., Consortium, W.M.H., et al.: The WU-Minn human connectome project: an overview. Neuroimage **80**, 62–79 (2013)
27. Zhang, Y., et al.: Segmentation of brain MR images through a hidden Markov random field model and the expectation-maximization algorithm. IEEE Trans. Med. Imaging **20**(1), 45–57 (2001)

Hybrid Graph Convolutional Neural Networks for Super Resolution of DW Images

Jiquan Ma and Hui Cui

Abstract Deep learning based super-resolution has been proved as a promising technique in increasing the resolution of Diffusion-weighted (DW) images, which is widely used in brain white matter analyses. Existing models using single type of convolutional neural networks (CNN) cannot support effective learning in heterogeneous data space composed of grid structure and wavevector domain. We propose a novel technique that employs a graph model in a deep network to improve the spatial resolution of DW image. The model is composed of a residual CNN to learn from spatial information in 3D grid structural domain and graph CNN (GCNN) to emphasize diffusion angular information in non-Euclidean domain. Given a low resolution DW image, for each direction of diffusion gradients, 3D convolutions and residual CNN are firstly performed to generate coarse-level super resolution image. Then the learning outputs from grid structure space are stacked and refined by the features in diffusion gradient space modelled by GCNN. We evaluate the proposed hybrid graph CNN model using real brain data from Human Connectome Project. Extensive experiments demonstrate improved results with richer fiber tracts that are closest to the ground truth. Our hybrid graph CNN model benefits the learning of spatial and angular features in complex or heterogeneous spaces of DWI.

J. Ma (✉)
Department of Computer Science and Technology, Heilongjiang University, Harbin 150080, Heilongjiang, China
e-mail: majiquan@hlju.edu.cn

H. Cui
Department of Computer Science and Information Technology, La Trobe University, Bundoora 3083, Victoria, Australia
e-mail: L.Cui@latrobe.edu.au

N. Gyori et al. (eds.), *Computational Diffusion MRI*, Mathematics and Visualization,
https://doi.org/10.1007/978-3-030-73018-5_16

1 Introduction

As a non-invasive technique for characterizing water molecule mobility in vivo, DW imaging (DWI) plays an important role in studying brain white matter [1–3]. However, DWI may suffer from the trade-off of resolution, SNR and acquisition time [4]. For instance, low resolution (LR) DWIs introduce a series of artifacts, such as partial volume effect (PVE), which may affect the sensitivity of downstream analysis including fiber orientation distribution function (ODF), fiber tractography , etc.

Super resolution (SR) techniques play an important role in increasing the resolution of DW images [5–7]. SR aims to establish a mapping from LR-DWI to its high resolution (HR) counterpart. Recovering HR from LR, however, is an inverse process and challenging problem. Single-image based SR (SISR) uses single LR image to restore the HR counterpart which is suitable for clinical practice. Recently, deep learning based SISR attracted extensive research interests and have shown remarkable performances [8]. For instance, CNN and recurrent neural networks (RNN) yielded state-of-the-art results in computer vision and speech recognition tasks. CNN models, however, are designed for 2D/3D data in Euclidean domain. It is difficult to be applied to the non-Euclidean domain or heterogeneous feature spaces, such as citation network, social network, and protein interaction network. This is also the case with DWI which is acquired from a combined 6D domain. For the applications on DW images, CNN based approaches are mostly constructed on the grid structural spatial domain. These methods, however, fail to incorporate the effect of diffusion wavevector domain.

Graph CNN (GCNN) extends the properties of CNN to graph-structured data, which provides possibilities to enhance the resolution of DWI by using deep neural networks. GCNN is a more general learning framework and can be applied in many fields, where data is organized by a graph directly or indirectly. GCNN was firstly proposed by [9] which was further improved by localized convolution operation [10]. To some extent, grid structure data can be considered as a unique graph with uniformed edge fields. A few applications have verified the capacity of CNN in classification and segmentation [11–13]. For DWI, the unified graph convolutional architecture, however, fails to take advantages of the grid structure in the spatial domain. In addition, spatial and angular information are not differentiated and adaptively fused.

To tackle this challenge, we propose a novel technique that employs graph-based operations in a deep CNN model to learn from heterogeneous data with both spatial and angular information. The first contribution is the new idea to employ a graph model, GCNN, to learn the similarity between voxels in diffusion gradient direction domain (q-space). In our model, q-space means the normalized version by $q/|q|$. We also construct a reference graph to fill in the gap between CNN and GCNN. The experimental results using the Human Connectome Project (HCP) dataset shows that GCNN further improved the learning outcome of global residual CNN in grid structural domain. Secondly, as a post-acquisition approach, our model is independent of imaging protocols which can be extended to other images. Besides, our research reveals the potential applications on the field of learning information from heterogeneous data domain.

2 Dataset

We collect DW images from HCP [14–16]. The images have voxel size of $1.25 \times 1.25 \times 1.25\text{mm}^3$, 90 gradient directions per shell, $b = 1000, 2000, 3000$ s/mm^2. The DW images with $b = 1000$ s/mm^2 were used for experiments. Since DW images are corrupted with heavy noise [17, 18], we use the LPCA filter [19] to remove noise before experiments. In total, there are 25 subjects where 8 of them are used for training, 2 of them are used for validation, and the rest 15 are used for testing. During the training process, 3D patches of sizes $11 \times 11 \times 11$ are extracted from each subject, generating around 30000 samples for each subject. LR DW images ($75 \times 90 \times 75$) were generated from the HR images ($150 \times 180 \times 150$) using a 3D mean filter with a reduction scale factor = 2 and voxel size of $2.5 \times 2.5 \times 2.5$ mm^3.

3 Methods

We formulate the SR process as an end to end learning process to learn the mapping from LR DW to HR DW images. The two major components in our model include CNN based coarse-level SR prediction in 3D grid structure space and GCNN module for refinement in diffusion gradient space. Overview of the proposed model is given in Fig. 1. Given an input LR DW image, for each direction of diffusion gradients, 3D convolutions and residual CNN are firstly performed in grid structure space to generate coarse SR. Then the learning output is refined by the features in diffusion gradient space modelled by GCNN. To fill the gap between CNN and GCNN, a reference graph is constructed. The predicted HR DW image is finally obtained by the output layer of convolutions.

3.1 Coarse SR Prediction in 3D Grid Structure Space

CNN has been proved of its superior performance over classical approaches especially using grid structure data. Because in each direction of diffusion gradients, DWI can be considered as a 3D grid structure in spatial domain, we exploit 3D CNN to predict coarse level SR from LR DWI.

For each direction of diffusion gradients, convolution operation is firstly performed to reduce the effects of low-resolution interpolation. The convolution module is composed of three 3D convolution layers, including two 3D filters with size $1 \times 1 \times 1$ at the first layer and last layer and one 3D filter with size $3 \times 3 \times 3$ at

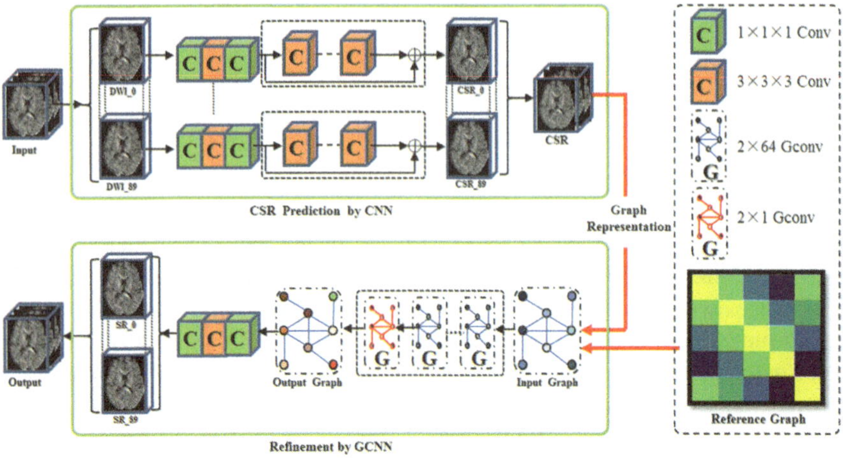

Fig. 1 Architecture of the proposed hybrid GCNN and CNN model. Firstly, residual CNN is used to generate coarse SR DW images. A Graph CNN is then used to refine the coarse-level predictions using the features learnt from diffusion space

the middle layer. After that, a global residual CNN based model, very deep super resolution (VDSR) [20] is exploited to predict coarse level SR. To extract the spatial information in LR DWI volumes, we change the 2D operations to 3D.

3.2 *Refinement by GCNN in Diffusion Gradient Space*

Generally, imaged voxels exhibit different signal intensities at different gradient directions, which means that rich information related to micro-structures in the brain could be encoded. We aim to improve the prediction performance by incorporating such information in diffusion gradient domain. In order to utilize the correlations across different diffusion gradient directions, GCNN is used to refine the quality of coarse-level prediction. Given $I^{CSR}_{X\times Y\times Z,n}$ denoting a predicted coarse-level SR 3D patch of size $X \times Y \times Z$ from one diffusion direction, a 4D volume $R^{CSR}_{X\times Y\times Z\times N}$ can be obtained by stacking all the 3D patches from N directions. Then a reference graph G is constructed according to the correlation between diffusion gradients to project 4D features to graph structure.

Reference Graph Construction Given $R^{CSR}_{X\times Y\times Z\times N}$, the reference graph $G = (V, E, W)$ is constructed with nodes V corresponding to the voxels in the volume with size $X = Y = Z = 64$, $N = 90$. W_{ij} is edge weight connecting nodes V_i and V_j, which is computed by the distance along gradient directions as

$$W_{ij} = Dist_s \times Dist_a \tag{1}$$

where $Dist_s$ and $Dist_a$ represent the distance between two voxels in terms of spatial domain and diffusion gradient direction domain. Respectively, $Dist_s$ is defined as:

$$Dist_s = e^{\frac{(\frac{|v_i, v_j|}{max})^2}{-\sigma_1}}, i \in [1, ..., 64], j \in [1, ..., 64] \tag{2}$$

where i and j are indices of the position of nodes, $|v_i, v_j|$ denotes the Euclidean distance between voxels v_i and v_j in $I^{CSR}_{X \times Y \times Z, n}$, max is the maximum distance. $Dist_a$ is defined as

$$Dist_a = e^{\frac{1.0-(g_m \cdot g_n)^2}{-\sigma_2}}, m \in [1, ..., 90], n \in [1, ..., 90] \tag{3}$$

were m and n are indices of diffusion gradient direction, $(g_m \cdot g_n)$ is the dot product of diffusion gradient directions g_m and g_n. σ_1 and σ_2 are kernel size for spatial domain and diffusion gradient direction domain. By adjusting kernel size σ_1 and σ_2, the reference graph G can be built with attention to a specific domain. σ_1 and σ_2 are fixed empirical parameters, which are used to control the bandwidths of two Gaussian kernels for x-space and q-space. For instance, when kernel size σ_1 is greater than σ_2, the graph edge weight will be dominated by spatial domain, otherwise by diffusion gradient direction domain.

An illustration is given in Fig. 2. For spatial domain (x-space), the edge weight is determined by the distance between voxels in the 3D patch. For diffusion gradient domain (q-space) , edge weight is related with the intersection angle between diffusion gradients. In xq-space, edge weight is co-determined by spatial domain and diffusion gradient domain. In order to make a clear instance, the illustration of graph construction is presented using a smaller 3D patch size.

GCNN and Optimization Given the constructed graph, we define the graph filter Following [21], as:

$$g_\theta(\Lambda) = \sum_{k=0}^{K-1} \theta_k T_k(\tilde{\Lambda}) \tag{4}$$

where the parameter $\theta \in \mathbb{R}^k$, $T_k(\tilde{\Lambda}) \in \mathbb{R}^{n \times n}$ and $\tilde{\Lambda} = 2\Lambda/\lambda_{max} - I_n$. Where I_n is a vector of polynomial coefficients, $\Lambda = diag([\lambda_0, \ldots, \lambda_{n-1}]) \in \mathbb{R}^{n \times n}$, $\{\lambda_l\}_{l=0}^{n-1}$ are ordered real non-negative eigenvalues associated with graph Laplacian. $\tilde{\Lambda}$ is a diagonal matrix of scaled eigenvalues that lie in [1, 1]. The j_{th} output feature map of the sample s is given by

$$y_{s,j} = \sum_{i=1}^{F_{in}} g_{\theta_{i,j}}(L) x_{s,i} \in \mathbb{R}^n, \tag{5}$$

and then error gradients for back propagation can be calculated as:

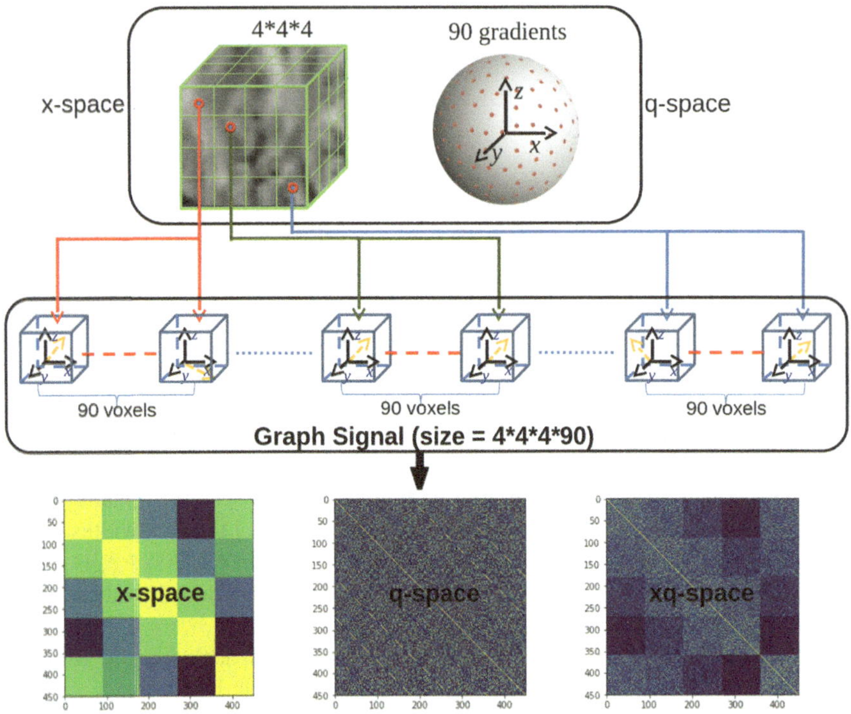

Fig. 2 Illustration of graph construction. 3D voxel patch size is 4 * 4 * 4 and the number of diffusion gradients is 90. Adjacency matrix (AM) of reference graph is visualized at the bottom. In order to make it clear, we just zoom in partial AM corresponding to the top five voxels (5 * 90 = 450)

$$\frac{\partial E}{\partial \theta_{i,j}} = \sum_{s=1}^{S} [\bar{x}_{s,i,0,\dots}, \bar{x}_{s,i,K-1}]^T \frac{\partial E}{\partial y_{s,j}} \tag{6}$$

3.3 Loss Function

For effective back forward propagation, we propose a combined loss function $Loss_{total}$ with voxel value deviation $Loss_v$, spectral graph deviation $Loss_g$. to evaluate the deviation in graph structure, and L_2 regulate item to avoid over-fitting. $Loss_{total}$ is defined as

$$Loss_{total} = \alpha \cdot Loss_v + \beta \cdot Loss_g + \lambda \cdot \frac{1}{2} \sum {\omega_{i,j}}^2 \tag{7}$$

where $Loss_v$ is defined as

Table 1 Comparison results in terms of PSNR and SSIM

	Linear	CNN	GCNN	PM
DWI	29.69, 0.90	30.49, 0.95	28.29, 0.94	32.65, 0.96
FA	30.11, 0.92	32.46, 0.96	30.35, 0.93	33.64, 0.98

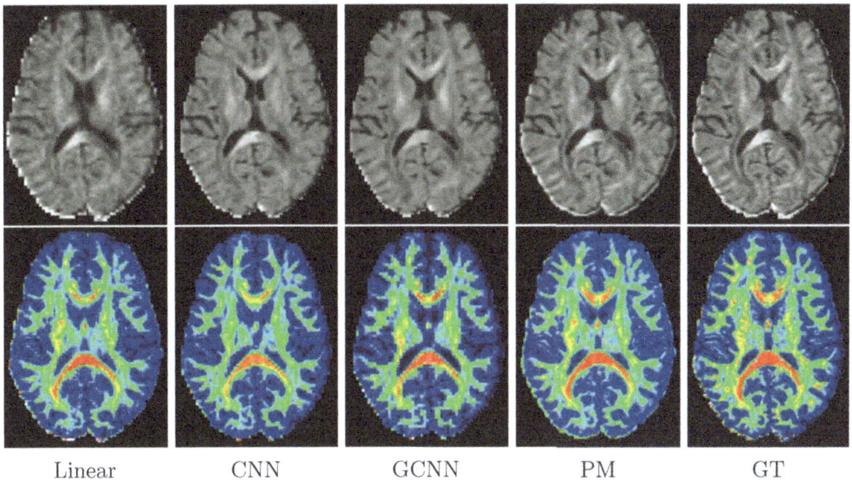

Fig. 3 Comparison of predicted HR-DWI and FA mapping using different models at axial view. The last column is the ground truth (GT)

$$Loss_v = \frac{\sum_{i=1}^{n} |V_{o_i} - V_{l_i}|}{n} \tag{8}$$

where n is the total number of the voxels in the sampled 3D patch, i is the i_{th} voxel. According to the spectral graph theory, a graph can be represented by the Laplacian matrix. Thus, $Loss_g$ is measured by Laplacian loss as

$$Loss_g = \frac{\sum (L \times V_p - L \times V_g)^2}{size(V)} \tag{9}$$

where V_p is the predicted graph, V_g is ground truth, L is Laplacian matrix of the reference graph, $size(V)$ is the total number of nodes in the graph, λ is the regulation coefficient. In our experiments, $\alpha + \beta = 1.0$, $\lambda = 0.0001$.

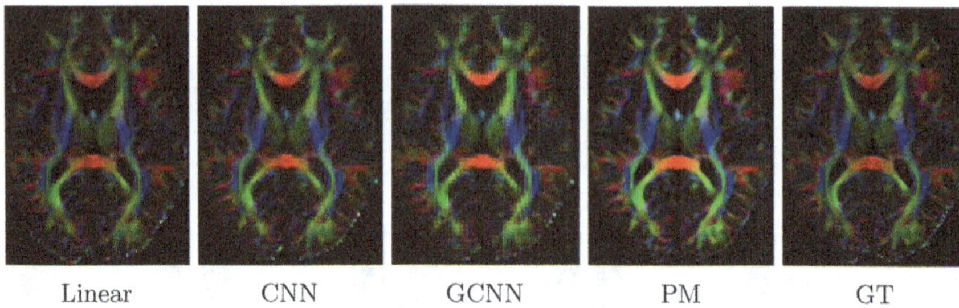

Fig. 4 Comparison of colored FA using different models at axial view

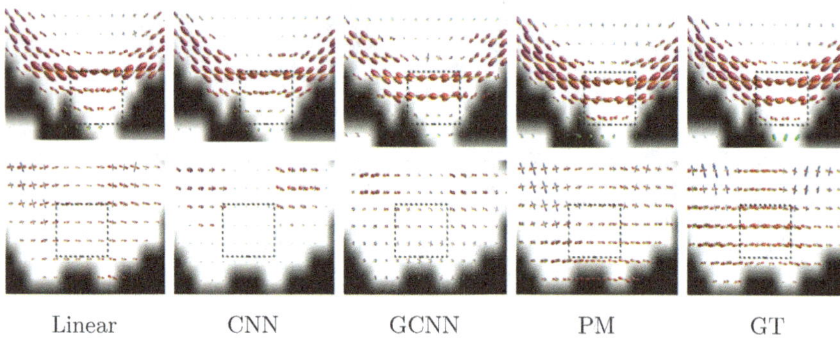

Fig. 5 Comparison of orientation distribution function (ODF)

4 Experiments

Comparison Methods and Parameter Settings We demonstrate the effectiveness of our network by comparison with linear interpolation, residual CNN, and graph CNN in terms of peak signal to noise ratio (PSNR) and structural similarity index (SSIM). It is noted that all the deep learning baseline methods share the same training settings with our proposed model. We also evaluate the quality of derived diffusion quantities, including fractional anisotropy (FA) images, colored FA and fiber tracts [22].

Our model was trained using an Adam optimizer with an initial learning rate of 10^{-4} and gradual decay every 5000 iterations. Other training settings included random sample step as 300 to control the number of training samples fed into the network in each iteration, batch size as 20, $\alpha = 0.5$, $\beta = 0.5$, max epoch=300. In order to crop the 3D voxel patch which has too many zero values, non-zero-ratio is set as 0.6. Our network was implemented using TensorFlow 1.4. It took about 12 hours for training using GPU GTX 1070oc.

Experimental Results and Discussion The quantitative results are given in Table 1. As shown, the proposed method (PM) achieved the best performance in terms of PSNR and SSIM.

The results in Fig. 3 indicate that PM improves the quality of reconstructed HR DW images and associated FA images significantly. Our model generated DW and FA

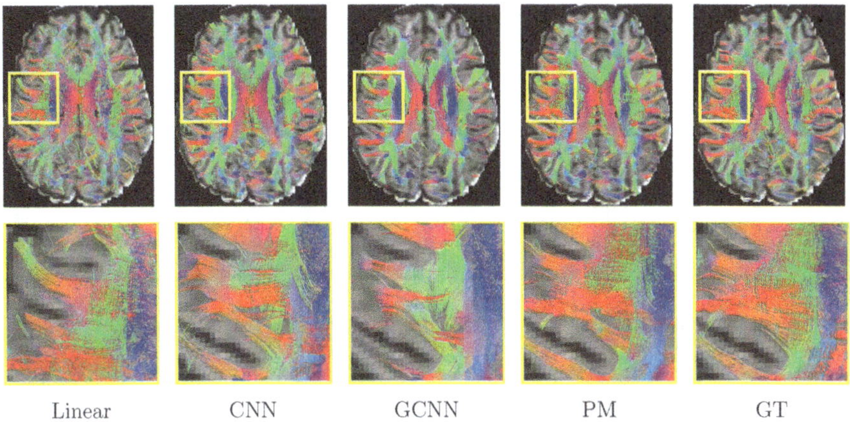

Fig. 6 Comparison of fiber tractography using different models at axial view

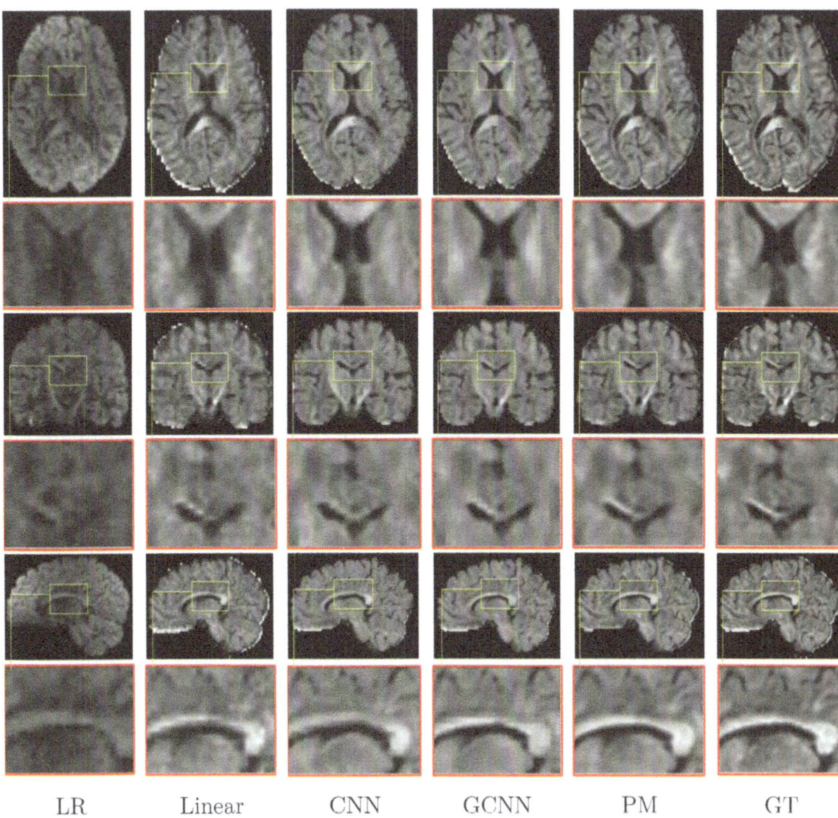

Fig. 7 Comparison of local zoom in views of DW images

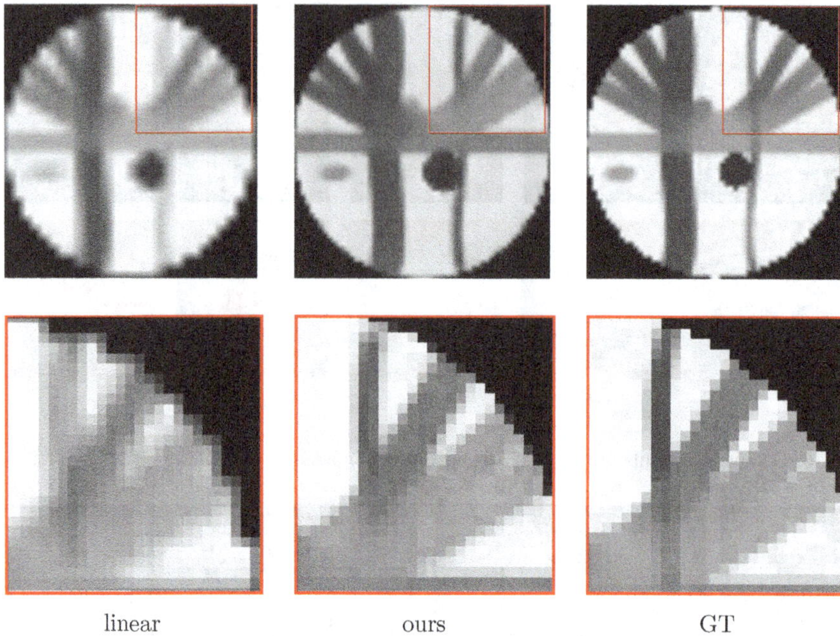

Fig. 8 Results of DWI SR on Phantomas data set from 32*32*32 to 64×64×64

images that are closest to ground truth. All the DWI and FA mappings are displayed using Mango Toolbox developed by Research Imaging Institute, UTHSCSA. The comparison of coloured FA in axial view is given in Fig. 4. Figure 5 shows the results of ODF fitting. Figure 6 is comparison of fiber tractography using different methods. Details of reconstructed DWI are compared in Fig. 7. In order to validate our model over cross phantom data sets, synthetic dMRI data are generated by Phantoms [23]. The comparison over Phantoms data is given in the Fig. 8. As demonstrated by the results, our hybrid model achieved the best SR results when compared with linear, CNN, and GCNN models. As can be observed, our model gives the best visual results with more structure details.

Experiments shown that low order Chebyshev polynomial is helpful to make an effective learning in graph. In this task, the best performance is obtained when the order of Chebyshev polynomial is 2.

5 Conclusion

We propose a domain directed hybrid graph based convolutional learning model for SR of low resolution DW images. Extensive experiments on HCP data demonstrated the improved performance in both qualitative and quantitative evaluations. Our model has the potential to be applied to other research where there is necessity to learning from heterogeneous data domains.

References

1. Johansen-Berg, H., Behrens, T.E.: Diffusion MRI: From Quantitative Measurement to in Vivo Neuroanatomy. Academic, New York, USA (2013)
2. Wu, E.X., Wu, Y., Tang, H., et al.: Study of myocardial fiber pathway using magnetic resonance diffusion tensor imaging. Mag. Reson. Imaging **25**(7), 1048–1057 (2007)
3. Essayed, W.I., Zhang, F., Unadkat, P., et al.: White matter tractography for neurosurgical planning: a topography-based review of the current state of the art. NeuroImage: Clinical. **15**, 659–672 (2017)
4. Chen, G., Dong, B., Zhang, Y., et al.: XQ-SR: Joint x-q space super-resolution with application to infant diffusion MRI. Med. Image Anal. **57**, 44–55 (2019)
5. Sánchez, I., Vilaplana, V.: Brain MRI super-resolution using 3D generative adversarial networks. arXiv preprint arXiv:1812.11440. (2018)
6. Greenspan, H., Oz, G., Kiryati, N., et al.: Mri inter-slice reconstruction using super-resolution. Mag. Reson. Imaging **20**(5), 437–446 (2002)
7. Alexander, D.C., Zikic, D., Zhang, J., et al.: Image quality transfer via random forest regression: applications in diffusion MRI. In: Golland P., Hata N., Barillot C., Hornegger J., Howe R. (eds.), International Conference on Medical Image Computing and Computer-Assisted Intervention. MICCAI 2014. LNCS, vol. 8675, pp. 225–232. Springer, Cham, Heidelberg (2014)
8. Dong, C., Loy, C.C., He, K., et al.: Image super-resolution using deep convolutional networks. IEEE Trans. Pattern Anal. Mach. Intell. **38**(2), 295–307 (2016)
9. Bruna, J., Zaremba, W., Szlam, A., et al.: Spectral networks and locally connected networks on graphs (2013). arXiv:1312.6203
10. Henaff, M., Bruna, J., LeCun, Y.: Deep convolutional networks on graph-structured data (20105). arXiv:1506.05163
11. Yan, X., Ai, T.: Analysis of irregular spatial data with machine learning: classification of building patterns with a graph convolutional neural network (2018). arXiv:1809.08196
12. Golrizkhatami, Z., Taheri, S., Acan, A.: Multi-scale features for heartbeat classification using directed acyclic graph cnn. Appl. Artif. Intell. **32**(7–8), 613–628 (2018)
13. Gusi T., Wei H., Zongming G., et al.: Rgcnn: Regularized graph cnn for point cloud segmentation (2018). arXiv:1806.02952
14. Andersson, J.L., Skare, S., Ashburner, J.: How to correct susceptibility distortions in spin-echo echo-planar images: application to diffusion tensor imaging. Neuroimage **20**(2), 870–888 (2003)
15. Andersson, J.L., Sotiropoulos, S.N.: Non-parametric representation and prediction of single- and multi-shell diffusion-weighted mri data using gaussian processes. Neuroimage **122**, 166–176 (2015)
16. Andersson, J.L., Sotiropoulos, S.N.: An integrated approach to correction for off-resonance effects and subject movement in diffusion mr imaging. Neuroimage **125**, 1063–1078 (2016)
17. Tristán-Vega, A., Aja-Fernández, S.: Dwi filtering using joint information for dti and hardi. Med. Image Anal. **14**(2), 205–218 (2010)
18. Basu, S., Fletcher, T., Whitaker, R.: Rician noise removal in diffusion tensor mri. In: Larsen, R., Nielsen, M., Sporring, J. (eds.), MICCAI 2006. LNCS, vol. 4190, pp. 117-125. Springer, Berlin, Heidelberg (2006). https://xs.scihub.ltd/, https://doi.org/10.1007/11866565_15
19. Manjón, J.V., Coupé, P., Concha, L., et al.: Diffusion weighted image denoising using overcomplete local PCA. PloS one **8**(9), e73021 (2013)
20. Kim, J., Kwon L.J., Mu L.K.: Accurate image super-resolution using very deep convolutional networks. In: Proceedings of the IEEE Conference on Computer Vision and Pattern Recognition, pp. 1646–1654. Las Vegas, NV, USA (2016)

21. Defferrard, M., Bresson, X., Vandergheynst, P.: Convolutional neural networks on graphs with fast localized spectral filtering. In: 30th International Proceedings on NIPS, pp. 3844–3852. Barcelona, Spain (2016)
22. Wang, Z., Bovik, A.C., Sheikh, H.R., et al.: Image quality assessment: from error visibility to structural similarity. IEEE Trans. Image Process. **13**(4), 600–612 (2004)
23. Caruyer E., Daducci A., Descoteaux M., Houde J., Thiran J., et al.: Phantomas: a flexible software library to simulate diffusion MR phantoms. In: International Society for Magnetic Resonance in Medicine, pp. 6406–6407. Milan, Italy (2014)

Manifold-Aware CycleGAN for High-Resolution Structural-to-DTI Synthesis

Benoit Anctil-Robitaille, Christian Desrosiers, and Herve Lombaert

Abstract Unpaired image-to-image translation has been applied successfully to natural images but has received very little attention for manifold-valued data such as in diffusion tensor imaging (DTI). The non-Euclidean nature of DTI prevents current generative adversarial networks (GANs) from generating plausible images and has mainly limited their application to diffusion MRI scalar maps, such as fractional anisotropy (FA) or mean diffusivity (MD). Even if these scalar maps are clinically useful, they mostly ignore fiber orientations and therefore have limited applications for analyzing brain fibers. Here, we propose a manifold-aware CycleGAN that learns the generation of high-resolution DTI from unpaired T1w images. We formulate the objective as a Wasserstein distance minimization problem of data distributions on a Riemannian manifold of symmetric positive definite 3×3 matrices SPD(3), using adversarial and cycle-consistency losses. To ensure that the generated diffusion tensors lie on the SPD(3) manifold, we exploit the theoretical properties of the exponential and logarithm maps of the Log-Euclidean metric. We demonstrate that, unlike standard GANs, our method is able to generate realistic high-resolution DTI that can be used to compute diffusion-based metrics and potentially run fiber tractography algorithms. To evaluate our model's performance, we compute the cosine similarity between the generated tensors principal orientation and their ground-truth orientation, the mean squared error (MSE) of their derived FA values and the Log-Euclidean distance between the tensors. We demonstrate that our method produces 2.5 times

This work was supported financially by the Réseau de Bio-Imagerie du Québec (RBIQ), the Research Council of Canada (NSERC), the Fonds de Recherche du Québec (FQRNT), ETS Montreal, and NVIDIA with the donation of a GPU.

B. Anctil-Robitaille (✉) · C. Desrosiers · H. Lombaert
Department of IT and Software Engineering, ETS Montreal, Montreal, Canada
e-mail: benoit.anctil-robitaille.1@ens.etsmtl.ca

C. Desrosiers
e-mail: christian.desrosiers@etsmtl.ca

H. Lombaert
e-mail: herve.lombaert@etsmtl.ca

N. Gyori et al. (eds.), *Computational Diffusion MRI*, Mathematics and Visualization,
https://doi.org/10.1007/978-3-030-73018-5_17

better FA MSE than a standard CycleGAN and up to 30% better cosine similarity than a manifold-aware Wasserstein GAN while synthesizing sharp high-resolution DTI.

1 Introduction

Unpaired image-to-image translation and image synthesis have been widely used in medical imaging [21]. Whether they are employed to generate missing modalities, normalize images or enhance images quality and resolution, generative adversarial networks (GANs) [4] have been proven effective in multiple challenging medical image analysis tasks. However, they have been mainly studied on real-valued images, thus impeding the development of applications for manifold-valued data such as diffusion tensor images (DTI). Despite the growing interest in the brain's structural connectivity, applications of GANs to DTI have been mostly limited to generating derived scalar maps like fractional anisotropy (FA) and mean diffusivity (MD), which ignore the fibers' orientation and provide limited insights on their structural organization.

Among the literature, [5] investigates the generation of diffusion MRI scalar maps from T1w images using a CycleGAN [23]. The authors show that the structural and diffusion spaces share a sufficient amount of information to be able to synthesize realistic FA and MD maps from downsampled T1w images. In [22], dual GANs and Markovian discriminators are used to harmonize multi-site FA and MD maps of neonatal brains. They demonstrate that using a GAN-like architecture can better capture the complex non-linear relations between multiple domains than standard normalization methods.

While the previously mentioned works present applications of GANs on DTI-derived metrics, they do not tackle the challenge of generating DTI. Being able to synthesize such images would unlock a vast amount of useful methods that are already well studied on real-valued modalities, while preserving all the geometrical information encoded in the diffusion tensors. However, DTI data is manifold-valued: the data of each voxel lies on a Riemannian manifold of symmetric positive definite 3×3 matrices, i.e., the SPD(3) manifold. The non-Euclidean nature of DTI prevents standard GANs from generating plausible images as there is no guarantee that the generated diffusion tensors lie on the SPD(3) manifold. A solution presented in [2] is to employ the Log-Euclidean metric to accurately process data on the SPD(n) manifold. By using the log and exp projections proposed in [2], one can apply Euclidean operations on tensors and guarantee that the resulting tensors will lie on such manifold. Those computationally efficient mapping operations form an interesting framework for manifold-valued data learning, and have been used in [7] to develop a deep neural network called SPDNet which learns discriminative SPD matrices. With the help of the matrix backpropagation of spectral layers defined in [9], they designed a network that learns data on SPD(n). Nonetheless, SPDNet [7] is limited to single SPD matrix learning and cannot help in learning multiple spatially-organized SPD matrices as it is the case with DTI.

Related to our work, [8] proposes a manifold-aware Wasserstein GAN for manifold-valued data generation, which leverages the aforementioned log and exp mappings. In their work, they generate plausible slices of DTI from noise vectors. By comparing the produced images of their network with those produced by a regular GAN, one can clearly see that the manifold mappings are necessary to produce credible diffusion tensors. However, the proposed manifold-aware GAN could not provide any additional clinical insights, nor help in understanding the brain's connectivity as the generated images are not conditioned by any real contextual information such as T1w images. Furthermore, [8] only focuses on the generation of 2D DTIs, which is of limited application for the assessment of the structural organization of the brain's fibers.

This paper presents a novel manifold-aware Wasserstein CycleGAN that generates high-resolution (HR) DTI from unpaired T1w images. Our method leverages the detailed structural information provided by T1w images while constraining the synthesized diffusion tensors to lie on the SPD(3) manifold using the Log-Euclidean metric. Specifically, the contributions of this work are as follows:

- We present the first CycleGAN model for the unpaired mapping between images and SPD(3) manifold-valued data.
- This is also the first deep learning model to generate DTI data from structural MRI. As mentioned before, previous approaches have focused on generating diffusion scalar maps like FA or MD, and not diffusion tensors as in this work.

Our proposed manifold-aware CycleGAN method is presented in the next section.

2 Method

Let X_{HR} be the domain of high-resolution structural images and Y_{HR} be the domain of high-resolution diffusion tensor images. Our goal is to learn mapping functions $G_Y : X_{\text{HR}} \mapsto Y_{\text{HR}}$ and $G_X : Y_{\text{HR}} \mapsto X_{\text{HR}}$ that translate the real-valued domain X_{HR} into the manifold-valued domain Y_{HR} and the other way around. However, as it is often the case, we do not have access to high-resolution DTI. Thus, we train our model with unpaired training samples $\{\mathbf{x}_i\}_{i=1}^N$ where $\mathbf{x}_i \in X_{\text{HR}}$ is a 3D structural image (e.g., T1w), and $\{\mathbf{y}_j\}_{j=1}^M$ where $\mathbf{y}_j \in Y_{\text{LR}}$ is a DT image with lower resolution. We employ the $\log_{\text{Id}}$ and $\exp_{\text{Id}}$ mapping to project the generated and the real DTI on the tangent plane at the 3×3 identity matrix, to ensure that $G_Y(\mathbf{x})$ lies on the SPD(3) manifold and to compare the manifold-valued data distributions as in [8]. Two discriminators, D_X and D_Y, assess the quality of the generated images $G_X(\mathbf{y})$ and downsampled generated DTI $\downarrow G_Y(\mathbf{x})$ with respect to their real data distributions $G_X(\mathbf{y}) \sim \mathbb{P}_{X_{\text{HR}}}$ and $\downarrow G_Y(\mathbf{x}) \sim \mathbb{P}_{\log_{\text{Id}}(Y_{\text{LR}})}$. We formulate the objective as a Wasserstein distance minimization problem on the SPD(3) manifold with adversarial and cycle-consistent losses. The adversarial portion of the objective helps G_X and G_Y generate images that match the target distribution. On the other hand, the cycle-consistent losses provide high-resolution gradients that are necessary to generate high-resolution DTI with a proper structure (Figs. 1 and 2).

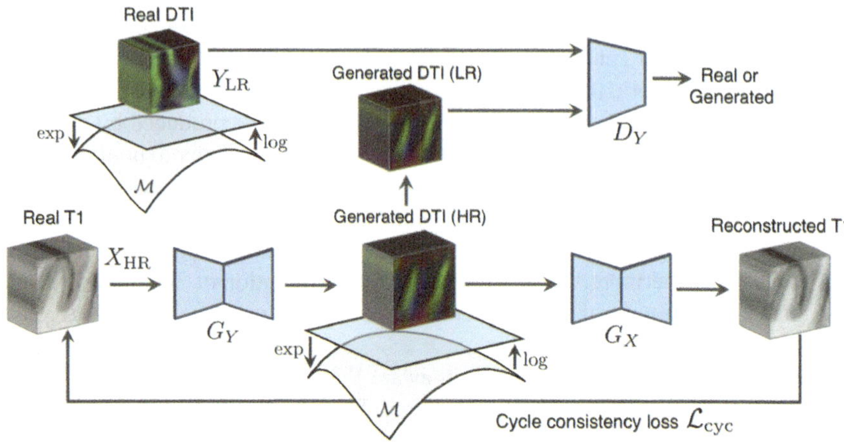

Fig. 1 The forward cycle of our manifold-aware CycleGAN: G_Y generates high-resolution DTIs on the SPD(3) manifold by projecting its prediction using the $\exp_{\mathrm{Id}}$ and $\log_{\mathrm{Id}}$ mappings consecutively. D_Y assesses the downsampled generated images and provides adversarial feedback to G_Y. G_X tries to reconstruct the original T1w images from $G_Y(\mathbf{x})$ and supplies high-resolution gradient information to G_Y

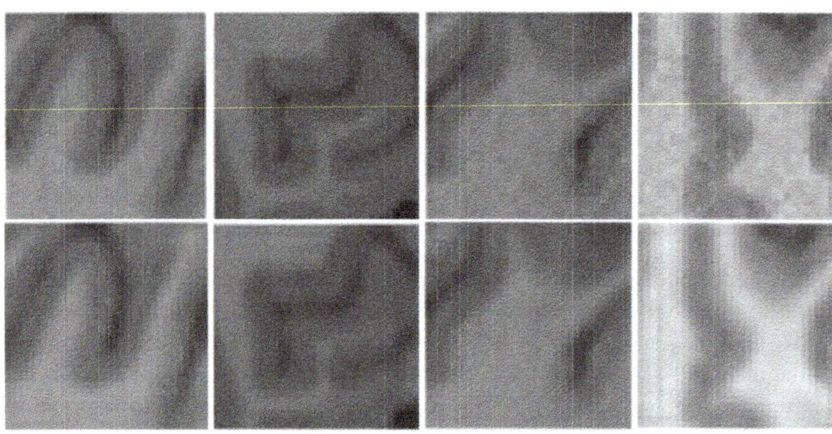

Fig. 2 (**top row**) Real high-resolution T1w patches and (**bottom row**) recovered T1w patches

2.1 Log-Euclidean Metric

Diffusion tensor matrices are well defined in the Log-Euclidean metric, where a matrix logarithm and exponential can be conveniently processed in one metric and always be mapped back to valid symmetric diffusion tensors [2]. Let $\mathbf{M} = \mathbf{U}\boldsymbol{\Sigma}\mathbf{U}^{\top}$ be the eigendecomposition of a symmetric matrix $\mathbf{M}$. The computation of the logarithm and the exponential of a tensor noted as $\log_{\mathrm{Id}}$ and $\exp_{\mathrm{Id}}$ are defined as follows:

$$\forall \mathbf{P} \in S^*_{++},\ \log_{\mathrm{Id}}(\mathbf{P}) = \mathbf{U}\log(\boldsymbol{\Sigma})\mathbf{U}^\top \in \mathcal{T}_{\mathrm{Id}} \tag{1}$$

$$\forall \mathbf{S} \in \mathcal{T}_{\mathrm{Id}},\ \exp_{\mathrm{Id}}(\mathbf{S}) = \mathbf{U}\exp(\boldsymbol{\Sigma})\mathbf{U}^\top \in S^*_{++} \tag{2}$$

We use these maps throughout our work to project the generated and real DTI on the SPD(3) manifold and on the tangent plane at the 3×3 identity matrix $\mathcal{T}_{\mathrm{Id}}$. Moreover, the Log-Euclidean distance between two tensors $\mathbf{P}_1$ and $\mathbf{P}_2$ is defined as:

$$dist(\mathbf{P}_1, \mathbf{P}_2) = \left\| \log_{\mathrm{Id}}(\mathbf{P}_1) - \log_{\mathrm{Id}}(\mathbf{P}_2) \right\|_2. \tag{3}$$

In our framework, we use this distance to measure the similarity between predicted and real DTI in the tangent plane.

2.2 *Adversarial Loss*

In a traditional GAN setup [4], a generator G and a discriminator D compete in a minimax game where G tries to generate data close to a true data distribution so that D cannot identify if the generated data is real or not. In [1], D is replaced by a discriminator that leverages the Wasserstein distance to estimate the similarity between the real and generated data distributions. The Wasserstein GAN (WGAN) architecture tends to stabilize the training as the Wasserstein distance never saturates, and thus always provides relevant gradients to G. The adversarial part of our objective follows the WGAN framework and is divided in two separate loss terms, $\mathcal{L}_{\mathrm{GAN}_X}$ and $\mathcal{L}_{\mathrm{GAN}_Y}$, respectively for structural images and DTI:

$$\begin{aligned}&\mathcal{L}_{\mathrm{GAN}_X}(G_X, D_X, Y_{\mathrm{LR}}, X_{\mathrm{HR}})\\&\quad = \mathbb{E}_{\mathbf{x}\sim\mathbb{P}_{X_{\mathrm{HR}}}}\big[D_X(\mathbf{x})\big] - \mathbb{E}_{\mathbf{y}\sim\mathbb{P}_{Y_{\mathrm{LR}}}}\big[D_X\big(G_X(\uparrow \log_{\mathrm{Id}}(\mathbf{y}))\big)\big]\end{aligned} \tag{4}$$

$$\begin{aligned}&\mathcal{L}_{\mathrm{GAN}_Y}(G_Y, D_Y, X_{\mathrm{HR}}, Y_{\mathrm{LR}})\\&\quad = \mathbb{E}_{\mathbf{y}\sim\mathbb{P}_{Y_{\mathrm{LR}}}}\big[D_Y(\log_{\mathrm{Id}}(\mathbf{y}))\big] - \mathbb{E}_{\mathbf{x}\sim\mathbb{P}_{X_{\mathrm{HR}}}}\big[D_Y\big(\downarrow G_Y(\mathbf{x})\big)\big]\end{aligned} \tag{5}$$

where $\uparrow$ and $\downarrow$ indicates trilinear up and downsampling.

In $\mathcal{L}_{\mathrm{GAN}_X}$, G_X generates 3D structural images from high-resolution DTI projected on the identity-based tangent plane using the $\log_{\mathrm{Id}}$ mapping. Since we only have samples from the low-resolution data distribution $\mathbf{y} \sim \mathbb{P}_{Y_{\mathrm{LR}}}$, the real DTI data is upsampled via trilinear interpolation before being fed to G_X. In the same loss, D_X measures the Wasserstein distance between the data distribution of generated and real structural images. Likewise, in $\mathcal{L}_{\mathrm{GAN}_Y}$, D_Y computes the Wasserstein distance between the distribution of downsampled generated DTI and real low-resolution DTI in the tangent plane, using the Log-Euclidean metric [2].

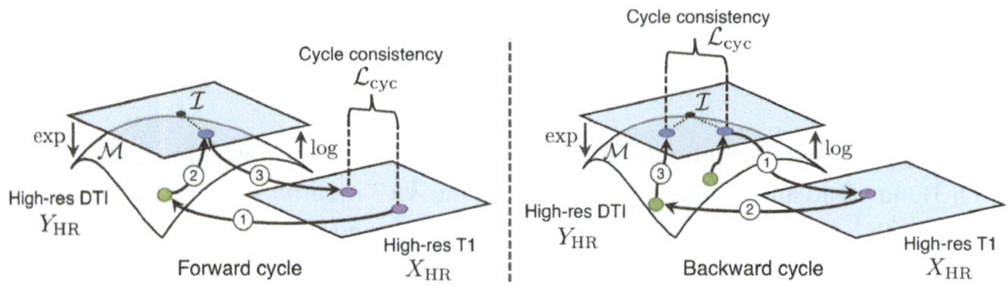

Fig. 3 On the left, our forward cycle: (1) A T1w image is translated into a high-resolution DTI where each voxel belongs to the SPD(3) manifold, (2) the tensors are projected to $\mathcal{T}_{\text{Id}}$ using the $\log_{\text{Id}}$ map and (3) the image is translated back to the T1w domain where $\mathcal{L}_{\text{cyc}}$ is computed. On the right, our backward cycle: (1) An upsampled DTI on $\mathcal{T}_{\text{Id}}$ is translated to the T1w domain, (2) the generated T1w image is translated back to DTI and (3) the recovered DTI is projected to $\mathcal{T}_{\text{Id}}$ where $\mathcal{L}_{\text{cyc}}$ is computed

2.3 Cycle Consistency Loss

The adversarial loss alone is not sufficient to drive the generation of high-resolution DTIs. Indeed, the discriminator D_Y only assesses downsampled DT images so its feedback cannot help G_Y improve beyond a certain level of detail. To mitigate this problem, we introduce a second loss that enforces the forward and backward cycle consistency of the network [23], and provides high-resolution gradient information. In our case, the forward cycle consistency ensures that, from the translated DTI $G_Y(\mathbf{x})$, we are able to reconstruct the corresponding original structural images $\mathbf{x} \sim \mathbb{P}_{X_{\text{HR}}}$ which are originally in high resolution. The backward cycle consistency ensures that we are able to reconstruct the original upsampled DTI $\mathbf{y} \sim \mathbb{P}_{\log_{\text{Id}}(Y_{\text{LR}})}$ from the translated structural images $G_X(\uparrow \log_{\text{Id}}(\mathbf{y}))$. The total cycle consistency loss is defined as

$$\begin{aligned}\mathcal{L}_{\text{cyc}}(G_Y, G_X) &= \lambda_{\text{cyc}_X} \mathbb{E}_{\mathbf{x}\sim\mathbb{P}_{X_{\text{HR}}}}\left[\|G_X(G_Y(\mathbf{x})) - \mathbf{x}\|_1 \right] \\ &+ \lambda_{\text{cyc}_Y} \mathbb{E}_{\mathbf{y}\sim\mathbb{P}_{Y_{\text{LR}}}}\left[\left\|G_Y(G_X(\uparrow \log_{\text{Id}}(\mathbf{y}))\right) - \uparrow \log_{\text{Id}}(\mathbf{y})\right\|_1 \right]\end{aligned} \quad (6)$$

Here, λ_{cyc_X} and λ_{cyc_Y} balance the contribution of the forward and backward cycles respectively and have been empirically set to a value of 3 and 1. Note that ℓ_1 norm is employed instead of ℓ_2 to make the loss less sensitive to large reconstruction errors (Fig. 3).

2.4 Manifold-Aware Wasserstein CycleGAN

Our full objective is

$$\begin{aligned}\mathcal{L}(G_X, G_Y, D_X, D_Y) &= \lambda_{\text{GAN}_X} \mathcal{L}_{\text{GAN}_X}(G_X, D_X, Y_{\text{LR}}, X_{\text{HR}}) \\ &+ \lambda_{\text{GAN}_Y} \mathcal{L}_{\text{GAN}_Y}(G_Y, D_Y, X_{\text{HR}}, Y_{\text{LR}}) + \mathcal{L}_{\text{cyc}}(G_Y, G_X)\end{aligned} \quad (7)$$

The adversarial parts $\mathcal{L}_{\text{GAN}}(G_X, D_X, Y_{\text{LR}}, X_{\text{HR}})$ and $\mathcal{L}_{\text{GAN}}(G_Y, D_Y, X_{\text{HR}}, Y_{\text{LR}})$ of our full objective guide the generators G_X and G_Y towards the synthesis of images close to their real data distributions using a Wasserstein distance on the SPD(3) manifold. The cycle consistency denoted as $\mathcal{L}_{\text{cyc}}(G_Y, G_X)$ gives fine-grained retro-action that helps generate HR DTI while preventing mode collapse. Once training is done, the high-resolution DTI $\mathbf{y}_{\text{HR}}$ of a structural image $\mathbf{x}$ can be obtained by applying the exponential map to the DTI generator output: $\mathbf{y}_{\text{HR}} = \exp_{\text{Id}}(G_Y(\mathbf{x}))$.

3 Experiments

Data We employ the pre-processed T1-weighted (T1w) and diffusion MRI (dMRI) data of 1,065 patients from the HCP1200 release of the Human Connectome Project [20] to evaluate our manifold-aware CycleGAN. The T1w (0.7 mm isotropic, FOV = 224mm, matrix = 320, 256 sagittal slices in a single slab) and diffusion (sequence = Spin-echo EPI, repetition time (TR) = 5520 ms, echo time (TE) = 89.5 ms, resolution = $1.25 \times 1.25 \times 1.25$ mm^3 voxels) data acquisition was done using a Siemens Skyra 3T scanner [18] and pre-processed following [3]. The diffusion tensors were fitted using DSI Studio toolbox [13]. Both T1w and DTI were decomposed in overlapping patches of 32^3 voxels centered on the foreground.

Experiments Setup We used 50,000 unpaired T1w and DTI patches randomly selected among 1,055 subjects as our training set. For the validation set and the test set, we took paired T1w and DTI patches covering the full brain of respectively 3 and 2 randomly chosen subjects. We compared our manifold-aware CycleGAN (MA-CycleGAN) method with two baselines: a manifold-aware Wasserstein GAN without cycle (MA-GAN) and a Wasserstein CycleGAN without the $\log_{\text{Id}}$ and $\exp_{\text{Id}}$ mappings. These baseline methods allow us to assess the impact of both the cycle consistency and the manifold mapping. We measure the quality of the generated HR DTI by computing three metrics: (1) the mean cosine similarity between the principal eigenvectors of the generated images and their ground-truth, (2) the mean squared error between the FA of the generated images and their ground-truth, and (3) the mean Log-Euclidean distance between the generated tensors and their ground-truth following Equation (3). Because the principal eigenvector's direction is more relevant at voxels with higher anisotropy, cosine similarity is measured at three different FA thresholds taken on the ground-truth images: FA ≥ 0 (all voxels), only voxels with FA ≥ 0.2, and only voxels with FA ≥ 0.5. An FA threshold near 0.2 is commonly used for tract-based analysis of white matter [19].

While the cosine similarity allows us to evaluate the precision of the predicted orientation of the generated tensors, the mean squared error on the derived FA highlights the network's ability to estimate local diffusion anisotropy. As for the Log-Euclidean distance, it takes into account both the orientation and the anisotropy of the tensors. Furthermore, a qualitative inspection of the generated tensors and FA is performed in Fig. 4.

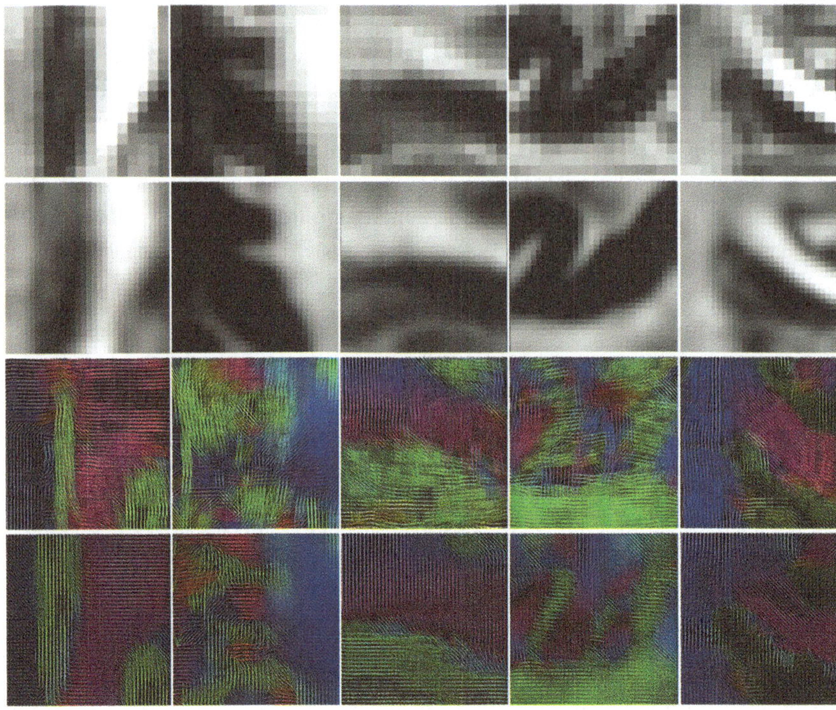

Fig. 4 (**top row**) Real low-resolution FA, (**second row**) generated high-resolution FA, (**third row**) real low-resolution color orientation tensors, and (**bottom row**) generated high-resolution color orientation tensors. Best viewed in color

Implementation Details Both generators are based on the Unet implementation from [16] where the convolutions have been changed to 3D convolutions. In addition, we changed the last activation layers to fit the scale of the generated data and the number of channels with respect to our inputs and outputs. For G_X, we use a sigmoid as the final activation function to generate values in the range [0,1]. In the case of G_Y, we use a hard hyperbolic tangent activation function. Furthermore, to guarantee that the synthesized tensors can be decomposed following Eqs. (1) and (2), we convert the 9-channels output of G_Y into a 3×3 matrix $\mathbf{Y}$ and make this matrix symmetric as follows: $\mathbf{Y}' = \frac{1}{2}(\mathbf{Y} + \mathbf{Y}^\top)$. For our discriminators, we employ a Resnet-18 architecture [6] where all convolutions have been changed to 3D convolutions. Both the generators and discriminators were trained for 30 epochs with the Adam optimizer [14] and a batch size of 8. A starting learning rate of 1×10^{-4} was used jointly with a reduce-on-plateau strategy. To stabilize the training of our network, we pre-trained the generators independently with 25,000 paired patches randomly sampled from 5 subjects during 10 epochs. The paired patches have been computed from aligned high-resolution structural T1w images and upsampled DTIs. The alignment was performed using FLIRT [10, 12] from FSL [11].

Table 1 Fractional anisotropy mean squared error (FA MSE), Log-Euclidean distance, and cosine similarity between principal eigenvectors of compared methods: manifold-aware Wasserstein GAN (MA-GAN), Wasserstein CycleGAN without the $\log_{\mathrm{Id}}$ and $\exp_{\mathrm{Id}}$ mappings (CycleGAN), and our manifold-aware Wasserstein CycleGAN (MA-CycleGAN). A smaller FA MSE and Log-Euclidean distance corresponds to a superior performance while a higher cosine similarity is better. Since the principal eigenvector's direction is more relevant at voxels with higher diffusivity, we report cosine similarity at increasing FA thresholds of 0, 0.2 and 0.5

			Cosine similarity		
Models	FA MSE	Log distance	FA ≥ 0	FA ≥ 0.2	FA ≥ 0.5
MA-GAN	0.0277	0.6067	0.4512	0.6369	0.6700
CycleGAN	0.0431	0.5699	0.5371	0.6717	0.7064
MA-CycleGAN (ours)	**0.0172**	**0.5515**	**0.5846**	**0.7217**	**0.8041**

Fibers Orientation Analysis To evaluate the predicted tensor orientation, we compute the cosine similarity between the principal orientation of each generated tensor and its ground-truth, for FA threshold 0.0, 0.2 and 0.5. Results in Table 1 show that our method performs better than the two baselines, yielding average improvements in cosine similarity (for FA $\geq$ 0.0, 0.2, 0.5) of 0.133, 0.085, 0.134 compared to the manifold-aware GAN without cycle and 0.048, 0.050, 0.100 compared to the CycleGAN without manifold mapping. As mentioned before, a good estimation of main diffusion orientation is generally more important at voxels with greater diffusivity. Hence, the cosine similarity for FA $\geq$ 0.2 and FA $\geq$ 0.5 is a better indicator of performance than for FA $\geq$ 0.0. We see in Table 1 that our model's estimation of fiber orientation improves with a higher FA threshold, reaching a similarity of 0.804 for voxels with FA $\geq$ 0.5. This can be observed in Fig. 6, which gives the FA MSE, Log-Euclidean distance and cosine similarity at each voxel of sagittal, coronal and axial slices from the same subject. As can be seen, orientations in regions with typically high FA, like the corpus callosum, are better predicted by our model than those in regions with lower FA.

FA Analysis Next, we evaluate the fractional anisotropy (FA) of generated HR DTI. FA is one of the most commonly used DT-derived metrics, thus an accurate prediction of this metric is critical. Table 1 shows the mean squared error (FA MSE) obtained by our method and the two baselines. Once more, our manifold-aware CycleGAN outperforms the manifold-aware GAN and the standard CycleGAN with an MSE of 0.172 compared to 0.0277 and 0.0431. Furthermore, we observe that the two methods using manifold mapping perform better than the standard CycleGAN. This performance gap is due to the fact that, without projecting the generated data on the SPD(3) manifold using the Log-Euclidean metric, there is no guarantee that the generated tensors lie on such manifold. Consequently, the tensors eigenvalues used in the computation of FA are not strictly positive, which increases the differences between the generated FA and the ground-truth. In addition, as seen in Figs. 4 and 5,

7. Huang, Z., Van Gool, L.: A riemannian network for SPD matrix learning. 31st AAAI Conference on Artificial Intelligence, AAAI 2017 (2017)
8. Huang, Z., Wu, J., Van Gool, L.: Manifold-valued image generation with wasserstein generative adversarial nets. In: Proceedings of the AAAI Conference on Artificial Intelligence, pp. 3886–3893 (2019)
9. Ionescu, C., Vantzos, O., Sminchisescu, C.: Matrix backpropagation for deep networks with structured layers. Technical Report (2015)
10. Jenkinson, M., Bannister, P., Brady, M., Smith, S.: Improved optimization for the robust and accurate linear registration and motion correction of brain images. NeuroImage (2002)
11. Jenkinson, M., Beckmann, C.F., Behrens, T.E.J., Woolrich, M.W., Smith, S.M.: Review FSL. NeuroImage (2012)
12. Jenkinson, M., Smith, S.: A global optimisation method for robust affine registration of brain images. Med. Image Anal. (2001)
13. Jiang, H., Van Zijl, P.C.M., Kim, J., Pearlson, G.D., Mori, S.: DtiStudio: Resource program for diffusion tensor computation and fiber bundle tracking (2005)
14. Kingma, D.P., Ba, J.L.: Adam: A method for stochastic optimization. In: 3rd International Conference on Learning Representations, ICLR 2015—Conference Track Proceedings. International Conference on Learning Representations, ICLR (2015)
15. Pfefferbaum, A., Sullivan, E.V.: Increased brain white matter diffusivity in normal adult aging: relationship to anisotropy and partial voluming. Mag. Reson. Med. Official J. Int. Soc. Mag. Reson. Med. (2003)
16. Ronneberger, O., Fischer, P., Brox, T.: U-net: Convolutional networks for biomedical image segmentation. In: Lecture Notes in Computer Science (including subseries Lecture Notes in Artificial Intelligence and Lecture Notes in Bioinformatics) (2015)
17. Schyboll, F., Jaekel, U., Weber, B., Neeb, H.: The impact of fibre orientation on t1-relaxation and apparent tissue water content in white matter. Mag. Reson. Mater. Phys. Biol. Med. **31**(4), 501–510 (2018)
18. Sotiropoulos, S.N., Jbabdi, S., Xu, J., Andersson, J.L., Moeller, S., Auerbach, E.J., Glasser, M.F., Hernandez, M., Sapiro, G., Jenkinson, M., Feinberg, D.A., Yacoub, E., Lenglet, C., Van Essen, D.C., Ugurbil, K., Behrens, T.E.: Advances in diffusion MRI acquisition and processing in the human connectome project. NeuroImage (2013)
19. Taoka, T., Morikawa, M., Akashi, T., Miyasaka, T., Nakagawa, H., Kiuchi, K., Kishimoto, T., Kichikawa, K.: Fractional anisotropy-threshold dependence in tract-based diffusion tensor analysis: evaluation of the uncinate fasciculus in alzheimer disease. Am. J. Neuroradiol. **30**(9), 1700–1703 (2009)
20. Van Essen, D.C., Smith, S.M., Barch, D.M., Behrens, T.E., Yacoub, E., Ugurbil, K.: The WU-minn human connectome project: an overview. NeuroImage (2013)
21. Yi, X., Walia, E., Babyn, P.: Generative adversarial network in medical imaging: a review. Med. Image Anal. **58** (2019)
22. Zhong, J., Wang, Y., Li, J., Xue, X., Liu, S., Wang, M., Gao, X., Wang, Q., Yang, J., Li, X.: Inter-site harmonization based on dual generative adversarial networks for diffusion tensor imaging: application to neonatal white matter development. BioMed. Eng. Online **19** (2020)
23. Zhu, J.Y., Park, T., Isola, P., Efros, A.A.: Unpaired image-to-image translation using cycle-consistent adversarial networks. In: Proceedings of the IEEE International Conference on Computer Vision (2017)

Diffusion MRI Applications

Beyond Lesion-Load: Tractometry-Based Metrics for Characterizing White Matter Lesions within Fibre Pathways

Maxime Chamberland, Mia Winter, Thomas A. W. Brice, Derek K. Jones, and Emma C. Tallantyre

Abstract In multiple sclerosis studies, lesion volume (or lesion load) derived from conventional T2 imaging correlates modestly with clinical assessment. Determining which specific white matter pathways are impacted by lesions may provide additional insights regarding task-specific clinical impairment. Using diffusion MRI, we introduce a set of tract-based metrics that go beyond traditional lesion load approaches and show how they relate to task performance (i.e., working memory, information processing and verbal fluency) in a cohort of 40 patients with multiple sclerosis.

1 Introduction

Lesion load (LL) is a volumetric index derived from structural MRI often used in clinical practice to characterise the degree of damage in the brains of patients with multiple sclerosis (MS). Focal lesions on T2-weighted brain imaging in MS reflect the permanent footprint of previous episodes of inflammation [1]. Although widely

M. Chamberland (✉) · M. Winter · D. K. Jones
Cardiff University Brain Research Imaging Centre (CUBRIC), School of Psychology, Cardiff University, Cardiff, UK
e-mail: chamberlandm@cardiff.ac.uk

M. Winter
e-mail: WinterM3@cardiff.ac.uk

D. K. Jones
e-mail: jonesd27@cardiff.ac.uk

M. Winter
Department of Clinical Neuropsychology, University Hospital of Wales, Cardiff, UK

T. A. W. Brice · E. C. Tallantyre
Division of Psychological Medicine and Clinical Neurosciences, Cardiff University School of Medicine, Cardiff, UK
e-mail: tallantyreec@cardiff.ac.uk

E. C. Tallantyre
Helen Durham Centre for Neuroinflammation, University Hospital of Wales, Cardiff, UK

N. Gyori et al. (eds.), *Computational Diffusion MRI*, Mathematics and Visualization,
https://doi.org/10.1007/978-3-030-73018-5_18

used in the diagnosis, prognosis and monitoring of people with MS [2], LL correlates only modestly with clinical measures of disability; resulting in the so-called clinical-radiological paradox [3, 4]. There are likely to be several explanations, including the occurrence of focal lesions in critical vs non-critical white matter (WM) pathways for a given cognitive function, and the presence of diffuse WM micro-structural damage that is not easily seen on conventional MRI. While LL is a convenient outcome measure in clinical studies of MS, there is a need for more advanced, and anatomically- and microstructurally-specific imaging metrics to allow clinicians and neuroscientists to disentangle the relationship between focal and diffuse pathology with clinical disability in MS [5].

A variety of MRI–based approaches that probe various physical properties of brain tissue have been shown to be sensitive to demyelination and axonal loss in MS [6, 7] (for review, see [8]). Diffusion MRI (dMRI) allows information about the structural architecture and tissue micro-structure to be obtained by probing the random motion of water molecules [9]. The ability to derive quantitative features such as fractional anisotropy (FA) or mean diffusivity (MD) from diffusion tensor imaging (DTI) [10] and to virtually reconstruct pathways with tractography [11] has led to an exponential growth of dMRI clinico-research studies. In MS, multiple groups have examined the relationship of DTI measures within the normal-appearing WM and cognitive function [12–15]. Most studies report significant associations, although these were not always stronger than the relationship between cognitive performance and LL [3, 4], potentially due to inconsistencies and limitations in tractography and associated microstructural metrics at the time.

Indeed, applying tractography to MS data can be problematic due to premature termination of streamlines within lesions [16, 17]. However, recent advances in local modeling and tractography such as multi-tissue multi-shell constrained spherical deconvolution (MSMT-CSD, [18]) and anatomically-based tractography [19, 20] allow the propagation of streamlines through lesions more reliably [21–27]. Furthermore, although tractography still faces significant challenges in the field in general [28–30], recent machine learning based approaches have shown promise in reproducible tract segmentation across subjects [31]. Based on these recent methodological breakthroughs, we propose a set of tract-based metrics to improve the link between WM lesions and clinical correlates. We demonstrate the utility of the proposed metrics in a cohort of 40 MS patients.

2 Theory and Methods

2.1 Clinical Assessment

40 subjects (27 women, mean age: 58 years, range: 44–78) with longstanding relapse-onset MS were recruited to this study (mean disease duration: 27 years, range: 15–47; median Expanded Disability Status Scale (EDSS) at clinical assessment: 2.5, range

0–6.0). The participants were cognitively assessed (see [32]) and the MS functional composite (MSFC, [33]) score was derived as a composite of walking speed, dexterity and information processing. Additionally, performance was assessed on three tasks involving (1) working memory (Letter-Number Sequencing, LNS [34]); (2) information processing speed (Speed of Information Processing adjusted for motor speed, SoIP [35]); and (3) verbal fluency (category switching, CF-Switching [36]). Performance on these type of tasks has been associated with compromise of widespread brain networks including fronto-temporo-parietal regions. This study was approved by the local ethics committee and all participants gave written informed consent.

2.2 Acquisition

Patients were scanned using dMRI within 12 months of their clinical assessment using a Siemens PRISMA 3T system using a 32-channel receive-only RF head coil. All participants underwent the following sequences: (1) 3D T2-weighted spin echo sequence (TR/TE: 3200/1408 ms; voxel size: $1 \times 1 \times 1$ mm^3), (2) 3D T2-FLAIR sequence (TR/TE: 5000/388 ms, TI: 1800 ms, voxel size = $0.5 \times 0.5 \times 0.5$ mm^3), (3) 3D T1 MPRAGE (TR/TE: 2300/3 ms; flip angle: 9°; voxel size: $1.0 \times 1.0 \times 1.0$ mm^3), (4) Diffusion-weighted spin-echo EPI with 14 b0 images, 30 directions at b = 1200 s/mm^2, 60 directions at b = 2400 s/mm^2 and $2 \times 2 \times 2$ mm^3 voxels.

2.3 Processing

Diffusion data were denoised [37] and corrected for subject motion and distortion [38, 39]. Next, apparent fibre density (AFD) maps were derived from fiber orientation distribution functions (fODFs) obtained from MSMT-CSD [18] using a single group response function. WM lesion masks were semi-automatically delineated using 3D T2 and FLAIR images by a trained technician (co-author TB, blinded to the purpose of the study) using NeuROI.[1]

For each dataset, automated WM tract segmentation was performed using TractSeg [31] to obtain the following task-relevant bundles of interest (identified by co-author MW): genu and splenium of the corpus callosum, cingulum (CG), inferior longitudinal fasciculus (ILF) and uncinate fasciculus (UF). For each bundle, 2000 streamlines were generated. The AFD was then averaged within each bundle of interest. A whole brain set of streamlines was also derived by concatenating all TractSeg outputs in each subject.

[1] www.nottingham.ac.uk/research/groups/clinicalneurology/neuroi.aspx.

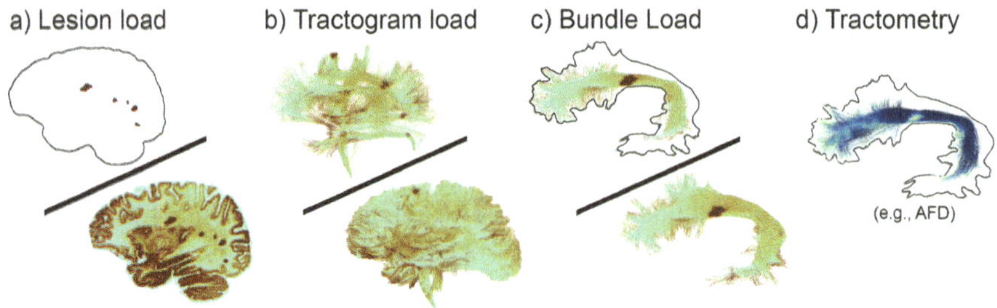

Fig. 1 Graphical overview of various lesion load metrics for an example subject. **a** Conventional voxel-based lesion load (normalized by head size). **b** Whole-brain tractogram load. **c** Topology-based bundle load (example bundle: arcuate fasciculus). **d** Lesion-based Tractometry (example bundle: arcuate fasciculus)

2.4 *Proposed Metrics*

LL (Fig. 1a) is typically defined as the total volume (mm^3) of the lesions (V_{les}), or its normalized variation:

$$LL = \frac{V_{les}}{V_{brain}}, \tag{1}$$

where V_{brain} is the intracranial brain volume (mm^3). A meta-analysis recently reported that only 5% of studies who calculate LL account for intracranial volume [4]. A simple extension of LL is the tractogram load (TL) metric (Fig. 1b), defined as the following ratio:

$$TL = \frac{T_{les}}{T}, \tag{2}$$

where T_{les} is the volume of all streamlines (voxelized) passing through all segmented lesions (i.e., the subset of streamlines) and T is the total volume of the whole-brain tractogram. Note that this metric implicitly integrates distal information about the entire streamlines as opposed to local lesion information only (V_{les}). Similarly, the bundle load (BL) metric (Fig. 1c) is a refined sub-case of TL and can be defined as the following ratio:

$$BL = \frac{B_{les}}{B}, \tag{3}$$

where, for a given bundle, B_{les} is the total volume of the subset of streamlines that traverse the lesion (i.e., not to be confounded with V_{les} which is the volume of the lesion alone) and B is the total volume of the current bundle-of-interest.

Finally, we also present a lesion-informed Tractometry approach termed here *lesionometry* (Fig. 1d) defined as:

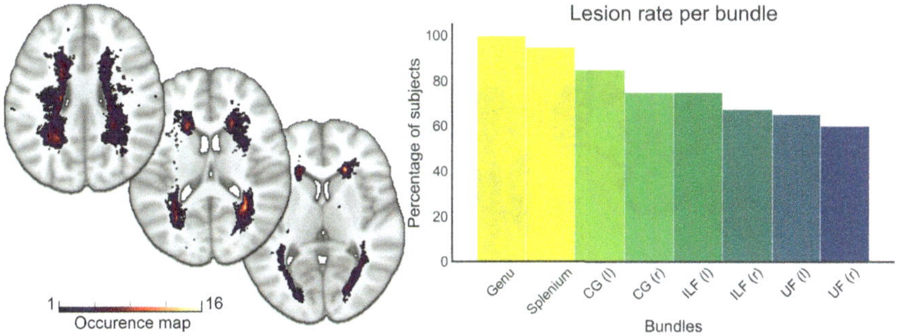

Fig. 2 Density map of all white matter lesions across 40 subjects. The group-average map shows voxels where a lesion was present in at least one of the patients (left). The lesion rate was also derived for the extracted bundles of interests (right)

$$\text{Lesionometry} = \frac{1}{nm}\sum_{i=1}^{m}\sum_{j=1}^{n}\mathcal{M}(s_{ij}), \tag{4}$$

where dMRI measures (e.g., $\mathcal{M}$ = [FA, MD, AFD, ...]) are sampled at each vertex j forming the streamline s_i that is traversing a lesion (B_{les}). If no lesion were present within a bundle, then conventional tract-average was used.

We hypothesize that having a more focused approach around lesions may result in stronger relationships between the metrics and their associated clinical scores. The anticipated directions were as follows: an increase in lesion load (LL, TL, BL) is associated with poorer task performance, and increased AFD is associated with better performance. Pearson correlations were used to assess correlation between the proposed metrics and clinical scores, after correcting for age and gender. Data visualization was done using FiberNavigator [40].

3 Results

3.1 Lesion Mapping

Figure 2 (left) shows that most of the lesions were located in deep WM and periventricular areas. Amongst the WM bundles that were extracted, 90% of the subjects had at least one lesion in the corpus callosum region (Fig. 2, right). Figure 3 qualitatively illustrates the complete reconstruction of the ILF in a single subject where a lesion occurred in the inferior occipital lobe (Fig. 3, green arrow). Streamlines traversing the lesion are color-coded using the anatomical FLAIR image.

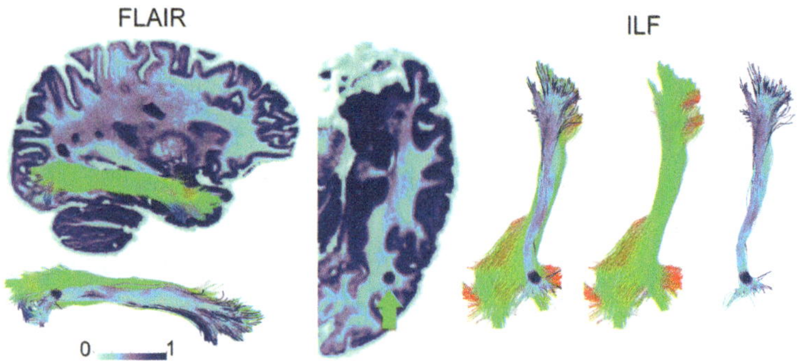

Fig. 3 Tractography of the inferior longitudinal fasciculus (green) in an individual with an MS lesion (green arrow). Streamlines successfully traversing the lesion (right, axial view) are indicated in blue. Color overlay: Intensity normalized FLAIR image

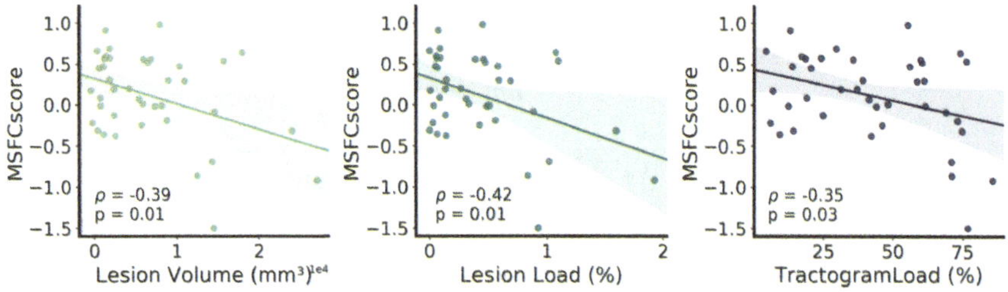

Fig. 4 Global correlations between MSFC and the 3 whole-brain measures (Lesion Volume, Lesion Load and Tractogram Load). As the load increases, the MSFC performance decreases. The correlation between Tracogram Load and MSFC appears to be less driven by outliers

3.2 *Volumetric Metrics*

At the whole-brain level, lesion volume (LV), lesion load (LL) and tractogram load (TL) showed similar negative correlations with MSFC (Fig. 4). Although TL exhibits a slightly lower correlation, the error margin appears less spread than in LV and LL. This could potentially be explained by the presence of outliers in the latter cases. TL also shows that on average, 42% of the underlying WM architecture can be indirectly affect by lesions. On the other hand, lesion load only affects 0.4% of the brain.

At the local level, bundle-specific loads (BL) showed stronger associations with tasks than the aforementioned global measures (Fig. 5). For example, the splenium showed stronger association with SoIP (Pearson's $r = -0.50$, $p = 0.001$) than typical LV (Pearson's $r = -0.35$, $p = 0.03$). In addition, the bilateral ILF showed greater association with CF-switching (Pearson's $r = -0.35$, $p = 0.01$) than conventional LL (Pearson's $r = -0.30$, $p = 0.05$).

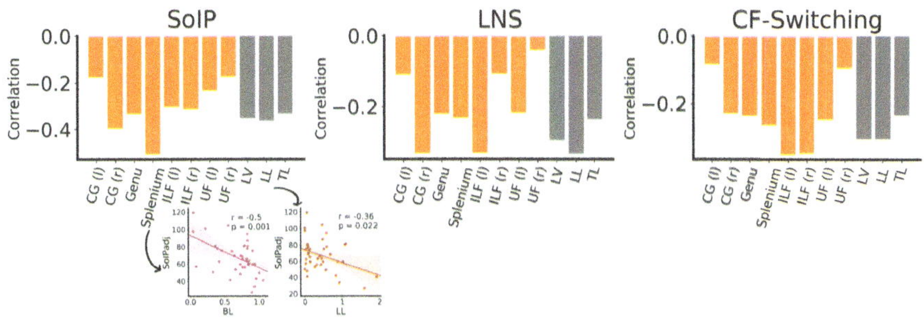

Fig. 5 Bundle load correlations across 3 cognitive tasks show dissociating patterns across bundles. For example, the ILF correlated with CF-switching but not as much with SoIP, whereas the splenium shows correlation with SoIP but not so strongly with LNS

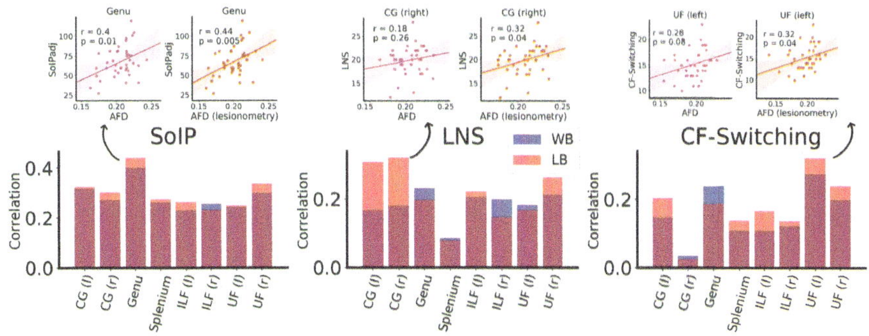

Fig. 6 Lesionometry comparison between whole-bundle (WB, blue) and lesioned-bundle (LB, orange) AFD averages. In general, the LB approach shows greater correlations with task than the conventional WB average. The overlapping bar alpha value was set to 0.5 (i.e., red color)

3.3 *Tractometry-Based Metrics*

Figure 6 shows results for the lesionometry approach across three tasks. In most cases, the lesion-based tract-averaging (orange) showed greater correlations than typical whole-tract averages (blue). In particular, the link between the left and right CG and LNS (Fig. 6, middle) almost doubled (e.g., from $r = 0.18$ to $r = 0.32$ and $r = 0.17$ to $r = 0.31$, respectively). As anticipated, AFD was positively associated with task performance.

4 Discussion and Conclusion

Focal WM lesions are the hallmark of MS, but inadequately explain disability, creating a clinico-radiological paradox [3, 4]. Neuropathology shows that axonal pathology is widespread in the brain in MS. The diffuse axonal damage in MS may be an independent process, or may be driven by anterograde and retrograde degeneration originating at the site of focal lesions [41]. Imaging techniques capable of unravelling the relationship between focal and diffuse pathology, and the correlates of MS disability, remain an unmet need. Diffusion MRI provides quantitative information about WM microstructure. Tractography in MS could provide valuable information about how lesion location relates to key WM bundles, and also inform on the microstructure within lesions and at distant but related sites. However, tractography has been relatively underutilised in MS, perhaps because of technical challenges [16, 17]. Here we demonstrate how dMRI and tractography can generate meaningful metrics relating to the burden of lesions, but also their interaction with the structural WM network.

Using lesion mapping, we were able to recreate the known predisposition of WM lesion location [42]. However, we also developed a novel metric by demonstrating the relationship between lesions and WM tracts, which reflected the proportion of the WM tracts that interacted with lesions (TL). Although TL did not improve the strength of correlations beyond LL with a composite measure of disability in people with MS, a reduction in the error margins suggests that this metric may be worthy of further investigation. Nonetheless, it provides a convenient way of visualising the structural network affected by lesions (see Fig. 1b) and also illustrates the high proportion of the WM fibres that interact with a lesion beyond conventional 2D slice-based visualizations.

At the local level, we were able to demonstrate that BL could provide valuable information beyond global metrics to explain task-specific performance. Such a targeted approach may therefore be preferred when trying to relate task performance to specific WM bundles. From a tractometry point-of-view, the lesionometry approach showed stronger association with task performance than typical whole-bundle averages of dMRI measures. Given that focal damage is suspected to extend beyond the visible site of the lesions (i.e., along WM tracts traversing the lesion), sampling dMRI measures selectively within the damaged portion of the bundle may allow more sensitivity to underlying changes in tissue microstructure.

It is already known that the location of the lesions is relevant to clinical disability [43–46]. Furthermore, profiling dMRI measures along WM bundles using the lesionometry approach introduced in this paper, will in theory result in tract profiles that are more sensitive to the underlying lesions. Finally, the metrics introduced in this paper were assessed using an exploratory approach to provide interested readers with an immediate application; undoubtedly, these features could be leveraged in a more advanced context using machine learning. In summary, we introduced a set of easy-to-use tract-based metrics to complement existing LL approaches to quantify the extent of brain damage associated with WM lesions in clinical applications.

References

1. Thompson, A.J., Banwell, B.L., Barkhof, F., Carroll, W.M., Coetzee, T., Comi, G., Correale, J., Fazekas, F., Filippi, M., Freedman, M.S., et al.: Diagnosis of multiple sclerosis: 2017 revisions of the McDonald criteria. Lancet Neurol. **17**(2), 162–173 (2018)
2. Vellinga, M., Geurts, J., Rostrup, E., Uitdehaag, B., Polman, C., Barkhof, F., Vrenken, H.: Clinical correlations of brain lesion distribution in multiple sclerosis. J. Mag. Reson. Imaging Official J. Int. Soc. Mag. Reson. Med. **29**(4), 768–773 (2009)
3. Barkhof, F.: The clinico-radiological paradox in multiple sclerosis revisited. Curr. Opin. Neurol. **15**(3), 239–245 (2002)
4. Mollison, D., Sellar, R., Bastin, M., Mollison, D., Chandran, S., Wardlaw, J., Connick, P.: The clinico-radiological paradox of cognitive function and MRI burden of white matter lesions in people with multiple sclerosis: A systematic review and meta-analysis. PloS One **12**(5), (2017)
5. Filippi, M.: Linking structural, metabolic and functional changes in multiple sclerosis. Eur. J. Neurol. **8**(4), 291–297 (2001)
6. Mallik, S., Samson, R.S., Wheeler-Kingshott, C.A., Miller, D.H.: Imaging outcomes for trials of remyelination in multiple sclerosis. J. Neurol. Neurosurg. Psychiatry **85**(12), 1396–1404 (2014)
7. Filippi, M., Preziosa, P., Rocca, M.A.: Microstructural MR imaging techniques in multiple sclerosis. Neuroimaging Clinics **27**(2), 313–333 (2017)
8. Rocca, M.A., Amato, M.P., De Stefano, N., Enzinger, C., Geurts, J.J., Penner, I.K., Rovira, A., Sumowski, J.F., Valsasina, P., Filippi, M., et al.: Clinical and imaging assessment of cognitive dysfunction in multiple sclerosis. Lancet Neurol. **14**(3), 302–317 (2015)
9. Stejskal, E.O., Tanner, J.E.: Spin diffusion measurements: spin echoes in the presence of a time-dependent field gradient. J. Chem. Phys. **42**(1), 288–292 (1965)
10. Basser, P., Mattiello, J., Lebihan, D.: Estimation of the Effective Self-Diffusion Tensor from the NMR Spin Echo (1994)
11. Conturo, T.E., Lori, N.F., Cull, T.S., Akbudak, E., Snyder, A.Z., Shimony, J.S., McKinstry, R.C., Burton, H., Raichle, M.E.: Tracking neuronal fiber pathways in the living human brain. Proc. Natl. Acad. Sci. **96**(18), 10422–10427 (1999)
12. Rovaris, M., Iannucci, G., Falautano, M., Possa, F., Martinelli, V., Comi, G., Filippi, M.: Cognitive dysfunction in patients with mildly disabling relapsing-remitting multiple sclerosis: an exploratory study with diffusion tensor MR imaging. J. Neurol. Sci. **195**(2), 103–109 (2002)
13. Rovaris, M., Riccitelli, G., Judica, E., Possa, F., Caputo, D., Ghezzi, A., Bertolotto, A., Capra, R., Falautano, M., Mattioli, F., et al.: Cognitive impairment and structural brain damage in benign multiple sclerosis. Neurology **71**(19), 1521–1526 (2008)
14. Akbar, N., Lobaugh, N.J., O'Connor, P., Moradzadeh, L., Scott, C.J., Feinstein, A.: Diffusion tensor imaging abnormalities in cognitively impaired multiple sclerosis patients. Canadian J. Neurol. Sci. **37**(5), 608–614 (2010)
15. Mesaros, S., Rocca, M., Kacar, K., Kostic, J., Copetti, M., Stosic-Opincal, T., Preziosa, P., Sala, S., Riccitelli, G., Horsfield, M., et al.: Diffusion tensor mri tractography and cognitive impairment in multiple sclerosis. Neurology **78**(13), 969–975 (2012)
16. Ciccarelli, O., Catani, M., Johansen-Berg, H., Clark, C., Thompson, A.: Diffusion-based tractography in neurological disorders: concepts, applications, and future developments. Lancet Neurol. **7**(8), 715–727 (2008)
17. Lipp, I., Parker, G.D., Tallantyre, E.C., Goodall, A., Grama, S., Patitucci, E., Heveron, P., Tomassini, V., Jones, D.K.: Tractography in the presence of multiple sclerosis lesions. NeuroImage **209**, (2020)
18. Jeurissen, B., Tournier, J.D., Dhollander, T., Connelly, A., Sijbers, J.: Multi-tissue constrained spherical deconvolution for improved analysis of multi-shell diffusion MRI data. NeuroImage **103**, 411–426 (2014)
19. Smith, R.E., Tournier, J.D., Calamante, F., Connelly, A.: Anatomically-constrained tractography: improved diffusion mri streamlines tractography through effective use of anatomical information. NeuroImage **62**(3), 1924–1938 (2012)

20. Girard, G., Whittingstall, K., Deriche, R., Descoteaux, M.: Towards quantitative connectivity analysis: reducing tractography biases. NeuroImage **98**, 266–278 (2014)
21. Gajamange, S., Raffelt, D., Dhollander, T., Lui, E., van der Walt, A., Kilpatrick, T., Fielding, J., Connelly, A., Kolbe, S.: Fibre-specific white matter changes in multiple sclerosis patients with optic neuritis. NeuroImage: Clinical **17** (2018) 60–68
22. Mito, R., Raffelt, D., Dhollander, T., Vaughan, D.N., Tournier, J.D., Salvado, O., Brodtmann, A., Rowe, C.C., Villemagne, V.L., Connelly, A.: Fibre-specific white matter reductions in alzheimer's disease and mild cognitive impairment. Brain **141**(3), 888–902 (2018)
23. Tur, C., Grussu, F., Prados, F., Charalambous, T., Collorone, S., Kanber, B., Cawley, N., Altmann, D.R., Ourselin, S., Barkhof, F., et al.: A multi-shell multi-tissue diffusion study of brain connectivity in early multiple sclerosis. Multi. Sclerosis J. 1352458519845105 (2019)
24. Charalambous, T., Tur, C., Prados, F., Kanber, B., Chard, D.T., Ourselin, S., Clayden, J.D., Wheeler-Kingshott, C.A.G., Thompson, A.J., Toosy, A.T.: Structural network disruption markers explain disability in multiple sclerosis. J. Neurol. Neurosurg. Psychiatry **90**(2), 219–226 (2019)
25. Dumont, M., Roy, M., Jodoin, P.M., Morency, F.C., Houde, J.C., Xie, Z., Bauer, C., Samad, T.A., Van Dijk, K.R., Goodman, J., et al.: Free water in white matter differentiates mci and ad from control subjects. Frontiers Aging Neurosci. **11**, 270 (2019)
26. Beaudoin, A.M., Rheault, F., Theaud, G., Whittingstall, K., Lamontagne, A., Descoteaux, M.: White matter tractometry correlates with fatigue severity in young adults with multiple sclerosis (1322). Neurology **94**(15 Supplement) (2020)
27. Storelli, L., Pagani, E., Preziosa, P., Filippi, M., Rocca, M.A.: Measurement of white matter fiber-bundle cross-section in multiple sclerosis using diffusion-weighted imaging. Multi. Sclerosis J. 1352458520938999 (2020)
28. Maier-Hein, K.H., Neher, P.F., Houde, J.C., Côté, M.A., Garyfallidis, E., Zhong, J., Chamberland, M., Yeh, F.C., Lin, Y.C., Ji, Q.: Others: The challenge of mapping the human connectome based on diffusion tractography. Nat. Commun. **8**, 1349 (2017)
29. Jones, D.K., Knösche, T.R., Turner, R.: White matter integrity, fiber count, and other fallacies: The do's and don'ts of diffusion MRI. NeuroImage **73**, 239–254 (2013)
30. Schilling, K.G., Nath, V., Hansen, C., Parvathaneni, P., Blaber, J., Gao, Y., Neher, P., Aydogan, D.B., Shi, Y., Ocampo-Pineda, M., et al.: Limits to anatomical accuracy of diffusion tractography using modern approaches. NeuroImage **185**, 1–11 (2019)
31. Wasserthal, J., Neher, P., Maier-Hein, K.H.: TractSeg—fast and accurate white matter tract segmentation. NeuroImage **183**, 239–253 (2018)
32. Tallantyre, E.C., Major, P.C., Atherton, M.J., Davies, W.A., Joseph, F., Tomassini, V., Pickersgill, T.P., Harding, K.E., Willis, M.D., Winter, M., et al.: How common is truly benign ms in a uk population? J. Neurol. Neurosurg. Psychiatry **90**(5), 522–528 (2019)
33. Fischer, J., Rudick, R., Cutter, G., Reingold, S., Force, N.M.S.C.O.A.T.: The multiple sclerosis functional composite measure (MSFC): an integrated approach to ms clinical outcome assessment. Multi. Sclerosis J. **5**(4) 244–250 (1999)
34. Weiss, L.G., Saklofske, D.H., Coalson, D., Raiford, S.E.: WAIS-IV Clinical Use and Interpretation: Scientist-Practitioner Perspectives. Academic (2010)
35. Coughlan, A., Oddy, M., Crawford, J.: Birt Memory and Information Processing Battery (bmipb). Brain Injury Rehabilitation Trust, London (2007)
36. Delis, D.C., Kaplan, E., Kramer, J.H.: Delis-Kaplan Executive Function System (2001)
37. Veraart, J., Novikov, D.S., Christiaens, D., Ades-Aron, B., Sijbers, J., Fieremans, E.: Denoising of diffusion MRI using random matrix theory. NeuroImage **142**, 394–406 (2016)
38. Andersson, J.L.R., Sotiropoulos, S.N.: An integrated approach to correction for off-resonance effects and subject movement in diffusion MR imaging. NeuroImage **125**, 1063–1078 (2016)
39. Andersson, J.L.R., Skare, S., Ashburner, J.: How to correct susceptibility distortions in spin-echo echo-planar images: application to diffusion tensor imaging. NeuroImage **20**(2), 870–888 (2003)
40. Chamberland, M., Whittingstall, K., Fortin, D., Mathieu, D., Descoteaux, M.: Real-time multi-peak tractography for instantaneous connectivity display. Frontiers Neuroinf. **8**, 59 (2014)

41. Filippi, M., Brück, W., Chard, D., Fazekas, F., Geurts, J.J., Enzinger, C., Hametner, S., Kuhlmann, T., Preziosa, P., Rovira, À., et al.: Association between pathological and mri findings in multiple sclerosis. Lancet Neurol. **18**(2), 198–210 (2019)
42. Holland, C.M., Charil, A., Csapo, I., Liptak, Z., Ichise, M., Khoury, S.J., Bakshi, R., Weiner, H.L., Guttmann, C.R.: The relationship between normal cerebral perfusion patterns and white matter lesion distribution in 1,249 patients with multiple sclerosis. J. Neuroimaging **22**(2), 129–136 (2012)
43. Filippi, M., Rocca, M.A., Martino, G., Horsfield, M.A., Comi, G.: Magnetization transfer changes in the normal appering white matter precede the appearance of enhancing lesions in patients with multiple sclerosis. Ann. Neurol. **43**(6), 809–814 (1998)
44. Thiebaut de Schotten, M., Dell'Acqua, F., Ratiu, P., Leslie, A., Howells, H., Cabanis, E., Iba-Zizen, M., Plaisant, O., Simmons, A., Dronkers, N., et al.: From phineas gage and monsieur leborgne to hm: revisiting disconnection syndromes. Cerebral Cortex **25**(12), 4812–4827 (2015)
45. Foulon, C., Cerliani, L., Kinkingnehun, S., Levy, R., Rosso, C., Urbanski, M., Volle, E., Thiebaut de Schotten, M.: Advanced lesion symptom mapping analyses and implementation as bcbtoolkit. Gigascience **7**(3), giy004 (2018)
46. Fox, M.D.: Mapping symptoms to brain networks with the human connectome. New England J. Med. **379**(23), 2237–2245 (2018)

Multi-modal Brain Age Estimation: A Comparative Study Confirms the Importance of Microstructure

Ahmed Salih, Ilaria Boscolo Galazzo, Akshay Jaggi, Zahra Raisi Estabragh, Steffen E Petersen, Karim Lekadir, Petia Radeva, and Gloria Menegaz

Abstract Brain age inferred from neuroimaging data could reveal important information about the evolution of structural and functional cerebral features across the life span. This has important implications for understanding healthy aging and for identifying Imaging-Derived Phenotypes (IDPs) that characterise age-related neurodegenerative illnesses, such as Alzheimer's and Parkinson's disease. The so-called brain age delta refers to the difference between image-derived brain age and chronological age. Accelerated aging (positive delta) or resilience to aging (negative delta) have been found to be useful correlates of factors such as disease and cognitive decline. Multiple studies have proposed prediction models using brain IDPs as predictor variables, mostly relying on simple linear regression. However, methodological and population heterogeneity in these studies precludes definitive conclusions regarding the most informative modelling methodologies or predictor IDPs. To provide first hints in this respect, in this paper we propose to address three questions. First, four different state-of-the-art models are ranked based on well-known performance indices (e.g., mean absolute error) using the UK Biobank brain MRI data in different single/multi-modal settings. Second, for the best model, the association

A. Salih (✉) · I. B. Galazzo · G. Menegaz
Department of Computer Science, University of Verona, Verona, Italy
e-mail: ilaria.boscologalazzo@univr.it

A. Jaggi · K. Lekadir · P. Radeva
Dept. de Matemátiques i Informàtica, University of Barcelona, Barcelona, Spain
e-mail: akshay.jaggi@ub.edu

K. Lekadir
e-mail: karim.jekadir@ub.edu

Z. R. Estabragh · S. E. Petersen
Queen Mary University of London, London, United Kingdom
e-mail: z.r.raisi-estabragh@qmul.ac.uk

S. E. Petersen
e-mail: s.e.petersen@qmul.ac.uk

N. Gyori et al. (eds.), *Computational Diffusion MRI*, Mathematics and Visualization,
https://doi.org/10.1007/978-3-030-73018-5_19

with individual IDPs are calculated to identify those that could play a prominent role in the aging process. Third, associations with non-brain variables are assessed as a first step towards a holistic approach. Our findings demonstrate a prominent role for dMRI IDPs in reducing the mean absolute error and rank high in the association study, dominating the first ten positions and being preceded only by three structural measures that are known to be related to the aging process. This provides evidence of the potential of dMRI IDPs as biomarkers of aging in health and disease.

1 Introduction

Neuroimaging data have been extensively used to assess brain changes during aging, under both healthy and disease conditions. Moreover, they can be exploited to predict "brain age" which is the apparent biological age of an individual and depends on several endogenous (subject-specific) as well as exogenous (environmental) factors. Metrics derived from various brain magnetic resonance imaging (MRI) sequences have been adopted to estimate brain age, either using raw data or handcrafted features. Brain age delta (or relative brain age) is calculated by subtracting chronological age from the estimated one. While a younger-appearing brain might be the result of a healthy life style [1], having an older-appearing brain has been previously associated with poor future outcomes [2] and with an increased likelihood to develop neurodegenerative illnesses such as Alzheimer's [3].

Statistical models for brain age estimation have been proven to be highly accurate, with prediction performance featuring high R^2 values and low mean absolute error (MAE) in the range of 4–5 years [2]. Most of the studies have investigated this aspect with features derived from a single brain MRI technique, most commonly, conventional T1-weighted structural images (sMRI). Morphometric measures from sMRI, such as volume and thickness of grey matter (GM) structures, should not be overlooked as they provide information on the individual degree of brain atrophy that encodes aging-induced degeneration [4]. However, more recently, diffusion MRI (dMRI), susceptibility weighted imaging (SWI), and resting-state fMRI have been explored for potentially providing a richer set of IDPs bringing complementary information [2, 5]. Thus, consideration of IDPs derived from multiple brain MRI sequences would be the most desirable approach allowing deeper phenotyping and more complete capturing of the different factors shaping the aging process.

Regarding modeling approaches, the performance accuracy depends on the statistical method utilised, as demonstrated by several authors. In a recent paper, Jonsson and colleagues (2019) applied deep learning as well as eight different regression methods to sMRI-based features extracted from three well-known databases, showing notable differences in the performance parameters across the different models [3]. Niu et al. [1] report similar variation in model performance in their study of brain age estimation with four regression models using several neuroimaging variables (sMRI, dMRI, and resting-state fMRI) in healthy controls and patients with anxiety disorders. In addition, the authors showed the potential for superior prediction accuracy with a multi-modal versus single-modal approach.

The recent availability of large imaging databases has provided new opportunities to exploit the importance of a multi-modal approach for brain age prediction. In this context, the UK Biobank (UKB) represents an important resource thanks to its comprehensive repository with genetic and phenotypic data for 500000 subjects aged between 40 and 69 (at recruitment). The UKB imaging study includes detailed MRI, providing high quality multi-modal neuroimaging data including sMRI, dMRI, SWI and fMRI [6]. These data are linked to detailed clinical, biological and lifestyle information. The availability of such a rich research resource has motivated many researchers to focus on brain age estimation with promising results [2, 3, 5, 7].

Smith et al. [5] estimated brain age using simulated and real data by applying simple linear regression. With regard to real data, 2641 IDPs covering sMRI, fMRI and dMRI were used for 19000 participants. The results, among others, attained MAE = 3.6 years.

In [2], phenotypes from six different MRI modalities were chosen to estimate brain age for 17461 subjects, running a Least Absolute Shrinkage and Selection Operator (LASSO) regression for each modality (MAE range = 3.897–5.928 years, where minimum e maximum were found for dMRI and task fMRI, respectively). When all the IDPs were combined, age was more accurately predicted (MAE = 3.515 years). Thirty-four IDPs were deemed informative for the prediction of the brain age after bootstrapping, and were predominantly from sMRI and dMRI.

Ning et al. [7] aimed to assessing the correlation between brain age delta and alcohol intake, smoking and genetic variations. To this end, 403 morphometric measurements from sMRI were chosen along with LASSO regression (MAE = 3.8 years). A significant association between brain age delta and the consumption of alcohol and smoking could be demonstrated.

Finally in [3], sMRI data for 12395 subjects were used to estimate brain age using transfer learning and 3-D Convolutional Neural Network (CNN). In this study, two sequence variants were identified having a strong relation with the brain age delta. The MAE of the model was 3.63 years.

The great variability in the number of subjects, IDPs, MRI modalities and statistical models precludes a straightforward comparison of all the studies. However, existing work suggests that: (i) sMRI provides relevant IDPs for estimation of brain age; (ii) dMRI-based phenotypes are similarly informative and need to be further investigated; and (iii) a multi-modal approach can improve, in general, the estimation accuracy. To the best of our knowledge, a systematic comparison among different statistical methods has not been addressed in existing literature. Therefore, in this paper we compared four regression methods in combination with different IDPs for brain age prediction, aiming at providing a balanced comparison across different single-modal and multi-modal approaches. In particular, we focused on Simple Linear Regression (SLR), LASSO, Support Vector Regression (SVR) and Bayesian Ridge Regression (BRR), while the handcrafted features were derived from sMRI, dMRI and SWI with a clear numeric prevalence of dMRI. Model performance was assessed using several parameters including MAE, R^2 and adjusted R^2.

Moreover, the associations between individual IDPs and brain age delta values were calculated for the best model. Finally, the association between brain age delta with selected biomedical and behavioral features was extracted to assess potential clinical/biological utility.

2 Data and Materials

Data were obtained from UKB. All the analyses here performed rely on the IDPs extracted centrally by researchers involved in the project [8]. Data were available from n = 16394 participants (age range = 40–70 years, n = 8652 females, n = 7742 males). This comprised a set of 714 IDPs for each subject, representing the summary metrics for sMRI, SWI and dMRI. From sMRI images, morphometric measures of brain volumes were reported as distinct IDPs, both normalised/not normalised for overall head size, in details: total brain volume (GM + white matter [WM]); volumes for WM, GM and cerebrospinal fluid (separately for each compartment); volume of peripheral cortical GM. Volume measures for subcortical structures were also calculated as further IDPs (e.g., thalamus, putamen, hippocampus), generally separated for left/right hemispheres and not normalised for head size. From SWI data, a T2* image was used and the median T2* value estimated as a separate IDP for each subcortical ROI identified from sMRI. Finally, several spatially-specific IDPs were extracted from dMRI data by following two different approaches. Indeed, nine dMRI-based indices derived from i) the diffusion tensor imaging (DTI), such as fractional anisotropy (FA) and mean diffusivity (MD), and ii) the neurite orientation dispersion and density imaging (NODDI) model, such as orientation dispersion (OD) and isotropic volume fraction (ISOVF), were calculated and averaged over specific areas/tracts. In the first approach, dMRI maps were aligned to a population-based WM tract skeleton and all the DTI/NODDI measures averaged over 48 regions defined using the Johns Hopkins University tract atlas [9]. In the second, probabilistic tractography was run for each subject and all the dMRI-based measures averaged within 27 distinct WM tracts. The final set of neuroimaging phenotypes included 25 IDPs from sMRI, 14 from SWI and 675 from dMRI. Full details on the acquisition protocols and image processing pipelines for the UKB brain data are available at https://biobank.ctsu.ox.ac.uk/crystal/crystal/docs/brain_mri.pdf.

The present analyses were conducted under data application number 2964. All participants provided formal consent, details on the UKB Ethics can be found at https://www.ukbiobank.ac.uk/the-ethics-and-governance-council.

3 Methods

3.1 Brain Age Estimation

Four different regression methods including LASSO [10], SLR [11], SVR [12] and BRR [13] were used to estimate the apparent brain age, all having chronological age as the dependent variable. All these models were implemented using Scikit-learn [14] library version 0.22.2 in Python 3.6.9.

In order to examine the impact of different imaging modalities, each of the four methods was run with single-modal and multi-modal brain IDPs, leading to seven different combinations per method. All the imaging features (independent variables) were normalized to zero mean and unit variance to account for the different measurement scales, while the actual age was demeaned only [5]. Gender and education were considered as confound variables and regressed out of all IDPs as in [6, 15]. Data were randomly split into training (80%, n = 13115) and testing (20%, n = 3279) sets, respectively. The test set was used to predict brain ages on unseen data.

Hyper-parameters for BRR, LASSO and SVR were tuned on the training data (further split on 80% for training and 20% for validation) with GridSearchCV and the optimal model was retained. After the parameters were optimized from training data, the optimal model was applied to estimate brain age in the test set. The performance of each model was assessed using the Coefficient of Determination (R^2) and the MAE. Adjusted R^2 was also calculated to account for the different number of predictors in each model.

Recent literature has demonstrated a proportional bias in brain age calculation, which might be caused by dilution bias of the prediction model [16, 17]. Moreover, this bias is also closely connected to the fact that brain age is overestimated in younger subjects and underestimated in older ones, while is more accurately predicted for participants whose actual ages are closer to the mean age of the training dataset [2, 5]. All these elements lead to a significant dependence of the brain age delta on chronological age, which resulted to be negatively correlated. Therefore, common practice is to apply a statistical age-bias correction procedure to overcome these limitations [5, 16] In this study, we adopted the procedure proposed by Beheshti et al. [16] that relies on a linear model given by the following equation:

$$D = \alpha * \Omega + \beta \tag{1}$$

where D is the brain age delta (estimated from training data), Ω is the chronological age of the training data, α and β represent the slope and the intercept. These two measures are subsequently used to correct the brain age predictions in the test set as described in Eq. 2:

$$CPBA = Predicted\ Brain\ Age - (\alpha * \Omega + \beta) \tag{2}$$

where CPBA stands for corrected predicted brain age.

After brain age was estimated and bias corrected in the test set, brain age delta was calculated for each subject. Pearson correlations for predicted brain age *vs* actual age (CPA) and brain age delta *vs* actual age (CBDA) were calculated twice for each model, before and after bias correction.

3.2 Associations with IDPs and Non-IDP Variables

For the best model results, Pearson correlations between brain age delta values and individual IDPs were calculated in order to identify the strongest associations, highlighting the features which contribute most to the modelling of the brain delta as suggested in [5]. The resulting p-values were Bonferroni-corrected for multiple comparison testing. Of note, the fully deconfounded versions of the IDPs were used in this step (including gender, education and age as confounds). As several studies demonstrated a significant association between brain and heart functionality, especially relying on brain volumetric measurements [18, 19], we also investigated whether a correlation between brain age delta and heart measures was present. In order to perform this analysis, five measures from Cardiac Magnetic Resonance (CMR) and eight Cardiovascular Risk Factors (CRFs) were considered. The correlation analysis was performed on a subgroup of the test set (n = 2730), as these measures were not available for all the test set subjects. CMR scans were performed on 1.5 T scanners using a standardised acquisition protocol [20]. The following indices derived for the left ventricle were retained: end-diastolic volume (LVEDV), end-systolic volume (LVESV), stroke volume (LVSV), mass (LVM), and ejection fraction (LVEF). Eight CRFs were also tested, covering biomedical and lifestyle measures: smoking status, material deprivation, body mass index, alcohol intake frequency, physical activity, diabetes diagnosis, presence of hypertension and high cholesterol. Smoking status and alcohol intake frequency were based on self-reports. Material deprivation was reported by UKB as the Townsend deprivation index. A continuous value for the amount of physical activity, measured in metabolic equivalent minutes/week, was calculated. Body mass index was derived from height and weight measures recorded at the baseline. Diabetes, hypertension, and hypercholesterolaemia were defined by cross-checking across self-report and blood biochemistry data. All the cardiac variables were initially normalized to zero mean and unit variance, and the main potential confounds (gender and age) regressed out from the data. Pearson correlation was finally computed between each of these measures and brain age delta values derived from the twenty-eight model combinations, and the results were Bonferroni-corrected to account for multiple comparison problems.

Table 1 Prediction performance of the four regression methods combined with different imaging features. Results are reported in terms of MAE values (years), and the optimal one for each IDPs combination is highlighted in bold

Mean absolute error					
IDPs	Number of features	BRR	SVM	SLR	LASSO
sMRI	25	4.509	**4.471**	4.506	4.509
SWI	14	6.026	6.0411	**6.024**	6.025
dMRI	675	**3.733**	3.758	3.761	3.738
sMRI+SWI	39	4.429	**4.393**	4.424	4.427
sMRI+dMRI	700	**3.498**	3.559	3.525	3.5
SWI+dMRI	689	**3.717**	3.74	3.741	3.719
All	714	**3.482**	3.526	3.512	3.483

4 Results

4.1 Brain Age Estimation

Results are summarised in Tables 1 and 2 reporting the overall performance of the four regression methods combined with the different IDPs. Table 1 reports the estimation performance for the test subjects in terms of MAE values before bias correction, as this represents the actual model performance. Results demonstrated that using all the 714 IDPs from the three imaging modalities provided the best model performance in terms of MAE for all regression methods. In particular, BRR gave the best results (MAE = 3.482 years), closely followed by LASSO (MAE = 3.483 years), while SVR performed less accurately among the four tested methods. When considering the different feature types, the performance of the models using SWI only was worst (MAE $\approx$ 6.0 years) compared to the other single-modal approaches that is sMRI (MAE $\approx$ 4.5 years) and especially dMRI (MAE $\approx$ 3.7 years). When considering the multi-modal models, adding dMRI phenotypes improved the accuracy of all methods.

These results were further confirmed by the R^2 and adjusted R^2 parameters (Table 2), for which the lowest value was reached using the SWI IDPs (R^2 = 0.075–0.085). When the IDPs from sMRI and dMRI were used jointly in the model, the performance was improved and very close to the one reached by using all the IDPs, and this finding held for all the four regression methods. For the sake of completeness, CPA and CBDA were calculated before and after bias correction, leading to the results summarised in Table 3. When using all the 714 IDPs, the correlation between brain age delta and actual age decreases towards zero after applying the bias correction steps. Conversely, CPA increased after bias correction in all four methods.

Table 2 Prediction performance of all the tested models in terms of R^2 and Adjusted R^2 values

	BRR		LASSO		SLR		SVR	
IDPs	R^2	Adj_R^2	R^2	Adj_R^2	R^2	Adj_R^2	R^2	Adj_R^2
sMRI	0.445	0.441	0.446	0.441	0.446	0.441	0.445	0.440
SWI	0.085	0.081	0.085	0.081	0.085	0.081	0.075	0.071
dMRI	0.613	0.512	0.612	0.511	0.606	0.504	0.604	0.501
sMRI+SWI	0.464	0.458	0.464	0.458	0.465	0.458	0.468	0.462
sMRI+dMRI	0.654	0.560	0.653	0.559	0.648	0.553	0.642	0.545
SWI+dMRI	0.618	0.516	0.61	0.515	0.611	0.507	0.609	0.505
All	0.658	0.562	0.657	0.562	0.652	0.555	0.650	0.553

Table 3 Correlation values between predicted brain age vs actual age (CPA) and between brain age delta vs actual age (CBDA), before and after bias correction

The model	Before correction		After correction	
	CPA	CBDA	CPA	CBDA
BRR	0.811	−0.592	0.903	−0.014
LASSO	0.810	−0.576	0.900	−0.018
SLR	0.807	−0.559	0.896	−0.026
SVR	0.806	−0.597	0.902	−0.015

4.2 Association with Brain IDPs

Considering that the BRR method combined with all the IDPs reached the lowest MAE and highest R^2/adjusted R^2r, here we report associations between individual IDPs and brain age delta values estimated from this model. In particular, Table 4 shows the first ten significant correlations (after correction for multiple comparisons), revealing a strong and significant association between these IDPs and the brain age delta. As further note, the association between brain age delta values and individuals IDPs were largely overlapped for the other regression methods, especially concerning the features in the top 10 positions. The order of the most significant features that are associated with the brain age delta is similar in the four methods, although the correlation values changed across them.

As it can be appreciated, the volumetric measurements from sMRI such as GM volume and volume of peripheral cortical GM (both normalised for head size) were negatively correlated with the brain age delta. Diffusion measures from DTI (such as MD, L1, L2 and L3) and from NODDI (such as ISOVF) in fornix were positively correlated with the brain age delta, while FA revealed an opposite pattern.

Table 4 Strongest associations between brain age delta values estimated from the winning model (BRR with all IDPs) and individual IDPs for test set subjects

IDPs	Correlation
Volume of grey matter (normalised for head size)	−0.5113
Volume of peripheral cortical grey matter (normalised for head size)	−0.4965
Volume of brain, grey+white matter (normalised for head size)	−0.4432
Mean ISOVF in fornix on FA skeleton	0.4092
Mean L1 in fornix on FA skeleton	0.4034
Mean MD in fornix on FA skeleton	0.4022
Mean L3 in fornix on FA skeleton	0.3969
Mean L2 in fornix on FA skeleton	0.3891
Mean FA in fornix on FA skeleton	−0.3887
Mean L2 in fornix cres+stria terminalis on FA skeleton (left)	0.3712

Table 5 Correlation between CMR, CRFs and brain age delta

Cardiovascular risk factors				Cardiac magnetic resonance			
Measure	Correlation	p-value	p_{FDR}	Measure	Correlation	p-value	p_{FDR}
Smoking	0.056	0.003	0.024	LVEDV	0.006	0.725	1
Deprivation	0.067	0	0.003				
Body Mass Index	0.053	0.005	0.040	LVESV	−0.004	0.795	1
Alcohol	0.038	0.046	0.369				
Exercises	0.001	0.920	1	LVSV	0.015	0.420	1
Diabetes	0.087	0	0				
Hypertension	0.066	0	0.004	LVM	0.044	0.021	0.107
High Cholesterol	0.056	0.003	0.025	LVEF	0.024	0.209	1

4.3 Association with Cardiac Variables

Table 5 reports Pearson correlations between brain age delta derived from the winning model (BRR with all IDPs) and CMR/CRFs measures (test set). For CMR, no significant associations were found after multiple comparison correction ($p_{FDR} > 0.05$), and only the correlation with the left ventricular mass (LVM) was significant before correction ($p = 0.021$). Regarding CRFs, all parameters except exercise and alcohol were significantly associated with the brain age delta ($p_{FDR} < 0.05$) .

5 Discussion

In this study, we investigated whether chronological age could be accurately predicted using brain MRI IDPs as predictor variables in various statistical models using data in the UKB. In particular, we focused on four well-known regression methods (SLR, SVR, LASSO and BRR) and considered measures from sMRI, SWI and dMRI as IDPs, either alone or in combination. Regarding the regression methods, overall, BRR achieved the highest accuracy as measured by MAE, R2 and adjusted R2 values. In particular, when dealing with a relatively small number of IDPs (< 50), for example in models with sMRI/SWI features only, better results were obtained using SVR and SLR. Conversely, in cases where a greater number of features was included, BRR reached the best performance, possibly because of its ability to handle multicollinearity between IDPs [21, 22].

Previous studies addressing modelling brain age using UKB data report MAE values between 3.5–3.8 years. Of note, Peng et al. [23] achieved the lowest MAE (2.14 years) although leveraging from deep CNN model, Simple Fully Convolutional Network, using sMRI from UKB for 14503 participants. In our study, the accuracy reached by BRR model in the different conditions was comparable (and even better in some cases) to such benchmarks, despite the generally lower number of subjects and MRI features.

Regarding the imaging predictors, models including all the 714 IDPs from the three brain MRI sequences had the best performance. However, when considering models with single-modal IDPs, dMRI reached the highest accuracy in terms of MAE values (MAE $\approx$ 3.7 years) compared to sMRI (MAE $\approx$ 4.5 years) and especially SWI (MAE $\approx$ 6.02 years), and this was further confirmed by the R^2/adjusted R^2 values. This might indicate that age-related alteration of brain can be better detected by dMRI, in agreement with literature findings [24]. A previous study also found similar results and further confirm that dMRI phenotypes are more informative than SWI IDPs in predicting brain age [2]. Phenotypes from sMRI and dMRI were generally the most informative for age prediction, as further supported by the correlation analysis between delta values and IDPs. Indeed, the strongest associations were found for features based on these modalities. In particular, our study revealed that brain age delta was negatively correlated with volumetric measures, while positively correlated with both ISOVF and diffusivities in the fornix.

GM volume was the most informative phenotype, in line with previous studies [2, 5]. This might be related to the fact that brain volume changes considerably over time and decreases during the aging process, causing atrophy [25] and macroscopic variations. Our analysis highlighted a prominent role of dMRI IDPs. Differences in diffusion properties across the life span have been demonstrated along specific WM tracts [24]. Diffusivity and FA values across the fornix spanned the first 10 ranking position, preceded only by atrophy measures in GM and WM. Noteworthy, the fornix is among those tracts that mature very early [26]. The IDPs that are present in the top 10 association ranking are, besides FA and MD, ISOVF, that is the isotropic volume fraction as estimated by the NODDI model, and the three tensor eigenvalues

L1, L2 and L3 that represent the axial (L1) and transversal (L2, L3) diffusivities. A reduction in the FA and an increase in diffusivity, as indicated by a positive correlation of ISOVF, MD, L1, L2 and L3 with delta age, could indicate impaired WM integrity. Moreover, myelin breakdown might be measured by radial diffusivity (L2 and L3) alterations, while increasing in apparent diffusivity value might be a sign of axonal disruptions [27]. Furthermore, AD (L1) and RD (L2 and L3) have been observed to increase in elderly people which may be a signal of deterioration of the WM fibers [28]. Noteworthy, ISOVF has been observed to increase in older people in most of the major tracts, pointing to a disrupted integrity [29]. Our results are in agreement with such findings reporting a negative association of FA and a positive association of L1 , L2, L3 and MD in fornix with brain delta age. Fornix tracts have a vital role in memory tasks, specially episodic memory. Alteration in diffusion measures during aging process might be good biomarkers for neurological diseases that are related to memory impairments [26]. This could indicate that such IDPs are more prone to alteration over the life span of an individual at least over specific WM tracts, making them potential biomarkers for the aging process in health and disease.

Regarding the associations with CMR measures, our study revealed a significant association with LVM, however there was loss of statistical significance after multiple comparison correction. A previous study reported association of increase in LVM with alterations in WM microstructure in elderly people [30]. In our study, the limited age range in the UKB did not permit consideration of relationship in very old individuals. Among the CRFs, all measures except exercise and alcohol were significantly correlated with brain age delta ($p_{FDR} < 0.05$), inline with what described by Cole et al [7, 31], despite using a different number of IDPs and subjects for estimating brain delta. In conclusion, results suggest that dMRI IDPs play a prominent role in reducing the MAE and rank high in the association study, providing evidence of the potential of dMRI IDPs as biomarkers of aging in health and disease. Future work will investigate the integration of other IDPs such as functional MRI, graph-based measures from brain connectomes as well as the genetic information to pursue the holistic path.

Acknowledgements This work was funded by INVITE program co-financed by the European Union within the Horizon 2020 Programme and by the Regione del Veneto.

References

1. Niu, X., et al.: Improved prediction of brain age using multimodal neuroimaging data. Hum. Brain Mapp. **41**(6), 1626–1643 (2020)
2. Cole, J.H.: Multi-modality neuroimaging brain-age in UK Biobank: relationship to biomedical, lifestyle and cognitive factors. Neurobiol. Aging (2020)
3. Atli Jónsson, B. et al.: Brain age prediction using deep learning uncovers associated sequence variants. Nat. Commun. **10**(1), 1–10 (2019)
4. Franke, K., Gaser, C.: Ten years of BrainAGE as a neuroimaging biomarker of brain aging: What insights have we gained? Frontiers Neurology **10**, 789 (2019)

5. Smith, S.M. et al.: Estimation of brain age delta from brain imaging. Neuroimage **200**, 528–539 (2019)
6. Elliott, L.T. et al.: Genome-wide association studies of brain imaging phenotypes in UK Biobank. Nature **562**(7726), 210–216 (2018)
7. Ning, K., et al.: Association of relative brain age with tobacco smoking, alcohol consumption, and genetic variants. Sci. Rep. **10**(1), 1–10 (2020)
8. UK Biobank.: About UK Biobank (2014). Available at https://www.uk-biobank.ac.uk/about-biobank-uk
9. Hua, K., et al.: Tract probability maps in stereotaxic spaces: analyses of white matter anatomy and tract-specific quantification. Neuroimage **39**(1), 336–347 (2008)
10. Tibshirani, Robert: Regression shrinkage and selection via the lasso. J. R. Stat. Soc. Ser. B (Methodological) **58**(1), 267–288 (1996)
11. Kenney, J.F., Keeping, E.S.: Linear regression and correlation. Math. Stat. **1**, 252–285 (1962)
12. Drucker, H. et al.: Support vector regression machines. Adv. Neural Inf. Process. Syst. 155–161 (1997)
13. MacKay, D.J.C.: Bayesian interpolation. Neural Comput. **4**(3), 415–447 (1992)
14. Pedregosa, F., et al.: Scikit-learn: machine learning in python. J. Mach. Learn. Res. **12**, 2825–2830 (2011)
15. Miller, K.L. et al.: Multimodal population brain imaging in the UK Biobank prospective epidemiological study. Nat. Neurosci. **19**(11), 15231536 (2016)
16. Beheshti, I. et al.: Bias-adjustment in neuroimaging-based brain age frameworks: a robust scheme. NeuroImage: Clin. **24**, 102063 (2019)
17. Le, T.T et al.: A nonlinear simulation framework supports adjusting for age when analyzing BrainAGE. Frontiers Aging Neuroscience **10**, 317 (2018)
18. van der Velpen, I.F. et al.: Impaired cardiac function and cognitive brain aging. Canadian J. Cardiol. **33**(12), 1587–1596 (2017)
19. Jefferson, A.L. et al.: Cardiac index is associated with brain aging: the framingham heart study. Circulation **122**(7), 690 (2010)
20. Petersen, S.E. et al.: UK Biobank's cardiovascular magnetic resonance protocol. J. Cardiovascular Mag. Reson. **18**(1), 8 (2015)
21. Assaf, A.G., Tsionas, M., Tasiopoulos, A.: Diagnosing and correcting the effects of multicollinearity: Bayesian implications of ridge regression. Tourism Manag **71**, 1–8 (2019)
22. Efendi, A., Effrihan.: A simulation study on Bayesian Ridge regression models for several collinearity levels. In: AIP Conference Proceedings, vol. 1913. 1, p. 020031. AIP Publishing LLC (2017)
23. Peng, H. et al.: Accurate brain age prediction with lightweight deep neural networks. BioRxiv (2019)
24. Yao, X. et al.: Evaluation of human brain aging via diffusion tensor imaging tract characteristics. In: 2017 13th International Conference on Natural Computation, Fuzzy Systems and Knowledge Discovery (ICNC-FSKD), pp. 2651–2655. IEEE (2017)
25. Driscoll, I., et al.: Longitudinal pattern of regional brain volume change differentiates normal aging from MCI. Neurology **72**(22), 1906–1913 (2009)
26. Douet, V., Chang, L.: Fornix as an imaging marker for episodic memory deffcits in healthy aging and in various neurological disorders. Frontiers Aging Neurosci. **6**, 343 (2015)
27. Liu, H., et al.: Aging of cerebral white matter. Ageing Res. Rev. **34**, 64–76 (2017)
28. Chen, D.Q. et al.: Age-related changes in diffusion tensor imaging metrics of fornix subregions in healthy humans. Stereotactic Funct. Neurosurg. **93**(3), 151–159 (2015)
29. Cox, S.R. et al.: Ageing and brain white matter structure in 3,513 UK Biobank participants. Nat. Commun. **7**(1), 1–13 (2016)
30. Moore, E.E., et al.: Increased left ventricular mass index is associated with compromised white matter microstructure among older adults. J. Am. Heart Assoc. **7**(13), e009041 (2018)
31. Cole, J.H. et al.: Body mass index, but not FTO genotype or major depressive disorder, inuences brain structure. Neuroscience **252**, 109–117 (2013)

Longitudinal Parcellation of the Infant Cortex Using Multi-modal Connectome Harmonics

Hoyt Patrick Taylor, Sahar Ahmad, Ye Wu, Khoi Minh Huynh, Zhen Zhou, Zhengwang Wu, Weili Lin, Li Wang, Gang Li, Han Zhang, and Pew-Thian Yap

Abstract Functional segregation and specialization of cortical regions is central to the significant changes that take place during early brain development. We present an automated scheme that harnesses local and long-range connectivity features of the cortex—derived from multiple imaging modalities—for longitudinal parcellation of the early developing human cortex. This is realized by using multi-modal connectome harmonics in a hierarchical agglomerative clustering (HAC) framework to produce group correspondent, individual-specific, and age-dependent parcellation maps for the first two years of development. We observed decreased regularity in cluster size and shape with increasing time, reflecting cortical specialization known to emerge during early development. Investigation of the modularity of structural connectivity defined by our parcellations suggests convergence toward adult resting-state networks as the brain develops.

1 Introduction

Dramatic, widespread, and highly impactful changes in cortical connectivity, myelination, and organization take place during the first two years of human brain development. These changes in large part lead to the remarkably consistent trends in the function and connectivity of cortical regions and the networks they comprise in healthy adults. Although the behavioral milestones that occur during the first two

H. P. Taylor (✉) · S. Ahmad · Y. Wu · K. M. Huynh · Z. Zhou · Z. Wu · W. Lin · L. Wang · G. Li · H. Zhang · P.-T. Yap
Department of Radiology and Biomedical Research Imaging Center (BRIC), University of North Carolina, Chapel Hill, USA
e-mail: hptaylor@live.unc.edu

H. Zhang
e-mail: hanzhang@med.unc.edu

P.-T. Yap
e-mail: ptyap@med.unc.edu

N. Gyori et al. (eds.), *Computational Diffusion MRI*, Mathematics and Visualization,
https://doi.org/10.1007/978-3-030-73018-5_20

years of development are well documented, the underlying development of functional segregation of the cortex has remained opaque despite rapid advancements in in-vivo imaging technology and analytic techniques seen in recent decades. The Baby Connectome Project [1] provides an opportunity to elucidate these gaps in understanding; longitudinal high-resolution anatomical, functional, and diffusion MR images of a large cohort of infants are made available for the first time. Simultaneously, novel complex network analysis techniques and multi-layer clustering routines detailed in [2] allow for utilization of this unique dataset to produce the first ever temporally dense, individually-specific, and group correspondent cortical maps, or parcellations, for infants in the first two years of development.

Previous work [2] demonstrates that the low-frequency harmonics of multi-layer vertex-wise structural connectivity graphs constructed using anatomical and diffusion MR images can be used to produce subject specific, group correspondent parcellation maps using hierarchical agglomerative clustering (HAC) that qualitatively agree with functionally-derived parcellations in the literature. Here, we extend the method detailed in [2] by utilizing multiple connectivity types: structural, functional, and local myelination similarity in order to obtain the multi-modal spectral coordinates. Further, we devise an iterative relaxation method that optimizes the contribution of each modality and subject to the group-level parcellation. Finally, we devise a method for determination of the optimal number of distinct cortical regions represented by the underlying connectivity data.

The present work achieves several advantageous features in cortical parcellation. First, it utilizes three separate imaging modalities: dMRI, fMRI, and T1/T2 MRI. Multi-modal parcellation schemes are highly advantageous as they are less susceptible to noise and provide more robust local and global connectivity information than single-modality schemes. Secondly, our multi-layer formulation and individual back-propagation ensure that for each cohort or time window, there is a one-to-one correspondence between group-consensus parcellations and individually back-propagated parcellations, which are computed based on each subject's multi-modal connectivity data. Third, our parcellation method includes a data-driven approach to the determination of the optimal number of parcels for a given cohort. This is particularly significant as it is often challenging to justify the choice for number of clusters in a parcellation routine. Finally, our approach provides a succinct and flexible solution to the 'missing data problem'. Often in staggered-cohort longitudinal studies such as the BCP, the number of subjects in equally-spaced time intervals is not constant, and there are often incomplete scans for one or more modalities for a given subject. Our approach accommodates for this, allowing for any number of modalities for each subject in each window, as well as allowing for variable number of subjects in each window.

2 Methods

Our multi-modal parcellation workflow has three major steps. First, modality and subject-specific connectivity matrices, A_x^{m}, are constructed for each modality (m) and each subject x (Fig. 1a, b). Second, temporally-weighted and degree-corrected multi-layer modality-specific adjacency matrices, $A_{\mathrm{ML}}^{\mathrm{m}}$, are constructed for each of seven 100-day windows from 0 to 700 days (Fig. 1c), and the first $N_{\mathrm{harm}} = 99$ non-trivial eigenvectors of the normalized Laplacian matrix of $A_{\mathrm{ML}}^{\mathrm{m}}$ are computed (Fig. 1d), yielding the modality-specific multi-layer connectome harmonics, ψ_k^m. Third, these connectome harmonics are then inputted to our unique iterative multi-modal parcellation routine (Fig. 1e–i), which yields multi-modal group-consensus parcellations and individually back-propagated multi-modal parcellations, which share one-to-one correspondence of parcels.

2.1 Data and Preprocessing

We used diffusion, functional, T1w, and T2w MRI data of 131 infant subjects enrolled as part of the UNC/UMN baby connectome project (BCP) [1], with subject ages

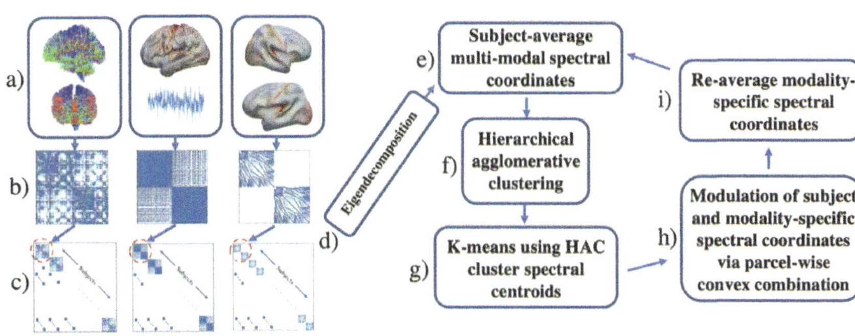

Fig. 1 **a** Structural connectivity, functional connectivity, and myelin content are computed from dMRI, fMRI, and T1/T2 data for each subject; **b** modality and subject specific connectome adjacency matrices are constructed; **c** temporal multilayer adjacency matrices are constructed for each modality; **d** eigendecomposition of (**c**) yields modality specific spectral coordinates; **e** averaging across subjects and concatenation across modalities yields consensus multi-modal spectral coordinates; **f** hierarchical agglomerative clustering followed by single-round k-means yields multi-modal consensus parcellation; **g** individual and modality-specific centroids of consensus parcels are used to initialize single round modality and individual k-means; **h** individual spectral coordinates are modulated via convex combination with spectral centroids of (**g**); **i** newly modulated modality and subject-specific spectral coordinates are re-averaged across subjects and concatenated across modalities, yielding the new consensus multi-modal spectral coordinates. **e** and **f** are iterated until the multi-modal consensus parcellations from consecutive iterations have normalized mutual information greater than 0.95

ranging from 16 and 700 days. Whole brain tractography specialized for infants was performed for each subject's 6-shell, 1.5 mm^3 isotropic dMRI data using the method described in [3]. Cortical surfaces were reconstructed using UNC infant preprocessing pipeline [4, 5]. The reconstructed white-matter surfaces (left/right) were mapped onto the unit sphere, and then registered using Spherical Demons [6]. The corresponding structural connectome adjacency matrix was constructed as described in [2]. Functional timeseries (TR = 0.8 s, 2 mm^3 isotropic) were mapped to the cortical surface vertices using the ribbon constrained volume-to-surface-mapping command from connectome workbench [7]. Correlations of timeseries of adjacent surface vertices and vertices connected via the structural connectome adjacency matrix were used to populate the functional connectome adjacency matrix. Myelin content was computed as the T1w/T2w ratio [8] and mapped onto the cortical surface. Each element of the myelin content adjacency matrix was computed as the inverse of the Euclidean distance between myelin content values of a pair of adjacent vertices.

2.2 Connectivity Matrices

In order to encode the information contained in the dMRI, fMRI, and T1/T2 MRI data, we formulate a connectome graph adjacency matrix for each modality and for each subject. It is important to note that all white matter surfaces contain an equal number of vertices ($N_{\text{vert}} = 81,924$), and are indexed the same. Thus, the surface adjacency matrix of subject x, A_x^{surf}, is identical to the surface matrix of all other subjects. For subject x, the structural connectivity matrix, A_x^{struc}, is constructed by the same method as described in [2]. The functional connectome adjacency matrix, A_x^{func} is constructed by computing the Pearson correlation coefficient of each vertex pair that share a connection in A_x^{struc}, as well as all vertex pairs who share a direct or once-removed connection in the polygonal connections that define the white matter surface. In other words, the nonzero elements of A_x^{func} are the correlations between the functional timeseries of vertices corresponding to nonzero elements of $A_x^{\text{struc}} + (A_x^{\text{surf}})^2$, where A_x^{surf} is the surface adjacency matrix. This approach was utilized in order to constrain functional connectivity using structural connectivity, as well as to reduce memory and computation costs. Nonzero elements of the myelin connectivity matrix of subject x, A_x^{my}, are given by

$$A_x^{\text{my}}(i, j) = \frac{1}{|M_i - M_j|}, \tag{1}$$

where M_i and M_j are the myelin values at vertex i and j. Myelin connectivity is computed for all nonzero index pairs of $(A_x^{\text{surf}})^2$, which correspond to all adjacent and once-removed adjacent vertex pairs.

2.3 Iterative Multi-modal Parcellation Via Connectome Harmonics

Parcellation was performed for each of seven 100-day windows in the first 2 years of life. For each window and each modality, a multi-layer graph adjacency matrix[9], $A^{\mathrm{m}}_{\mathrm{ML}}$, is constructed, linking the modality-specific adjacency matrices of subjects x and y, A^{m}_{x} and A^{m}_{y}, within the window using interlayer identity arcs, I^{inter}_{xy} with weight w_{xy}, such that $I^{\mathrm{inter}}_{xy} = \hat{I} w_{xy}$ given by

$$w_{xy} = c \exp - \frac{|t_x - t_y|^2}{2\sigma}, \tag{2}$$

where t_x and t_y are the scan times (in days) for subjects x and y respectively, $\sigma = 50$ days, and $\hat{I}$ is the square identity matrix of size $N_{\mathrm{vert}} \times N_{\mathrm{vert}}$. Additionally, c is proportional to $\frac{N_{\mathrm{intra}}}{N_{\mathrm{inter}}}$ in order to normalize the degree of interlayer connections with respect to the degree of intra-subject connections. Note that this multi-layer formulation provides strong interlayer connections between subjects scanned at similar times and weak connection between subjects at disparate times within a given window. Eigenvectors of the normalized Laplacian matrix $L^{m}_{\mathrm{ML}} = L(A^{\mathrm{m}}_{\mathrm{ML}}) = D^{1/2} A^{\mathrm{m}}_{\mathrm{ML}} D^{1/2}$ of each modality are computed, yielding modality-specific connectome harmonics, ψ^{m}_{k}, such that $L^{m}_{\mathrm{ML}} \psi^{m}_{k} = \lambda_k \psi^{m}_{k}$, $k \in \{0, ..., N_{\mathrm{harm}}\}$. These $N_{\mathrm{vert}} \times N_{\mathrm{subjects}}$ eigenvectors are then partitioned into N_{subjects} length N_{vert} subject-specific modality specific connectome harmonics, averaged across subjects for each modality, and concatenated across modalities, yielding the multi-modal cohort-average connectome harmonics, $\bar{\psi}_k$ $k \in \{0, ..., N_{\mathrm{harm}} N_{\mathrm{modalities}}\}$. Here, $N_{\mathrm{harm}} = 99$ and $N_{\mathrm{modalities}} = 3$. Note that the multi-layer formulation enforces similarity in subject-specific modality specific connectome harmonics for subjects scanned at nearby timepoints. An initial multimodal consensus parcellation is computed via HAC [10] on the multi-modal cohort-average connectome harmonics. The number of clusters outputted by HAC is determined by identifying the most costly cluster-merging step (Sect. 2.4) [11]. Next, the modality and subject-specific parcel-mean spectral coordinates are computed by averaging spectral coordinates in each parcel defined by the multi-modal consensus parcels and are used to initialize the k-means algorithm for parcellation based on the individual modalities using their eigenspectra. To achieve agreement across modalities, the modality-specific spectral coordinates are then smoothed by convex combination with the spectral centroid of the parcel. These smoothed modality-specific spectral coordinates are then used to produce a new multimodal parcellation for each subject, and re-averaged and concatenated to produce a new group-modality-consensus parcellation. This process is iterated until the multimodal consensus parcellation stabilizes. Fig. 1 graphically illustrates the pipeline. The goal of this iterative refinement of the multi-modal consensus parcellation is to achieve balance between the contribution from each modality to the overall multi-modal parcellation. Further, the strategy of concatenation of the modality-specific group-average connectome harmonics allows for variable number of subjects in each window with each scan type, i.e., for many subjects in this study, only one or two imaging modalities were present, however the group consensus multi-modal data still shaped the resultant individually-specific parcellation for those subjects.

2.4 Optimal Cluster Number Determination

In clustering or parcellation tasks, the number of clusters is often arbitrarily selected prior to clustering. A better solution is obtained by determining the optimal number of clusters based on the underlying connectivity data. For each time window, we compute the cluster merging cost (the ward metric) of each merging step of the HAC dendrogram of the cohort-average multi-modal spectral coordinates and the surface adjacency matrix as a connectivity constraint. This ensures that cluster mergers can only occur between adjacent surface vertices. The optimal number of clusters for a given window is then taken by identifying the knee point of the plot of merging cost versus cluster number.

3 Results

3.1 Homogeneity

In order to evaluate the degree to which our identified parcels exhibit similar connectivity profiles, we formulate a unique method of computing the parcel-wise homogeneity of connectivity based on the method presented in [12]. For each parcel, full principal component analysis of the size $N_{\text{parcelvert}} \times N_{\text{vert}}$ connectivity profile is conducted. The explained variance ratio of the first principal component is taken as the homogeneity score for that parcel. We apply this process to the structural connectivity matrices of each subject and average across subjects in each time window using the multi-modal, multi-layer iterative consensus parcellation outputted by our method as well as a parcellation constructed via spectral clustering of the concatenated multi-modal spectral coordinates of the full-group (0–700 days) average connectivity matrices. We display this result in Fig. 3.

3.2 Community Detection

A relevant application of connectivity-based parcellation is to investigate modularity and network-structure of parcels with respect to the underlying connectivity data. To demonstrate this, we first construct the parcel-wise weighted structural connectivity matrix for each subject using the individually back-propagated multi-modal parcel-

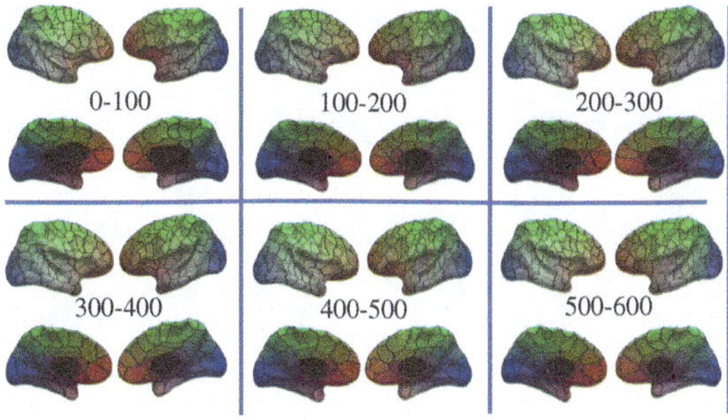

Fig. 2 Consensus parcellations for each time window (in days), color-mapped onto the group-average white matter surface from medial and lateral views. Regions are colored according to their relative distance to the frontal (red), visual (blue), and parietal (green) cortices. As age increases, regularity of cluster shapes and sizes decreases, particularly in the lateral/temporal lobes. Parcels in sensorimotor regions show convergence toward their shape and location in adult parcellations

lation labels. We then average the subject-specific parcel-wise weighted structural connectivity matrices—which is possible by virtue of the one-to-one correspondence between the subject-specific and group-consensus parcellations of each time window—and apply the Gen-Louvain community detection algorithm [13] to this average connectivity matrix. This algorithm identifies the community partition of parcels such that the within-community modularity is maximized. This result is displayed in Fig. 4. This represents just one possible application of our parcellation results.

4 Discussion

Figures 3 and 5 demonstrate that our iterative, multi-modal, multi-layer parcellation method achieves higher average parcel homogeneity than a simpler averaging approach, and that our individually back-propagated parcellations exhibit higher average structural connectivity homogeneity than both the other tested methods. Connectivity-based parcellations should identify regions with homogeneous connectivity, and therefore this underscores an important advantage of our method. Our multi-layer approach allows for one-to-one correspondence between parcels across all subjects, while allowing for variation in parcel boundaries across subjects, resulting in individually tailored, more highly homogeneous parcellations than simpler approaches.

Our temporally-weighted multi-layer multi-modal approach allows for arbitrarily many imaging modalities as long as a vertex-wise connectivity matrix can be constructed from each modality. Further, for a given temporal window for which a group multi-modal parcellation is carried out, it need not be the case that every subject in that window possesses data for each imaging modality in order for every subject to

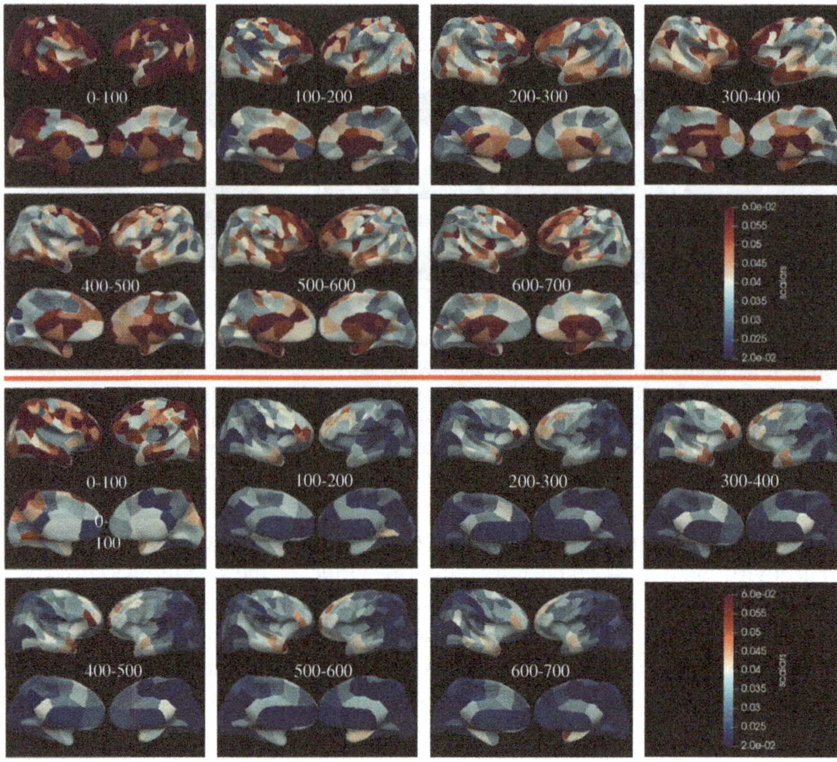

Fig. 3 Time-window average homogeneity of structural connectivity for **Top**: Multi-modal, multi-layer, iterative parcellation method and **Bottom**: Full-cohort (0–700 days) subject-average multi-modal parcellation

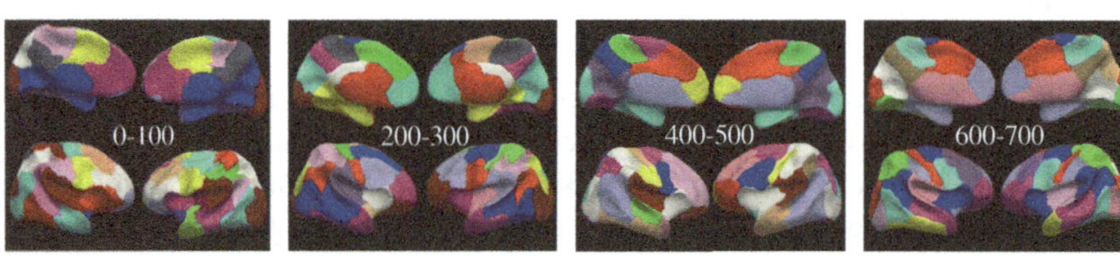

Fig. 4 Large-scale parcel modules identified using the Gen-Louvain community detection algorithm for four separate time windows (listed in days)

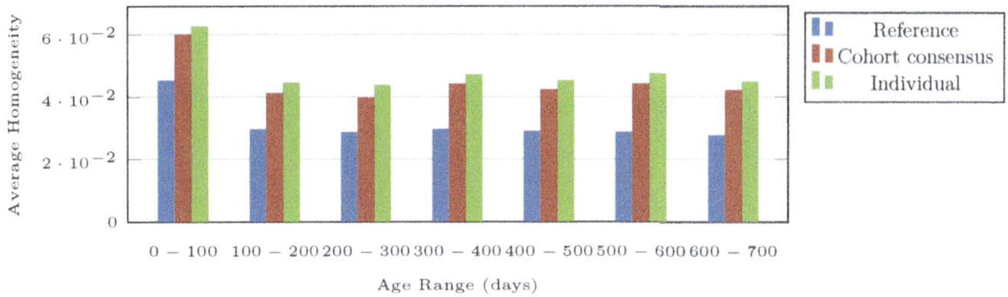

Fig. 5 Window and parcel-average homogeneity for 2-year average 'reference' parcellation from Fig. 3 (bottom), cohort consensus multi-modal iterative parcellation, and individually back-propagated multi-modal iterative parcellation

benefit from the group's multi-modal data. This is a simple solution to the common 'missing data problem' in longitudinal staggered cohort study designs.

All homogeneity results show a significant decrease in homogeneity between 0 and 100 days and 100–200 days, with relatively constant homogeneity between 100–700 days. This is aligned well with expectation, as we do not expect a high degree of regional specialization to occur until after at least 100 days. One possible explanation for the relatively constant average homogeneity after 100 days is that the specialization and segregation of function that occurs in this period takes place on a very small scale and does not involve major changes in structural connectivity.

Figure 3 also shows that the posterior temporal and parietal lobes are among regions that undergo the largest decrease in homogeneity. This is aligned with our expectation, as we know that many functions involving these regions develop during the first two years of life [14]. Interestingly, the primary motor, auditory, and visual cortices do not demonstrate higher average homogeneity than other high-level cognitive regions. This suggests that the parcel size in these regions may be disproportionately large in comparison to other regions, or that the structural connectivity estimates in these regions are more noisy.

Figure 2 displays the group-consensus results of our novel parcellation routine for each of seven 100 day windows in the first two years of development. These parcellations show large-scale qualitative similarity to one another, but develop in several important ways. As age increases, regularity of cluster size and shape decreases, reflecting regional specialization of cortical function. In particular, the lateral and temporal lobes demonstrate uniform clusters for 0–100 days, small clusters for 100–600 days, and eventual stabilization for 600–700 days. Several parcels also show distinct evolution toward their shape and position in adult parcellations such as the parcels in the sensorimotor cortices around the central sulcus, which become increasingly elongated and confined to specific strips.

Figure 4 demonstrates a convergence toward adult resting-state networks as time increases [15]. This is of particular note as the community detection was carried out using only structural connectivity information on the reduced parcel-wise graph using the multi-modal consensus parcellations. This result demonstrates that our parcellation defines a physiologically meaningful division of the cortex for network-level analysis of structural connectivity.

5 Conclusion

Our parcellation results confirm expected trends for cortical organization in early development. We demonstrate that our multi-modal parcellation routine can be used to produce biologically relevant parcellations. Further, our routine presents a flexible solution to the common 'missing data problem' in longitudinal neuroimaging studies, allowing for individualized results while maintaining group-correspondence of parcel labels, and successfully synthesizes data from multiple imaging modalities.

Acknowledgements This work was supported in part by NIH grants (NS093842, EB008374, EB022880, MH116225, MH117943, MH104324, and MH110274) and the efforts of the UNC/UMN Baby Connectome Project Consortium.

References

1. Howell, B.R., Styner, M.A., Gao, W., Yap, P.T., Wang, L., Baluyot, K., Yacoub, E., Chen, G., Potts, T., Salzwedel, A., Li, G., Gilmore, J.H., Piven, J., Smith, J.K., Shen, D., Ugurbil, K., Zhu, H., Lin, W., Elison, J.T.: The UNC/UMN Baby Connectome Project. BCP), An overview of the study design and protocol development (2019)
2. Taylor IV, H.P., Wu, Z., Wu, Y., Shen, D., Zhang, H., Yap, P.T.: Automated Parcellation of the Cortex Using Structural Connectome Harmonics. In: Lecture Notes in Computer Science (including subseries Lecture Notes in Artificial Intelligence and Lecture Notes in Bioinformatics) (2019)
3. Wu, Y., Hong, Y., Feng, Y., Shen, D., Yap, P.T.: Mitigating gyral bias in cortical tractography via asymmetric fiber orientation distributions. Med. Image Anal. (2020)
4. Li, G., Nie, J., Wang, L., Shi, F., Lin, W., Gilmore, J.H., Shen, D.: Mapping region-specific longitudinal cortical surface expansion from birth to 2 years of age. Cerebral Cortex (2013)
5. Li, G., Nie, J., Wang, L., Shi, F., Lyall, A.E., Lin, W., Gilmore, J.H., Shen, D.: Mapping longitudinal hemispheric structural asymmetries of the human cerebral cortex from birth to 2 years of age. Cerebral Cortex (2014)
6. Yeo, B.T., Sabuncu, M.R., Vercauteren, T., Ayache, N., Fischl, B., Golland, P.: Spherical demons: Fast diffeomorphic landmark-free surface registration. IEEE Trans. Med. Imaging (2010)
7. Marcus, D.S., Harwell, J., Olsen, T., Hodge, M., Glasser, M.F., Prior, F., Jenkinson, M., Laumann, T., Curtiss, S.W., Van Essen, D.C.: Informatics and data mining tools and strategies for the human connectome project. Frontiers Neuroinf. (2011)
8. Glasser, M.F., van Essen, D.C.: Mapping human cortical areas in vivo based on myelin content as revealed by T1- and T2-weighted MRI. J. Neurosci (2011)
9. Zhang, H., Shen, D., Lin, W.: Resting-state functional MRI studies on infant brains: a decade of gap-filling efforts (2019)
10. Murtagh, F., Contreras, P.: Algorithms for hierarchical clustering: An overview. Data Mining Knowl. Discov. Wiley Interdisciplinary Reviews (2012)

11. Salvador, S., Chan, P.: Determining the number of clusters/segments in hierarchical clustering/segmentation algorithms. In: Proceedings—International Conference on Tools with Artificial Intelligence, ICTAI (2004)
12. Gordon, E.M., Laumann, T.O., Adeyemo, B., Huckins, J.F., Kelley, W.M., Petersen, S.E.: Generation and evaluation of a cortical area parcellation from resting-state correlations. Cerebral Cortex (2016)
13. De Meo, P., Ferrara, E., Fiumara, G., Provetti, A.: Generalized Louvain method for community detection in large networks. In: International Conference on Intelligent Systems Design and Applications, ISDA (2011)
14. Wen, X., Zhang, H., Li, G., Liu, M., Yin, W., Lin, W., Zhang, J., Shen, D.: First-year development of modules and hubs in infant brain functional networks. NeuroImage (2019)
15. Buckner, R.L., Krienen, F.M., Castellanos, A., Diaz, J.C., Thomas Yeo, B.T.: The organization of the human cerebellum estimated by intrinsic functional connectivity. J. Neurophysol. (2011)

Automatic Segmentation of Dentate Nuclei for Microstructure Assessment: Example of Application to Temporal Lobe Epilepsy Patients

Marta Gaviraghi, Giovanni Savini, Gloria Castellazzi, Fulvia Palesi, Nicolò Rolandi, Simone Sacco, Anna Pichiecchio, Valeria Mariani, Elena Tartara, Laura Tassi, Paolo Vitali, Egidio D'Angelo, and Claudia A. M. Gandini Wheeler-Kingshott

Abstract Dentate nuclei (DNs) segmentation is helpful for assessing their potential involvement in neurological diseases. Once DNs have been segmented, it becomes possible to investigate whether DNs are microstructurally affected, through analysis of quantitative MRI parameters, such as those derived from diffusion weighted imaging (DWI). This study developed a fully automated segmentation method using the non-DWI (b0) images from a DWI dataset to obtain DN masks inherently registered with parameter maps. Three different automatic methods were applied to healthy subjects: registration to SUIT (a spatially unbiased atlas template of the cerebellum and brainstem), OPAL (Optimized Patch Match for Label fusion) and CNN (Convolutional Neural Network). DNs manual segmentation was considered the gold

M. Gaviraghi (✉) · G. Castellazzi
Department of Electrical, Computer and Biomedical Engineering, University of Pavia, Pavia, Italy
e-mail: marta.gaviraghi01@universitadipavia.it

G. Castellazzi
e-mail: gloria.castellazzi@unipv.it

G. Savini · F. Palesi · P. Vitali
Neuroradiology Unit, IRCCS Mondino Foundation, Pavia, Italy
e-mail: fulvia.palesi@unipv.it

P. Vitali
e-mail: paolo.vitali@grupposandonato.it

G. Castellazzi · C. A. M. G. Wheeler-Kingshott
Department of Neuroinflammation, Faculty of Brain Sciences, Queen Square MS Centre, UCL Queen Square Institute of Neurology, University College London, London, United Kingdom
e-mail: c.wheeler-kingshott@ucl.ac.uk

G. Castellazzi · E. D'Angelo · C. A. M. G. Wheeler-Kingshott
Brain Connectivity Center (BCC), IRCCS Mondino Foundation, Pavia, Italy
e-mail: egidiougo.dangelo@unipv.it

F. Palesi · N. Rolandi · A. Pichiecchio · E. D'Angelo · C. A. M. G. Wheeler-Kingshott
Department of Brain and Behavioral Sciences, University of Pavia, Pavia, Italy
e-mail: anna.pichiecchio@mondino.it

N. Gyori et al. (eds.), *Computational Diffusion MRI*, Mathematics and Visualization,
https://doi.org/10.1007/978-3-030-73018-5_21

standard. Results show that SUIT results have a Dice Similarity Coefficient (DSC) of 0.4907±0.0793 between automatic and gold standard masks. Comparing OPAL (DSC = 0.7624±0.1786) and CNN (DSC = 0.8658±0.0255), showed that a better performance was obtained with CNN. OPAL and CNN were optimised on high spatial resolution data from the Human Connectome Project. The three methods were then used to segment DNs of subjects with Temporal Lobe Epilepsy (TLE) from a 3T MRI research study with DWI data acquired with a coarser resolution. In TLE, SUIT performed similarly, with a DSC = 0.4145±0.1023. OPAL performed worse than using HCP data with a DSC of 0.4522±0.1178. CNN was able to extract the DNs without need for retraining and with a DSC = 0.7368±0.0799. Statistical comparison of quantitative parameters from DWI analysis, as well as volumes, revealed altered and lateralised changes in TLE patients compared to healthy controls. The proposed CNN is a viable option for accurate extraction of DNs from b0 images of DWI data with different resolutions and acquired at different sites.

1 Introduction

Cerebellar nuclei (CNs) have a fundamental role in the central nervous system; they are the main output channels of the cerebellum towards the supratentorial brain and the spinal cord [1]. The dentate nuclei (DNs) are the CNs with the largest volume (measuring about 2 cm in the anterior-posterior direction and 1 cm in transverse plane and coronal plane) [2]. Histologically, the DNs have the shape of an irregularly pleated grey foil, very thin and with a longitudinal section appearing as a curved line

S. Sacco
Department of Neurology, UCSF Weill Institute for Neurosciences, University of California, San Francisco, CA, USA
e-mail: saccosimone88@gmail.com

Department of Clinical Surgical Diagnostic and Pediatric Sciences, University of Pavia, Pavia, Italy

A. Pichiecchio
Neuroradiology Unit, IRCCS Mondino Foundation, Pavia, Italy

V. Mariani · L. Tassi
C. Munari Centre for Epilepsy Surgery, Grande Ospedale Metropolitano Niguarda, Milan, Italy
e-mail: valeria.mariani@osepedaleniguarda.it

L. Tassi
e-mail: laura.tassi@osepedaleniguarda.it

V. Mariani
Institute of Neuroscience, Italian National Research Council (CNR), Parma, Italy

E. Tartara
Epilepsy Centre, IRCCS Mondino Foundation, Pavia, Italy
e-mail: elena.tartara@mondino.it

that contains white matter inside. The DNs are known mainly for their involvement with the sensorimotor system, although recently studies are suggesting a role in procedural memory, emotional and cognitive functions [3].

Several studies have shown that DN morphological properties can be altered in different neurological pathologies [4, 5]. In human, there are general reports of cerebellar atrophy in Temporal Lobe Epilepsy (TLE) patients [6], while animal models have shown a direct involvement of the DNs: in particular, experimental studies have shown that electrical stimulation of the DNs shortened and inhibited the onset of seizures [7–9].

T1-weithed (T1-w) images are structural scans generally used for segmenting brain regions. The DNs, unfortunately, do not show contrast on T1-w scans, while they are visible on T2-weighted (T2-w) images [10]. Currently, DNs manual segmentation is still considered the gold standard [11–13], but it is time-consuming and suffers from inter- and intra-rater variability. A fully automatic segmentation is therefore desirable.

A recently published pilot study [14] proposes a fully automatic method using DWI, requiring time-consuming information from tractography. Another piece of work [15], proposes a deep learning approach using as input multiple data including T1-w, T2-w images and Fractional Anisotropy (FA) maps. Using quantitative maps such as FA, though, introduces a circular bias and should be avoided.

In reference [16] the authors propose a fusion technique based on explicit shape modelling, starting from high-resolution 7T quantitative susceptibility mapping (QSM) of the cerebellum. In a recent piece of work [17] a multi-atlas method was developed to segment iron-rich deep grey matter nuclei (including the DNs). However, QSM is not standard acquired in clinical settings.

The purpose of this study is to segment the DNs for microstructure quantification of metrics acquired using the EPI readout as for DWI data. Segmentation masks of the DNs can be used to extract average values of quantitative metrics to be compared between populations of subjects, to assess correlations with clinical scores or to monitor disease progression over time. Among the most interesting metrics there are parameters derived from clinically feasible Diffusion Tensor Imaging (DTI) or from advanced methods including Diffusion Kurtosis Imaging (DKI) [18] and Neurite Density and Orientation Dispersion Imaging (NODDI) [19]. Given the typical resolution of DWI scans at 3T ($2 \times 2 \times 2$ mm^3) and the low number of voxels included in segmentation masks of small structures such as the DNs, it is highly desirable to reduce the data manipulation due to post-processing steps (e.g. registration) and to have region segmented directly in DWI-space. It is essential that any automatic method is applicable with good performance to images of different quality and acquired with different scanners.

Here we developed a method to automatically segment DNs from non-diffusion weighted (b0) images, acquired as part of DWI scans. We specifically investigated three different approaches using high-resolution data derived from the Human Connectome Project [20]: (1) atlas registration; (2) patch-matching; (3) a deep learning network-based method. Masks obtained with each of these methods were compared to the gold standard manual segmentation of DNs. The methods were tested in a second dataset of subjects from a TLE study. The resulting best approach was employed to compare DN volumes and average DWI metrics between patients and healthy controls (HC), in view of future clinical studies.

2 Methods

2.1 Subjects

HCP dataset Pre-processed images of 100 healthy subjects scanned for the Human Connectome Project (HCP) were downloaded [20]. 24 subjects were discarded for cerebellar artefacts. The 76 remaining subjects (43 Females, 29.41±3.62 years) were used to develop the automatic DNs segmentation.

TLE dataset A second dataset of 84 subjects, recruited for an Italian multi-centre research project on TLE, were selected as clinical test data: 34 HC (16 Females, 31.97±7.73 years), 21 patients with left TLE (LTLE; 13 Females, 33.294±11.68 years) and 29 with right TLE (RTLE; 17 Females, 37.97±9.86 years).

2.2 MRI Protocol

HCP dataset MR images were acquired on a Siemens 3T Connectome Skyra scanner (diffusion: Gmax = 100 mT/m), a 32-channel receive head coil and standard shim coils. DWI data had minimal pre-processing, co-registered with T1-w data at a resolution of $1.25 \times 1.25 \times 1.25$ mm^3 and matrix size of $145 \times 174 \times 145$ [21]. Data included 18 volumes with b = 0 s/mm^2.

TLE dataset MR images were acquired using a Siemens 3T MAGNETOM Skyra scanner with standard gradients and a 32-channel receive coil.

DWI: spin-echo EPI, 90 volumes with b-value = 1000/2000 s/mm^2 (45 DW gradient directions per b-value) and 9 volumes with b = 0 s/mm^2; spatial resolution = $2.24 \times 2.24 \times 2$ mm^3 and matrix size of $100 \times 100 \times 66$.

T1-w: high-resolution 3DT1-w (T1w) volume with spatial resolution = $1 \times 1 \times 1$ mm^3.

2.3 DWI Processing

For each subject, the mean of the b0 volumes was calculated ($\overline{b0}$). For TLE subjects, quantitative metrics were extracted using DESIGNER (https://github.com/NYU-DiffusionMRI/DESIGNER): Axial Diffusivity (AD), Radial Diffusivity (RD), Mean

Diffusivity (MD) and FA from DTI fitting [22] and Axial Kurtosis (AK), Radial Kurtosis (RK) and Mean Kurtosis (MK) from DKI fitting [18].

2.4 DNs Segmentation

$\overline{b0}$ images of HCP subjects were used for developing the DN segmentation method. Manual segmentation was used as ground truth (GT). Automatic DN masks, from three different automatic segmentation methods, were compared to the GT masks and applied to the TLE dataset. Performance against GT was assessed by calculating three scores: Dice Similarity Coefficient (DSC), True Positive Rate (TPR) and Positive Predictive Value (PPV) (see paragraph 2.6).

Ground Truth (GT)—manual segmentation $\overline{b0}$ images of HCP subjects were manually segmented by rater 1 using Mango (http://ric.uthscsa.edu/mango/mango.html). In order to assess the automatic methods' performance against inter-raters variability, a second rater, rater 2, segmented the same data using Jim (http://www.xinapse.com/j-im-8-software). DSC scores were calculated first between manual segmentation masks from raters 1 and 2 for each HCP subject and then averaged over all 76 HCP subjects. 6 subjects were also segmented twice rater 1 on different days to calculate the intra-rater variability. For the TLE dataset, rater 1 manually segmented the $\overline{b0}$ of 18 subjects (6 for each group) to have a GT (GT_{TLE}) for this independent dataset.

Atlas-based method: SUIT The toolbox SUIT (A spatially unbiased atlas template of the cerebellum and brainstem) is an open source extension of SPM (Statistical Parametric Mapping, https://www.fil.ion.ucl.ac.uk/spm/) available for Matlab (The MathWorks, Inc., Natick, MA, United States of America).

SUIT [10] is an atlas-based method for cerebellar segmentation that performs a non-linear registration between a template (standard space) and the image to segment. The resulting transformation is then applied to an atlas defined in standard space and its labels are warped into the subject space. One of the labels is for the DNs. SUIT requires registering T1w images of each subject to the template; the inverse transformation is then used to warp DN labels from standard-space to subject-space. As the T1w images of the HCP dataset are already co-registered with the respective DWI, the DN segmentations obtained with SUIT are already in DWI space.

Pre-processing (OPAL and CNN) In order to segment DNs with OPAL and CNN we applied two pre-processing steps: (1) Intensity normalization: mean signal intensity and standard deviation were calculated for each subject's $\overline{b0}$ volume, considering only brain voxels, to obtain zero mean and standard deviation equal to 1 for all subjects; (2) Cropping: to reduce the computational time, images were cropped around the cerebellum reducing axial slices to 86x71 voxels.

Patch-matching method: OPAL OPAL (Optimized Patch Match for Label fusion) [23] joins information from different templates to obtain the desired segmentation. OPAL is an evolution of the Patch Match algorithm [24], implemented in C++ (https://github.com/KCL-BMEIS/NiftySeg/).

We built up a database of 46 subjects providing: $\overline{b0}$ images, the corresponding masks of the cerebellum and the DN GTs. This database was intended as a collection of reference templates. The DNs segmentation of each new subject was performed by dividing images into patches and comparing each patch with those from the reference templates, looking for the most locally similar match. The output is a probabilistic map of the DNs. We divided the remaining 30 subjects into validation and testing sets. We used the validation set to select the probability threshold (0.1, 0.2, 0.3, 0.4, 0.5) for binarizing the DN masks, where a lower threshold corresponds to larger DN masks. For each threshold and for each validation subject we calculated the DSC between the DN masks and the GTs. We selected the threshold that maximised the mean DSC and we assessed the performance of OPAL on the remaining 15 test subjects for an unbiased performance estimate.

Deep-learning method: CNN A CNN (Convolutional Neural Network) was implemented with Matlab19a using the Deep Learning Toolbox.

CNN architecture—The architecture used here was inspired to the one used for segmenting the spinal cord grey matter [25]. This architecture was based on dilated convolutions and on removal of pooling layers, responsible for information loss. This type of convolution expanded receptive fields without increasing the number of parameters [26]. The network implemented required as input a two-dimensional (2D) image, oriented in the axial plane. The architecture is shown in Fig. 1. All convolutional layers have a zero-padding of type "same" [26]. Therefore, the dimensions of

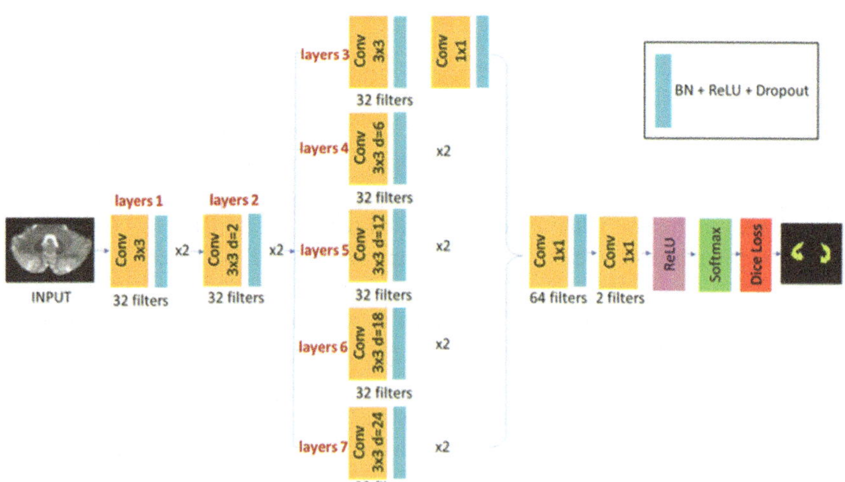

Fig. 1 Scheme of the CNN architecture adopted here

Table 1 Range of parameters used for the transformations of data augmentation step. For each slice, with 0.5 probability, a random number within this range was assigned to each transformation. For elastic deformation: α represents the scale factor and σ the standard deviation of the Gaussian filter

Transformation	Parameter range
Rotation	$[-4.6°, 4.6°]$
Shift	$[-3, 3]$ in x and y direction
Scaling	[0.98, 1.02] with bicubic interpolation
Elastic deformation	$\alpha = 4$ and $\sigma = 30$

each layer's output do not differ from those of the layer's input. For each layer the neurons are activated by the ReLU (Rectifier Linear Unit) function [27].

The architecture of the CNN is the following: Input layer (INPUT) treating each voxel of input images as a neuron; two layers of standard convolution (layers 1); two layers of dilated convolution with dilatation factor d = 2 (layers 2); five branches in parallel, each branch with two convolution layers. In the first branch there is a standard convolution for the first layer the kernel dimension is 3×3 while for the second it is 1×1 (layers 3); the remaining four branches have dilated convolution respectively with d = 6, 12, 18, 24 (layers 4, 5, 6, 7).

Each output of these parallel branches is concatenated in the third dimension and followed to: a convolution layer that uses 64 filters of dimensions 1×1; a convolution layer that uses 2 filters of dimensions 1×1; a Softmax layer [28] that represents the activation function for classification; a Loss layer.

The convolutional layers have 32 filters with dimension 3×3 except for the second layer of layers 3, which is 1×1, and the last two layers. Except for the last 1×1 convolution, each convolution layer is followed by batch normalization [28] and dropout [26]. Due to the imbalance between the class of belonging to the DN and the non belonging class (i.e. background), we decided to use the Dice Loss as loss function, based on the DSC and robust to class imbalance [29]. We used the Adam optimizer [30] with a small learning rate of $\eta = 0.001$ for setting the weights of the CNN.

Training—To reduce overfitting, data augmentation was applied. Four different transformations were considered: rotation, translation, scaling and elastic deformation. These transformations were applied to input ($\overline{b0}$)—desired output (GT) pairs. Data augmentation was applied independently on each slice with a probability of 0.5 for each transformation. The parameters used are reported in Table 1. The original $\overline{b0}$ images plus those from data augmentation and the corresponding GT masks were provided as input to the CNN for training. To speed up training, however, only slices containing the DN (on average 8 per subject) were automatically included as selected from the GT masks. The hyperparameters that must be chosen a priori before training were the batch size, the dropout and the number of epochs. For tuning these hyperparameters we tried a number of combinations (45 in total), using batch size (8, 16, 24, 32, 64), dropout (0.2, 0.3, 0.4) and epochs (30, 50, 100).

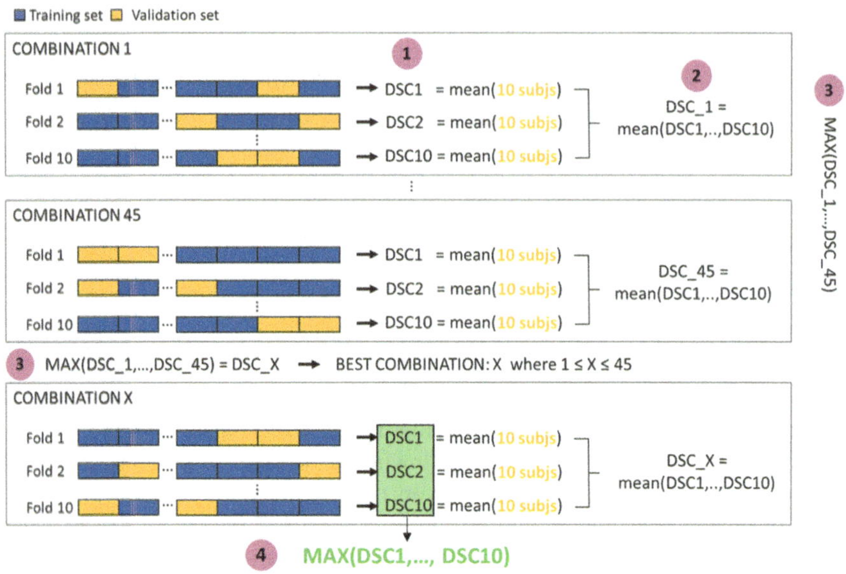

Fig. 2 Steps followed for hyperparameters optimization and CNN training

For each combination of hyperparameters, a Monte Carlo 10-folds cross validation was performed: firstly, we randomly extracted 6 of the 76 subjects as test set. Then, the remaining 70 subjects were randomly split into 60 subjects for training and 10 subjects for validation; this step was repeated for each of the 10 folds. The Monte Carlo 10-folds cross validation randomly selects subjects for the training and the validation set, therefore it is possible that a subject is never included or can be used more than once in the validation set. Steps used for CNN training are shown in Fig. 2: (1) for each fold of each combination of hyperparameters we calculated the DSC for the subjects included in the validation set (10 subjects); (2) we calculated the mean DSC for each hyperparameters combination by averaging the DSCs of the 10 folds; (3) we chose the combination of hyperparameters that maximized the average DSC; (4) among the 10 CNN that were trained with the best hyperparameters combination, we chose the one with the maximum DSC. Set the hyperparameters, we used the 6 test subjects for an unbiased estimate of the CNN performance. Subsequently, to check that the network did not overfit on the GTs of rater 1 used for training the scores for the 6 test subjects were calculated comparing the segmentations obtained with CNN and the masks from rater 2.

2.5 Post Processing for OPAL and CNN

Both OPAL and CNN labeling identified a number of false positive (FP) voxels as belonging to the DNs located in different brain regions, sometimes very distant from the DNs themselves. In order to remove these FP voxels, an automated post processing step was implemented: the DN masks obtained with SUIT were dilated twice and used to mask the DN masks generated by OPAL and CNN.

2.6 Quantitative Evaluation

For each method, performance was tested by comparing automatic DNs against GT masks using three scores [31].

DSC i.e. the overlap between two binary masks:

$$DSC = \frac{2\,TP}{2\,TP + FP + FN} \tag{1}$$

where TP indicates True Positive and FN False Negative. DSC ranges [0–1].

Sensitivity or TPR:

$$TPR = 100\,x\,\frac{TP}{TP + FN} \tag{2}$$

TPR ranges [0–100] with low TPR indicating a bias towards under-segmentation.

Precision or PPV:

$$PPV = 100\,x\,\frac{TP}{TP + FP} \tag{3}$$

PPV ranges [0–100] with low PPV indicating a bias towards over-segmentation.

Specificity or True Negative Rate (TNR) was not considered because the two classes (DN and background) are unbalanced, causing high and non-informative TNR values.

2.7 Comparison of Automatic Methods

We calculated DSC, TPR and PPV for each automated method. For OPAL and CNN we calculated these scores, on the validation and test sets, before and after post processing. Since SUIT is an atlas-based method we calculated these scores on the whole dataset, while for OPAL we exclueded the 46 subjects used as template. Regarding CNN, the scores were calculated for the validation (10 subjects) and test (6 subjects) sets for each of the 10 folds corresponding to the optimal set of hyperparameters. For each method we calculated the group average of these scores.

For the CNN we calculated two average values: the first one by averaging between the 10 folds corresponding to the best combination of hyperparameters, while the second one by averaging only results obtained with the network chosen as the final CNN (the one with the best perfomance) among the 10 networks.

2.8 Clinical Application to TLE Data

TLE data pre-processing and DNs segmentation The spatial resolution of the TLE $\overline{b0}$ images was lower than that of the HCP dataset, so TLE $\overline{b0}$ images were resampled to match the HCP resolution using FSL FLIRT (FMRIB's Linear Image Registration Tool) before applying each segmentation method. In order to remove the FPs we exploited the segmentation masks resulting from SUIT, which were moved from T1w space to b0 space using a rigid registration computed with SPM. Resulting DN masks were resampled to their original spatial resolution for quantitative analysis of parameter maps by applying the inverse of the roto-translation matrix. (GT_{TLE}) segmentations were used to assess performance of the three methods. We selected the best automatic DNs segmentation method based on the performance on both datasets (HCP and TLE). The best method was then applied to all TLE subjects to extract quantitative DNs parameters from DWI.

DN structural and microstructural characteristics in TLE patients For each DN (right and left DN independently), thew following quantitative measures were extracted: (1) volume; (2) average value of DTI metrics (AD, RD, MD and FA); (3) average value of DKI metrics (AK, RK and MK). Lateralization of volumes and metrics values was investigated using an Asymmetry Index (AI), with range [−2; 2] where 0 indicates perfect symmetry [32]:

$$AI = \frac{mean(DN\ left) - mean(DN\ rigth)}{\frac{mean(DN\ left)+mean(DN\ rigth)}{2}} \quad (4)$$

We considered a total of 24 measures for each subjects. Statistically significant differences between HC, RTLE and LTLE were investigated using SPSS (IBM, Armonk, NY, United States of America) as exploratory work.

Age and gender were compared and included in the statistical comparison. A general linear model (GLM) univariate analysis was implemented using as covariates those variables not homogeneous between groups. 24 GLM univariate comparisons, with =5%, were performed to explore which variables could significantly differentiate the three groups. Subsequently GLM univariate analysis was repeated for each metric in pairwise group comparisons.

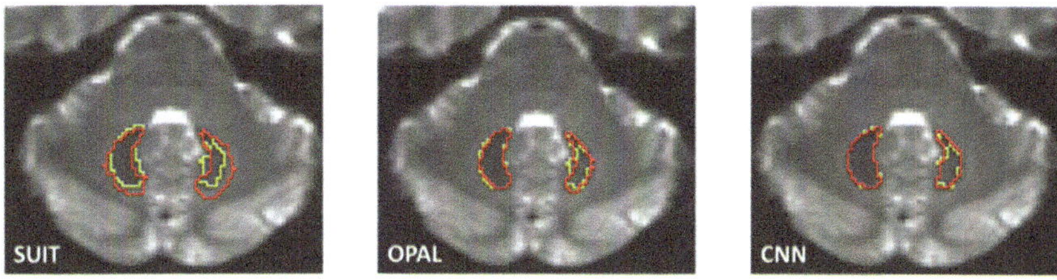

Fig. 3 Segmentation masks obtained with the three methods for a randomly selected subject (SUIT, OPAL and CNN). Each image shows the overlap of the segmentation obtained with the respective automated method (red) overlaid with the GT (yellow)

3 Results

The inter-rater variability of the manual segmentations resulted in a DSC = 0.8066±0.0575. Intra-rater variability produced a DSC = 0.7927±0.0369. In Fig. 3 DN masks of a randomly selected subject are displayed. OPAL probability threshold was set to 0.4. The Monte Carlo 10-folds cross validation of the CNN provided the best results with hyperparameters: batch size = 24, dropout = 0.2 and number of epochs = 100.

3.1 Comparison of the Three Automatic Methods

Table 2 reports DSC, TPR and PPV scores (mean±standard deviation) for the three methods.The best performance was achieved by CNN (DSC = 0.8658±0.0255) followed by OPAL (DSC = 0.7624±0.1786). SUIT performed worst, with the lowest scores (DSC = 0.4907±0.0793).

The scores between the segmentations obtained with CNN and rater 2 were: DSC = 0.8208±0.0371, TPR = 74.3759±5.6519, PPV = 91.7158±4.9100.

3.2 Application to TLE Dataset

Table 3 reports DSC, TPR and PPV scores between (GT_{TLE}) and the segmentation obtained with each automatic method. For OPAL it was necessary to reset the probability threshold to 0 as 0.4 (set for the HCP data) eliminated TP. Overall scores were: DSC = 0.1322±0.1512, TPR = 7.7931±9.2878 and PPV = 55.2716±50.8794. CNN outperformed the other methods with a DSC = 0.7368±0.0799.

Statistical comparisons showed that age was not homogeneous between the three groups of the TLE study (p-value = 0.017) while gender was matched (p-value = 0.491). Therefore, we included age as a GLM covariate in the DWI metric analysis.

Table 2 SUIT, OPAL and CNN performances. For CNN, two sets of scores are reported: (1) average scores from the 10 networks with the chosen hyperparameters; (2) metrics from results obtained with the CNN network chosen as the best performer. For DSC, in bracket we reported the values before the post processing step to remove false positives. For each score the best value is indicated in boldface

	DSC	TPR	PPV
SUIT	0.4907±0.0793	86.3444±6.6154	34.9475±7.6264
OPAL—validation set	0.7434 ± 0.2168 (0.7427 ± 0.2164)	73.4617 ± 24.0014	76.9896± 22.3599
OPAL—test set	0.7624 ± 0.1786 (0.7602 ± 0.1780)	76.3791 ± 23.1454	83.2686 ± 9.3198
CNN—validation set (10 networks)	0.8519±0.0144 (0.7607±0.0311)	**86.7444±2.7735**	84.5275±1.0535
CNN—validation set (1 network)	0.8366±0.0579 (0.7916±0.0602)	83.8757±9.9464	84.4935±8.0567
CNN—test set (10 networks)	0.8650±0.0067 (0.7943±0.0323)	84.6590±1.2522	88.6746±0.8117
CNN—test set (1 network)	**0.8658±0.0255** (0.8440±0.0270)	84.5150±4.0032	**88.9238±3.8065**

Table 3 Comparison of SUIT, OPAL and CNN against GT on 18 TLE subjects

	DSC	TPR	PPV
SUIT	0.4145±0.1023	84.3647±8.4051	27.9597±8.6905
OPAL	0.4522±0.1178	84.3277±16.0649	28.6451±12.1937
CNN	**0.7368±0.0799**	**88.6787±4.5745**	**65.7410±10.6841**

We found significant differences between the three groups: AD of the left DN (p-value = 0.024), MD of the left DN (p-value = 0.039) and volume of the right DN (p-value = 0.014). The first row of Fig. 4 shows boxplots of these metrics for each group. Pairwise comparisons between two of the three groups showed that: AD of the left DN is significantly different between LTLE and RTLE patients (p-value = 0.004), MD of the left DN is significantly different between LTLE and RTLE patients (p-value = 0.016), the volume of the right DN is significantly different between HC and LTLE patients (p-value = 0.049) and between HC and RTLE patients (p-value = 0.010). Moreover from pairwise comparisons other metrics resulted significantly different: volume of the left DN between HC and RTLE patients (p-value = 0.027) and RD of the left DN between HC and LTLE patients (p-value = 0.044) (second row of Fig. 4).

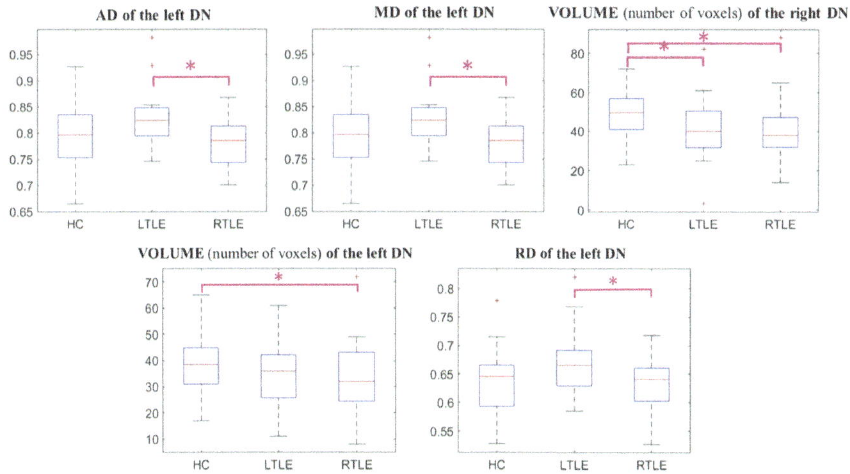

Fig. 4 In the first row: boxplots of the measures that resulted statistically different ($p<0.05$) between the three groups: AD of the left DN, MD of the left DN and volume of right DN. Significant pairwise comparisons are highlighted with asterisks. In the second row: boxplots of the measures that resulted statistically different ($p<0.05$) from pairwise comparisons (highlighted with an asterisk): volume of the left DN and RD of the left DN

4 Discussion

In this work we proposed an automatic DNs segmentation method that uses $\overline{b0}$ images from a DWI dataset. Specifically, analysis of DSC scores highlighted performances comparable with inter- and intra-raters segmentation (DSC>0.7). The use of $\overline{b0}$ images, inherently co-registered with DWI data, instead of high resolution T1w structural scans, allows the user to apply the masks directly to microstructural parameter maps obtained for clinical research studies.

On HCP data, segmentation masks obtained with OPAL and CNN were more accurate than the over-segmented DNs obtained with SUIT. Furthermore, the scores average values were superior for segmentations using CNN compared to OPAL.

OPAL applied to TLE data had worse performance (even after changing the threshold). This indicates that OPAL, which here used a reference database constructed on HCP data, cannot segment images acquired on a different scanner and with a worse resolution. Possibly, to improve the performance of OPAL, one would need to build a more appropriate database of reference templates.

Therefore, the implemented CNN outperforms OPAL and can be considered the best automated segmentation method of DWI images among the ones tested here (the code for the CNN is publicly available at https://github.com/marta-gaviraghi/segmentDN).

One further major advantage of CNN over OPAL lies in its greater transferability across sites and users. Indeed, OPAL requires that the database of $\overline{b0}$s and associated GTs is available to segment the DNs of new subjects. Conversely, CNN needs a database of images and GTs only for training, but after the network has learnt the association between images and segmentations, the reference images are no longer needed. One could question also the dependency of the method on the geometrical acquisition parameters, but here we demonstrated that the method worked well (DSC>0.73) also on a completely different dataset, acquired on a standard clinical 3T scanner and with a much coarser voxel resolution.

The CNN was applied to the $\overline{b0}$ data of the TLE dataset to segment the DNs and study their microstructural properties. While understanding the DNs involvement in TLE requires a dedicated study comparing regions from the entire brain, it was very interesting to see that the DN masks obtained from the $\overline{b0}$ images could be easily applied to DTI and DKI metrics and be used for some very preliminary assessment. The statistical comparison showed that the right DN volume is reduced in both RTLE and LTLE with respect to HC. The volume reduction of the right DN in TLE patients could indicate atrophy of this cerebellar nucleus, but to understand the source of such alteration one should also consider what happens to the underlying microstructure and hence assess parameters from, for example, DTI or DKI fitting of the data as it was performed here. From our exploratory comparisons, AD and MD seem to be the most affected metrics, which might simply relate to a different proportion of white and grey matter structures captured by the masks in different groups. To disentangle the source of such changes, though, future studies should consider advanced microstructural models that probe more specific biophysical properties such as neuronal density, orientation dispersion and soma compartments [19]. These preliminary results support the hypothesis that DNs might be involved in TLE, consistently with previous studies in animal models of epilepsy [7–9]. The extent of such involvement must be explored further within a dedicated clinical study that correlates DN alterations with that of other brain regions, considering also clinical/anamnestic data such as comorbidities and treatment [33].

Methodologically, given the coarse resolution of DWI data, a potential limitation of using $\overline{b0}$ images is that it is not possible to extract the convoluted surface of the DNs and to specifically extract their grey matter. Current structural scans used for the segmentation of small regions (T1w scans) do not show contrast in the CN areas. If a detailed reconstruction of the DNs shape and size is considered a fundamental aspect for a specific study, a dedicated sequence with optimized contrast (e.g. based on T2 or T2* properties or QSM) and image resolution (e.g. to achieve sub-millimetre voxel size) should be considered, at the expense of longer acquisition times. For the purpose of our study, $\overline{b0}$ images served the purpose of achieving a significant improvement over the SUIT segmentation without resorting to additional MR sequences and longer acquisition time. Furthermore, the demonstrated translation of the CNN from the HCP to a clinical scanner DWI data is very encouraging and makes this CNN possibly viable for other applications that use EPI-readouts; future work could therefore investigate transferability of the proposed CNN to study functional MRI activations of the DNs in relation to their microstructure characteristics. Future work could explore other architectures (such as U-Net) in order to find the best one for this application. In order to remove FPs, morphological operations could be implemented as an alternative post-processing step to SUIT masking.

5 Conclusion

We proposed an automatic segmentation of the DNs using an automated method. The CNN implemented here can segment images with a spatial resolution and acquisition protocol different from the training set. By using the proposed CNN on a cohort of subjects affected by TLE we detected asymmetric microstructural changes within the DNs, which should be further investigated in dedicated studies. Future work could consider multimodal datasets including as input images with different MRI contrasts and an expanded GT database for training.

References

1. Sure, D.R., Culicchia, F.: Duus' Topical Diagnosis in Neurology. Thieme (2005)
2. Cattaneo, L.: Anatomia del sistema nervoso centrale e periferico dell'uomo. Monduzzi Editore (1989)
3. Habas, C.: Functional imaging of the deep cerebellar nuclei: A review. Cerebellum **9**, 22–28 (2010). https://doi.org/10.1007/s12311-009-0119-3
4. Solbach, K., et al.: Cerebellar pathology in Friedreich's ataxia: Atrophied dentate nuclei with normal iron content. NeuroImage Clin. **6**, 93–99 (2014). https://doi.org/10.1016/j.nicl.2014.08.018
5. Fukutani, Y., et al.: Cerebellar dentate nucleus in Alzheimer's disease with myoclonus. Dement. Geriatr. Cogn. Disord. **10**, 81–88 (1999). https://doi.org/10.1159/000017106
6. Hermann, B.P., et al.: Cerebellar atrophy in temporal lobe epilepsy. Epilepsy Behav. **7**, 279–287 (2005). https://doi.org/10.1016/j.yebeh.2005.05
7. Babb, T.L., et al.: Fastigiobulbar and dentatothalamic influences on hippocampal cobalt epilepsy in the cat. Electroencephalogr. Clin. Neurophysiol. **36**, 141–154 (1974). https://doi.org/10.1016/0013-4694(74)90151-5
8. Krook-Magnuson, E., et al.: Cerebellar directed optogenetic intervention inhibits spontaneous hippocampal seizures in a mouse model of temporal lobe epilepsy. eNeuro. **1** (2014). https://doi.org/10.1523/ENEURO.0005-14.2014
9. Kros, L., et al.: Cerebellar output controls generalized spike-and-wave discharge occurrence. Ann. Neurol. **77**, 1027–1049 (2015). https://doi.org/10.1002/ana.24399
10. Diedrichsen, J.: A spatially unbiased atlas template of the human cerebellum. Neuroimage. **33**, 127–138 (2006). https://doi.org/10.1016/j.neuroimage.2006.05.056
11. Acosta-Cabronero, J., et al.: The whole-brain pattern of magnetic susceptibility perturbations in Parkinson's disease. Brain. **140**, 118–131 (2017). https://doi.org/10.1093/brain/aww278
12. Lindig, T., et al.: Pattern of Cerebellar Atrophy in Friedreich's Ataxia: Using the SUIT Template. Cerebellum **18**, 435–447 (2019). https://doi.org/10.1007/s12311-019-1008-z
13. Akram, H., et al.: Connectivity derived thalamic segmentation in deep brain stimulation for tremor. NeuroImage Clin. **18**, 130–142 (2018). https://doi.org/10.1016/j.nicl.2018.01.008
14. Ye, C., et al.: Fully automatic segmentation of the dentate nucleus using diffusion weighted images. **1128–1131** (2012)
15. Bermudez Noguera, C., et al.: Using deep learning for a diffusion-based segmentation of the dentate nucleus and its benefits over atlas-based methods. J. Med. Imaging. **6**, 1 (2019). https://doi.org/10.1117/1.jmi.6.4.044007
16. Bazin, P.-L., et al.: Automated Segmentation of Cerebellar Nuclei from Ultra-High-Field Quantitative Susceptibility maps with multi-atlas shape fusion. Proc. Jt. Annu. Meet. ISMRM-ESMRMB, Paris, Fr. 695 (2018)

17. Li, X., et al.: Multi-atlas tool for automated segmentation of brain gray matter nuclei and quantification of their magnetic susceptibility. Neuroimage **191**, 337–349 (2019). https://doi.org/10.1016/j.neuroimage.2019.02.016
18. Jensen, J.H., Helpern, J.A.: MRI quantification of non-Gaussian water diffusion by kurtosis analysis. NMR Biomed. **23**, 698–710 (2010). https://doi.org/10.1002/nbm.1518
19. Zhang, H., et al.: NODDI: Practical in vivo neurite orientation dispersion and density imaging of the human brain. Neuroimage **61**, 1000–1016 (2012). https://doi.org/10.1016/j.neuroimage.2012.03.072
20. Van Essen, D.C., Smith, S.M., Barch, D.M., Behrens, T.E.J., Yacoub, E., Ugurbil, K.: The WU-Minn Human Connectome Project: An overview. Neuroimage **80**, 62–79 (2013). https://doi.org/10.1016/j.neuroimage.2013.05.041
21. WU - Minn Consortium Human Connectome Project: WU-Minn HCP 1200 Subjects Data Release: Reference Manual. 2017, 1-169 (2017). www.humanconnectome.org/documentation/S900/
22. Alexander, A.L., et al.: Diffusion Tensor Imaging of the Brain. Neurotherapeutics **4**, 316–329 (2007). https://doi.org/10.1021/jf505777p
23. Giraud, R., et al.: An Optimized PatchMatch for multi-scale and multi-feature label fusion. Neuroimage **124**, 770–782 (2016). https://doi.org/10.1016/j.neuroimage.2015.07.076
24. Barnes, C., et al.: PatchMatch: A randomized correspondence algorithm for structural image editing. ACM Trans. Graph. **28** (2009). https://doi.org/10.1145/1576246.1531330
25. Perone, C.S., et al.: Spinal cord gray matter segmentation using deep dilated convolutions. Sci. Rep. (2018). https://doi.org/10.1038/s41598-018-24304-3
26. Khan, S., et al.: A Guide to Convolutional Neural Networks for Computer Vision. Morga Claypool (2018)
27. Aylward, et al.: Deep Learning for Medical Image Analysis. Elsevier (2017)
28. Ioffe, S., Szegedy, C.: Batch Normalization: Accelerating Deep Network Training by Reducing Internal Covariate Shift (2015). arXiv:1502.03167
29. Fidon, L., et al.: Generalised wasserstein dice score for imbalanced multi-class segmentation using holistic convolutional networks (2018). arXiv:1707.00478v4
30. Kingma, D.P., et al.: Adam: A Method for Stochastic Optimization (2017). arXiv:1412.6980
31. Prados, F., et al.: Spinal cord grey matter segmentation challenge. Neuroimage (2017). https://doi.org/10.1016/j.neuroimage.2017.03.010
32. Bonekamp, D., et al.: Diffusion tensor imaging in children and adolescents: Reproducibility, hemispheric, and age-related differences. Neuroimage **34**, 733–742 (2007). https://doi.org/10.1016/j.neuroimage.2006.09.020
33. Mavroudis, I.A., et al.: Dendritic, axonal, and spinal pathology of the purkinje cells and the neurons of the dentate nucleus after long-term phenytoin administration: A case report. J. Child Neurol. **28**, 1299–1304 (2013). https://doi.org/10.1177/0883073812455694

Two Parallel Stages Deep Learning Network for Anterior Visual Pathway Segmentation

Siqi Li, Zan Chen, Wenlong Guo, Qingrun Zeng, and Yuanjing Feng

Abstract The segmentation of the anterior visual pathway(AVP) from MRI sequences is challenging because of the thin long architecture, structural variations along the path, and poor contrast with adjacent anatomic structures. The AVP plays a critical role in many devastating pathological conditions (e.g., pituitary tumors and craniopharyngiomas). However, most of the existing methods segment AVP on T1w images merely and often fail to achieve good results that cannot meet clinical needs. In this work, we introduced fractional anisotropy(FA) images into the training data set and proposed a deep learning network with two parallel stages for AVP segmentation. On an MRI dataset consisting of 102 subjects selected from the Human Connectome Project (HCP), we demonstrate that the proposed framework consistently improves the accuracy of AVP segmentation.

1 Introduction

Magnetic resonance imaging (MRI) is a non-invasive technology, which widely is used for studying and characterizing diseases of AVP such as optic neuritis, optic nerve hypoplasia, and optic pathway glioma [1]. Generally, the tiny structure of AVP requires an experienced neurosurgeon to accurately mark on the MRI from the scanned patients. Furtherly, due to the different judgment standards between

Supported by the National Natural Science Foundation of China [Grant Nos. 61976190, 61903336, and 61703369]; Key Research & Development Project of Zhejiang Province [Grant No. 2020C03070]; Major Science & Technology Projects of Wenzhou [Grant No. ZS2017007]; Science & Technology Innovation Projects for College Students of Zhejiang Province (New-shoot Talents Porgram) [Grant No. 2020R403079].

S. Li · Z. Chen · W. Guo · Q. Zeng · Y. Feng (✉)
Institute of Information Processing and Automation, College of Information Engineering, Zhejiang University of Technology, Hangzhou 310023, China
e-mail: fyjing@zjut.edu.cn

Z. Chen · Y. Feng
Zhejiang Provincial United Key Laboratory of Embedded Systems, Hangzhou 310023, China

N. Gyori et al. (eds.), *Computational Diffusion MRI*, Mathematics and Visualization,
https://doi.org/10.1007/978-3-030-73018-5_22

neurosurgeons, the performance of marked results is unstable, which seriously affects the repeatability of the treatment plan. The manual segmentation of AVP is a time-consuming and labor-intensive process along with unstable factors. Automated AVP segmentation will allow the extraction of quantitative features associated with optic pathway morphology from medical image slices. Therefore, the rapid and high-precision automatic segmentation of the AVP is of clinical and guiding significance for the formulation of neurosurgery planning.

The AVP consists of a pair of optic nerves also known as the cranial nerve II (CNII), transmitting visual signals from the retina to lateral geniculate nucleus(LGN). Due to its thin long architecture, poor contrast, and lack of obvious boundary with the surrounding tissues, the AVP is challenging to be segmented automatically [2]. For the segmentation of the AVP, many methods based on atlas and shape models have been proposed, which emphasize the use of statistical shape or statistical appearance because of the tortuous and slender structure. Bekes et al. [3] proposed a geometric model-based method for semi-automatic segmentation of the eyeballs, lenses, optic nerves, and chiasm with minimal user interaction. Noble et al. [4] presented a method based on atlas-navigated optimal medial axis and deformable model. Dolz et al. [5] proposed a method using a support vector machine for optic nerves.

Some machine learning-based methods have been proposed to segment visual structures since the increasing artificial intelligence application. Dolz et al. [6] proposed a deep learning classification scheme based on augmented enhanced features to segment organs at risk on the optic region in patients with brain cancer. Mansoor et al. [7] proposed an AVP segmentation method that uses deep learning to quickly locate the optic nerve shape model while combining T1-weighted(T1w), T2-weighted, and FLAIR MRI sequences. Ren et al. [8] introduced 3D Convolutional Neural Networks (3D-CNNs) to segment small tissues, including the optic nerve and chiasm automatically. Most previous works segmented the AVP on single modal data with one exception by Mansoor et al. [7]. They proposed an AVP segmentation mechanism steered by deep learning features and multiple MRI sequences (i.e., T1-weighted (T1w), T2-weighted (T2w), and FLAIR), reporting a DSC of 0.78 ± 0.12 on 165 AVP cases. However, these MRI sequences belong to structural imaging and more information ought to be obtained by using other modal data.

Clinically, T1w images are better for depicting normal anatomy and fractional anisotropy (FA) images are employed to characterize the structural anisotropy in the brain. Figure 1 shows the T1w images and FA images of No.100206 subject in HCP data. Figure 1a, d show the region of eyeballs to the chiasm (EC) on the T1w image and FA image respectively in the 34th slice. Figure 1b, e show the region of chiasm on the T1w image and FA image respectively in the 42nd slice. Figure 1c, f show the region of chiasm to the LGN (CL) on the T1w image and FA image respectively in the 50th slice. All the above slices are obtained from axial. As Fig. 1a, d shown, the edge of the EC on the T1w image can be easily obtained while the edge is missing on the FA image. On the contrary, the edge of the CL can be easily obtained on the FA image while the edge is blurred on the T1w image in Fig. 1c, f. In Fig. 1b, we can obtain the edge of chiasm easily both on T1w images and FA images. In conclusion, we can easily obtain the EC structure on T1w images and easily obtain

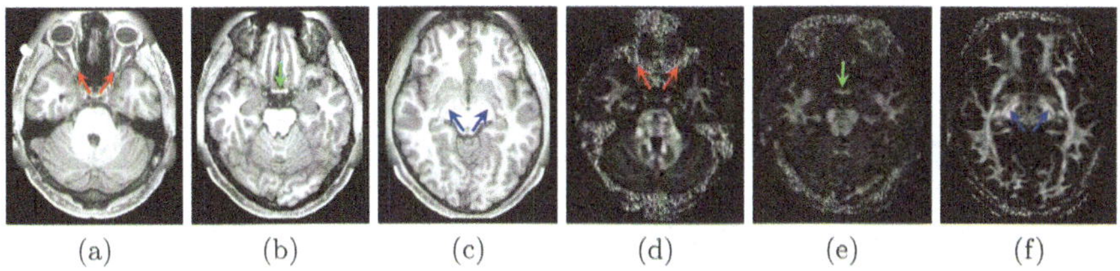

Fig. 1 A typical axial slice view of the T1w image and FA image from the No.100206 subject displays three parts of the AVP. **a** and **d** show respectively the region of EC on T1w and FA image in the 34th slice; **b** and **e** show respectively the region of chiasm on T1w and FA image in the 42nd slice; **c** and **f** show respectively the region of CL on T1w and FA image in the 50th slice

the CL structure on FA images. So we incorporate FA images and segment the AVP structure by combining T1w and FA images.

In this work, we propose a two parallel stages network (TPSN) for AVP segmentation. The main contributions of this work are as follows. First, we introduce FA images to extract the structural anisotropy in the brain for AVP segmentation. Second, we propose a two parallel stages network for the EC and CL feature extraction to form the AVP feature. Third, we propose a data fusion model to segment the complete AVP structure. Extensive experiments with our TPSN have been conducted, and the results demonstrate that our method provides consistent and noticeable performance improvement attributing to the fusion training of T1w images and FA images.

2 Methods

2.1 Data Preprocessing

Cropping: The overview of our architecture is shown in Fig. 2. We crop the HCP data with a spatial resolution of $145 \times 174 \times 145$ voxels to $128 \times 160 \times 128$ by removing as many zero background as possible to make them fit our network input size. This processing not only can effectively improve the calculation efficiency, but also retain the original image information as much as possible. In the end, we normalize the T1w images from 0 to 1.

Data Enhancement: To enlarge the training dataset, multiple data augmentation strategies have been utilized, including flipping and changing the color property randomly (i.e., brightness, contrast, and hue). We flip the input image horizontally according to probability 0.5. The following transformations were applied to each training sample and the intensity of each transformation was varied uniformly.

1. Contrast augmentation $X_{\text{augmented}} = X * \gamma$ with $\gamma \sim U[0.7, 1.3]$.
2. Brightness augmentation $X_{\text{augmented}} = X * \gamma$ with $\gamma \sim U[0.5, 1.5]$.
3. Hue augmentation $X_{\text{augmented}} = X * \gamma$ with $\gamma \sim U[-0.5, 0.5]$.

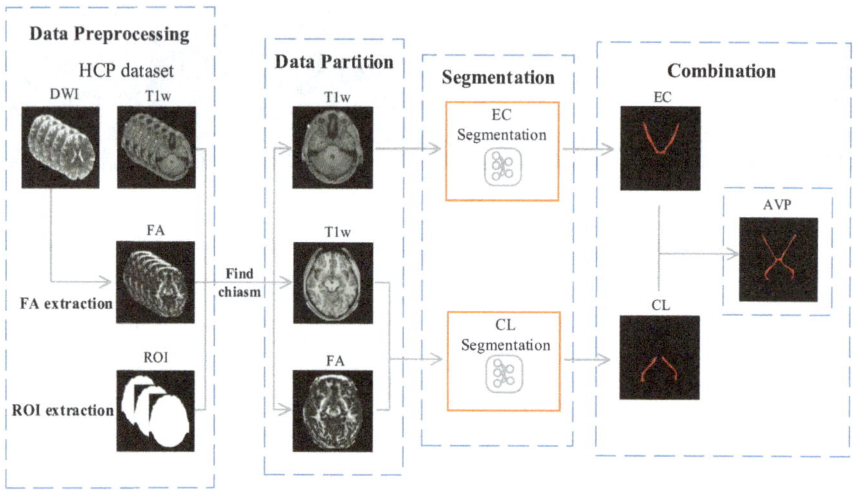

Fig. 2 Overview of the proposed two parallel stages deep learning network

FA Image Generation: In this work, we introduce the FA images to form a new blended feature extraction approach for the localization of AVP. The FA index is a widely used parameter of diffusion tensor imaging (DTI) for representing the motional anisotropy of water molecules, being sensitive to the presence and integrity of white matter fibers [9]. In DTI, a set of orthogonal vectors known as eigenvectors, which define the orientation of the principal axes of a diffusion ellipsoid in space, are calculated from the diffusion tensor. The length of each vector is represented by corresponding eigenvalues. An FA value is calculated using the following formula, based on eigenvalues in the diffusion tensor [10, 11]:

$$FA=\sqrt{\frac{3[(\lambda_1-\overline{\lambda})^2+(\lambda_2-\overline{\lambda})^2+(\lambda_3-\overline{\lambda})^2]}{2(\lambda_1{}^2+\lambda_2{}^2+\lambda_3{}^2)}}, \quad (1)$$

where $\overline{\lambda}$ is the average of λ_1, λ_2, and λ_3. For each volume, the FA images are registered to the T1w images using FLIRT (FMRIB's linear registration tool) in FSL (version 5.0.8) [12].

ROI Extraction: To solve the imbalance of AVP classes, we have delineated an ROI to extract training data. The distribution of AVP within a subject data is skewed as only 16% of brain slices involve AVP. To solve this problem, we first calculate the number of the AVP ground truth label slices for each training sample and then expend half of the number slices up and down respectively. The operation is performed on the axial slices. We extract the whole brain ROI by the following four steps. First, we extract images slice by slice for each volume. Second, we process the median filter on the images. Third, we binarize the images using threshold processing. Finally, we fill the holes for each image to find the whole brain ROI. To filter out unwanted information, we only deal with the data within the ROI. Noted that the whole brain will be set as our ROI during the testing phase.

2.2 Two Parallel Stages Network Architecture

As shown in Fig. 1d, the FA image has a poor contrast of EC with a lot of interference noise, which makes our fitting network worse. Based on the observation that the CL region in FA images generally has higher contrast than the CL region in T1w images as shown in Fig. 1c, f, we propose a two parallel stages network (TPSN) for AVP segmentation. In this work, T1w images are used to train for EC segmentation, while T1w and FA images are used to train for CL segmentation. The designed architecture is built upon the U-Net architecture [13], which has shown outstanding performance in various medical segmentation tasks [14, 15]. The details of our architecture are shown in Fig. 3.

We adopt the U-Net model as the base model, which consists of an encoding and a decoding part. The encoding part repeatedly applies two 3×3 convolutional layers with stride 1, each followed by a batch normalization (BN), a rectified linear unit (ReLU), and a 2×2 max-pooling operation with stride 2 on 4 levels. At each down-sampling step, the dimensions of the input image are reduced by half and the number of feature channels is doubled. The bottom level includes two 3×3 convolutional layers, each followed by a BN without pooling layer. The decoding part recovers the original dimensions of the input images by up-sampling the feature map, a concatenation with the corresponding feature channels from the contractive path and two 3×3 convolutional layers, each followed by a BN and a ReLU. The final layer is a 1×1 convolution followed by a sigmoid operation for mapping the feature vector to the binary prediction (i.e., AVP vs. non-AVP). Last but not least, the binary predictions of EC-T1w, CL-T1w, and CL-FA were generated through the base model.

In Fig. 3, our framework mainly consists of two stages: EC segmentation and CL segmentation. In the first stage, we set the axial slices of the T1w images as training data and use the U-Net model to extract the prediction P_{EC} from the input I_{EC}, which is defined as follows:

$$F_{\text{EC}} = H_U(I_{\text{EC}}), \tag{2}$$

where $H_U(\cdot)$ denotes the U-Net model, I_{EC} denotes the T1w images of EC, and F_{EC} denotes the feature of EC on T1w images, which is then used for the extraction of EC prediction.

In the second stage, CL Segmentation is responsible for the prediction of the CL structure on T1w and FA data. In this stage, axial slices of the T1 and FA images are respectively used as the input of U-Net, and two outputs are obtained. Inspired by [16], we perform a data fusion module, which fuses the AVP features of different

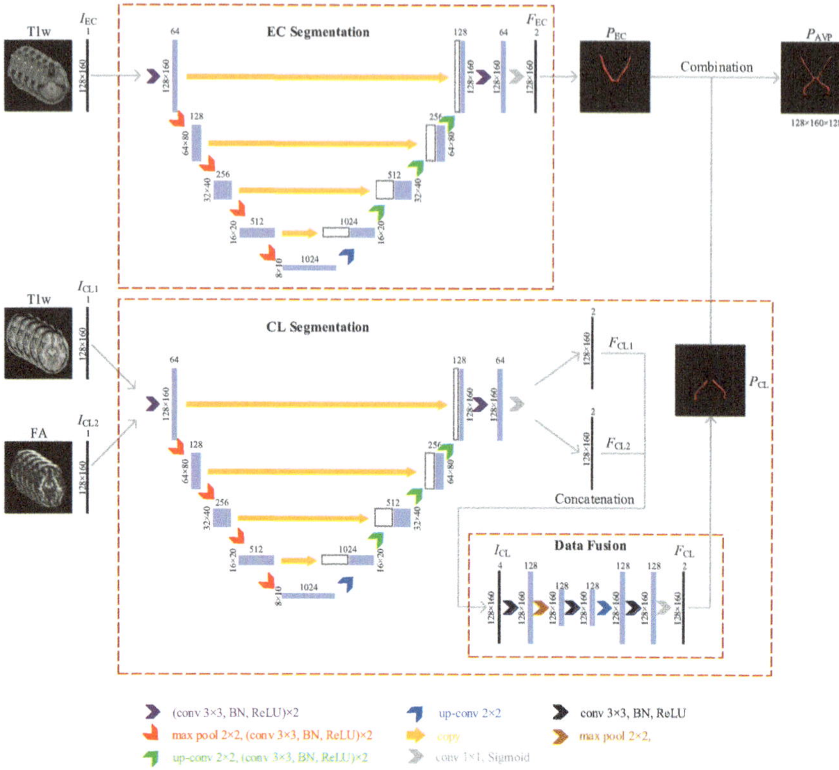

Fig. 3 The designed architecture of our two parallel stages network (TPSN). Blue boxes represent multi-channel feature maps. White boxes show copied feature maps. The number on top of each box gives the number of channels. Network operations are represented by differently colored arrows

modal data (i.e., T1w and FA). Then the two outputs are concatenated and used as the input of the data fusion module to get the final CL prediction.

Specifically, the second stage can be divided into two steps. First, we set the T1w and FA images of CL as training data separately and perform the same operation as the Eq. 2 shown. So we have $F_{\mathrm{CL1}} = H_U(I_{\mathrm{CL1}})$ and $F_{\mathrm{CL2}} = H_U(I_{\mathrm{CL2}})$, where I_{CL1} and I_{CL2} denote respectively the T1w and FA images of CL versus F_{CL1} and F_{CL2} denote respectively the CL features of T1w and FA images. Second, we concatenate the F_{CL1} feature and the F_{CL2} feature to form the fusion feature I_{CL}, which is then used to obtain the prediction of CL structure (i.e., F_{CL}). So we can further have

$$F_{\mathrm{CL}} = H_F(I_{\mathrm{CL}}) \,, \tag{3}$$

where $H_F\,(\cdot)$ denotes the data fusion net module. In this second stage, we can also combine the T1 and FA images as the input to the base model, but the combined data of this simple front-end fusion usually contains a lot of redundant information.

Therefore, we choose the back-end fusion method here to extract CL features. The advantage of this is that the errors of the fusion model come from different classifiers, and the errors from different classifiers are often uncorrelated and do not affect each other, and will not cause further accumulation of errors.

Finally, we obtain the EC prediction from EC and CL features, i.e., $P_{\mathrm{EC}} = H_M(F_{\mathrm{EC}})$ and $P_{\mathrm{CL}} = H_M(F_{\mathrm{CL}})$, where P_{EC} and P_{CL} denote the prediction of EC and CL structure; $H_M(\cdot)$ denotes the torch.max function in PyTorch. The EC and CL prediction are followed by a combination operation to form the complete AVP prediction,

$$P_{\mathrm{AVP}} = H_C(P_{\mathrm{EC}}, P_{\mathrm{CL}}) , \tag{4}$$

where $H_C(\cdot)$ denotes the combination operation and P_{AVP} denotes the prediction of complete AVP structure. The specific step of the combination operation is to input each axial slice of the test data into the trained model in sequence, and then combine the outputs from the axial slice into a complete volume according to the sequence.

3 Experiments

3.1 Dataset

In this work, we used the T1-weighted structural MRI and pre-processed diffusion MRI (dMRI) data of 102 unrelated subjects (age 22–35) from the *HCP* 1200 *Subject Release* of the Human Connectome Project (HCP). HCP subjects were scanned at Washington University in St. Louis on a customized Siemens Skyra 3T scanner (Siemens AG, Erlangen, Germany). The corresponding dMRI data, which was multi-shell data with 18 b = 0 and 270 gradient directions distributed equally over 3 shells with b-values of 1000, 2000, and 3000 s/mm^2, was acquired using a spin-echo planar imaging (EPI) sequence (repetition time (TR) = 5520 ms, echo time (TE) = 89.5 ms, matrix size = 145 × 174 × 145, resolution = 1.25 × 1.25 × 1.25 mm^3 voxels). The spatial resolution of T1w data was the same as that of dMRI data. We used the dMRI data that had already been processed by the minimal preprocessing pipeline (e.g. distortion correction, motion correction, registration to MNI space and brain extraction had already been completed) [17]. We divide them into the training set, validation set, and testing set according to the ratio of 8:1:1. Accuracies were measured on the validation set in all experiments.

The AVP ground truths of the 102 subjects HCP dataset were well described manually by experienced neurosurgeons. On the selected subjects, two neurosurgeons used 3D Slicer [18] to mark the AVP ground truths on T1 structural images. Each of the neurosurgeons labeled 51 volumes. Boundary definitions for the AVP ground truths were obtained according to the signal intensity differences in the T1w images.

3.2 Implementation Details

Our method had been implemented in PyTorch for network building. The system was running on Windows 10 with Cuda 10.1. We performed our experiment on an NVIDIA TitanX GPU and it took approximately 24 hours to complete all the training models upon all patches with 200 epochs. For each model, we set the training with a batch size of 64, and the weights were updated by Adam optimizer [19] with a learning rate of 0.002. The loss function used by the network was dice. To prevent over-fitting, an early stopping strategy was likewise utilized in the work if no improvement arose in the validation loss of the validation set after 10 epochs.

3.3 Results

To evaluate the segmentation performance, three metrics are used to evaluate the proposed method: Dice similarity coefficient (DSC) [20], Hausdorff distance (HD) [21], and average symmetric surface distance (ASD). The DSC also called the overlap index or F1 score, is the most used metric in the evaluation of medical volume segmentation, which compares volumes based on their overlap between segmentation and the ground truth. The HD and ASD are computed to measure the accuracy of the segmentation boundary. DSC is defined as follows:

$$\mathrm{DSC}\left(V_{\mathrm{gt}}, V_{\mathrm{seg}}\right) = \frac{2|V_{\mathrm{gt}} \cap V_{\mathrm{seg}}|}{|V_{\mathrm{gt}}| + |V_{\mathrm{seg}}|}, \tag{5}$$

where V_{gt} and V_{seg} represent the ground truth and the automatic segmentation result, respectively. The HD and ASD between two volumes are defined by:

$$\mathrm{HD} = \max(h(A, B), h(B, A)), \tag{6}$$

$$\mathrm{ASD} = mean(h(A, B), h(B, A)), \tag{7}$$

where A and B denote two finite point sets and $h(A, B)$ is called the directed HD and given by:

$$h(A, B) = \max_{a \in A} \min_{b \in B} \| a - b \|, \tag{8}$$

where $\| \cdot \|$ denotes the euclidean distance. Finally, the performance is assessed as an average of the measures of all AVP segmentation in the test-set. The value of DSC refers to the same measurement as Dice per volume in the testing set. For the HD and ASD evaluation metrics, the smaller the value is, the better the segmentation result. The quantitative and qualitative results of our method are as follows.

Quantitative results by TPSN: Figure 4 and Table 1 show the quantitative evaluation metrics of our designed architecture in comparison with the U-Net and U-Net++ [22]

Table 1 Summary of DSC, HD, and ASD metrics evaluated on 10 testing subjects for AVP segmentation

Methods	Input	DSC	HD(mm)	ASD(mm)
U-Net	T1	0.822	2.963	0.200
U-Net	FA	0.679	5.912	0.422
U-Net	T1, FA	0.833	2.913	0.197
U-Net++	T1	0.827	2.970	0.193
U-Net++	FA	0.691	5.545	0.420
U-Net++	T1, FA	0.843	3.066	0.168
TPSN(ours)	T1, FA	**0.855**	**2.330**	**0.162**

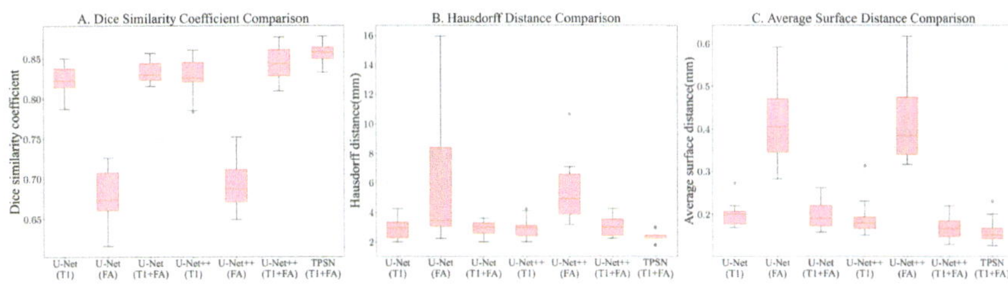

Fig. 4 Quantitative results comparing our proposed method with the other methods

framework. From Table 1, An average DSC of 0.855 for TPSN is obtained, showing a significantly improved performance by TPSN over the other methods ($p < 0.05$, paired t-test). Additionally, our method segment the AVP with 2.330 mm average HD and 0.162 mm average ASD which are less than the other methods, showing a better performance. Figure 4 shows the box plot of the metrics proving the effectiveness of our method. Although it was not possible in this work to use the same datasets as those used in previous studies, the higher DSC we achieved as indicated by the summary in Table 2 suggested that our method outperformed previously approaches for optic pathway segmentation.

Qualitative results by TPSN: Figure 5 shows some qualitative examples generated by different networks. We can see that the region from the chiasm to LGN of our segmentation results have less red and label versus larger blue label demonstrating the superior performance of TPSN over the other two networks. Our method mainly improved the segmentation performance of the region from the chiasm to LGN by introducing FA images and performing fusion training. Time-wise, our method can obtain the AVP segmentation of 10 test sets in 42 s on the premise that FA images have been generated.

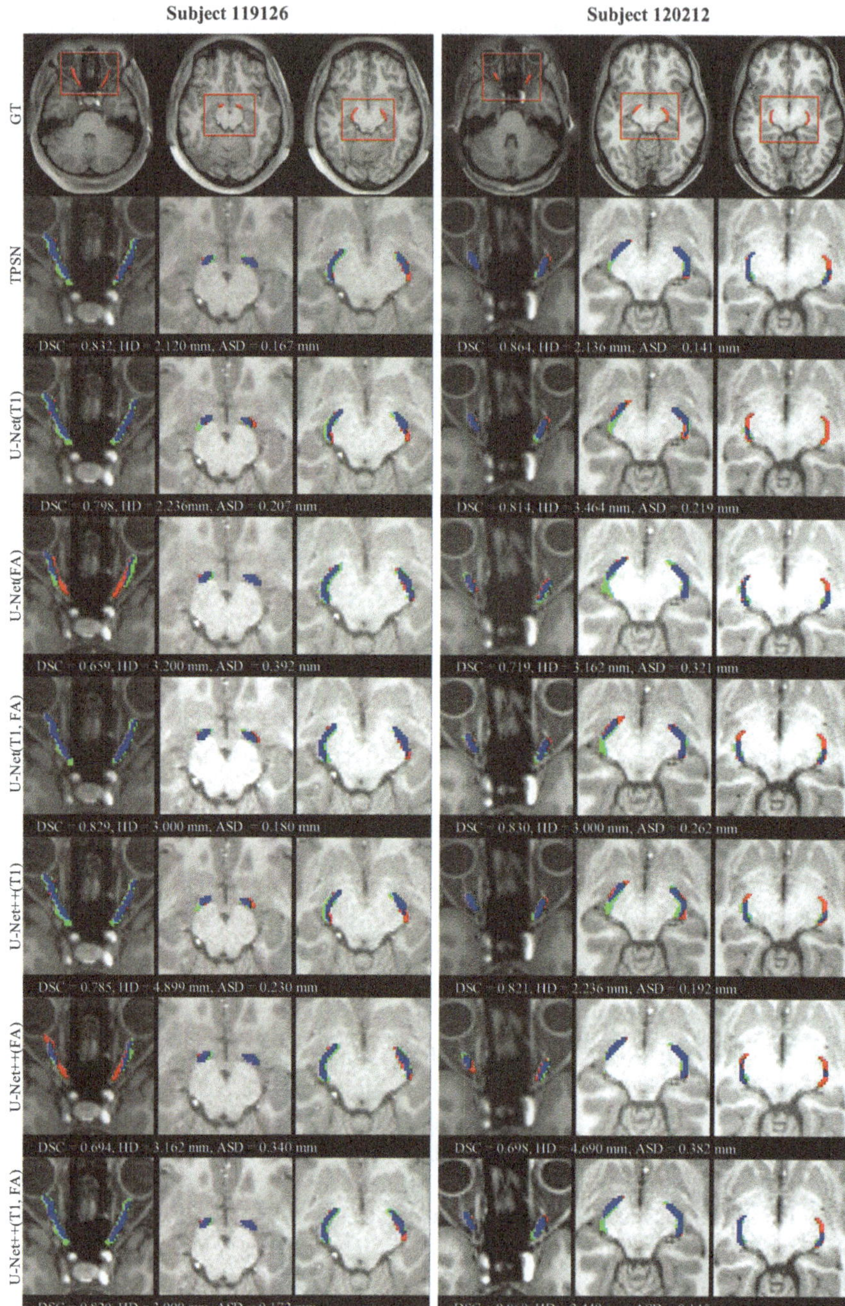

Fig. 5 Qualitative comparison of two representative testing set. The Blue label shows the overlap area of manual and automated segmentation, the red label shows the manual label, and the green label shows the automated segmentation

Table 2 Comparison of our method with the methods reported in the literature for visual pathway related segmentation

Method	Region of segmentation	Number of subject	DSC
Noble et al. [4]	Nerve+chiasm	10	0.80
Asman et al. [23]	Nerve	31	[0.48, 0.88]
Harrigana et al. [24]	Nerve	30	0.77
Yang et al. [25]	AVP	17	0.73±0.04
Dolz et al. [5]	Nerve	10	0.76
Mansoor et al. [7]	AVP	165	0.78±0.12
Ren et al. [8]	Nerve	48	0.71±0.08
Our method	AVP	102	**0.85±0.02**

4 Conclusion

Automated anterior visual pathway (AVP) segmentation is difficult due to its thin size, structural variation along the path, and non-discriminative contrast to the adjacent anatomic structures. The AVP is a white matter bundle composed of medullated fibers. For providing the fiber anatomical connectivity information in the brain, we generate the fractional anisotropy (FA) images from DWI data and introduce them into our framework. In this work, we designed a partitioned training framework with two parallel stages for the AVP automatic segmentation of MRI data, which is formed by two parallel stages, i.e., EC segmentation and CL segmentation. Our two parallel stages network (TPSN) improves the performance of AVP segmentation and we achieve an average DSC of 0.855 in the experiment conducted on HCP data, which demonstrates the effectiveness of our approach.

References

1. Chan, J.: Optic Nerve Disorders, pp. 130–131 (2007)
2. Salmela, M.B., Cauley, K.A., Andrews, T., Gonyea, J.V., Tarasiewicz, I., Filippi, C.G.: Magnetic resonance diffusion tensor imaging of the optic nerves to guide treatment of pediatric suprasellar tumors. Pediatric Neurosurg. **45**(6), 467–471 (2009)
3. Bekes, G., Máté, E., Nyúl, L.G., Kuba, A., Fidrich, M.: Geometrical model-based segmentation of the organs of sight on ct images. Med. Phys. **35**(2), 735–743 (2008)
4. Noble, J.H., Dawant, B.M.: An atlas-navigated optimal medial axis and deformable model algorithm (nomad) for the segmentation of the optic nerves and chiasm in mr and ct images. Med. Image Anal. **15**(6), 877–884 (2011)
5. Dolz, J., Leroy, H.A., Reyns, N., Massoptier, L., Vermandel, M.: A fast and fully automated approach to segment optic nerves on mri and its application to radiosurgery. In: 2015 IEEE 12th International Symposium on Biomedical Imaging (ISBI), pp. 1102–1105. IEEE (2015)

6. Dolz, J., Reyns, N., Betrouni, N., Kharroubi, D., Quidet, M., Massoptier, L., Vermandel, M.: A deep learning classification scheme based on augmented-enhanced features to segment organs at risk on the optic region in brain cancer patients (2017). arXiv:1703.10480
7. Mansoor, A., Cerrolaza, J.J., Idrees, R., Biggs, E., Alsharid, M.A., Avery, R.A., Linguraru, M.G.: Deep learning guided partitioned shape model for anterior visual pathway segmentation. IEEE Trans. Med. Imaging **35**(8), 1856–1865 (2016)
8. Ren, X., Xiang, L., Nie, D., Shao, Y., Zhang, H., Shen, D., Wang, Q.: Interleaved 3d-cnn s for joint segmentation of small-volume structures in head and neck ct images. Med. Phys. **45**(5), 2063–2075 (2018)
9. Assaf, Y., Pasternak, O.: Diffusion tensor imaging (dti)-based white matter mapping in brain research: a review. J. Molecular Neurosci. **34**(1), 51–61 (2008)
10. Basser, P.J., Pierpaoli, C.: Microstructural and physiological features of tissues elucidated by quantitative-diffusion-tensor mri. J. Mag. Reson. (1996)
11. Pierpaoli, C., Basser, P.J.: Toward a quantitative assessment of diffusion anisotropy. Mag. Reson. Med. **36**(6), 893–906 (1996)
12. Jenkinson, M., Bannister, P., Brady, M., Smith, S.: Improved optimization for the robust and accurate linear registration and motion correction of brain images. Neuroimage **17**(2), 825–841 (2002)
13. Ronneberger, O., Fischer, P., Brox, T.: U-net: Convolutional networks for biomedical image segmentation. In: International Conference on Medical Image Computing and Computer-assisted Intervention, pp. 234–241. Springer, Berlin (2015)
14. Çiçek, Ö., Abdulkadir, A., Lienkamp, S.S., Brox, T., Ronneberger, O.: 3d u-net: learning dense volumetric segmentation from sparse annotation. In: International Conference on Medical Image Computing and Computer-assisted Intervention, pp. 424–432. Springer, Berlin (2016)
15. Dong, H., Yang, G., Liu, F., Mo, Y., Guo, Y.: Automatic brain tumor detection and segmentation using u-net based fully convolutional networks. In: Annual Conference on Medical Image Understanding and Analysis, pp. 506–517. Springer, Berlin (2017)
16. Yan, Z., Yang, X., Cheng, K.T.: A three-stage deep learning model for accurate retinal vessel segmentation. IEEE J. Biomed. Health Inf. **23**(4), 1427–1436 (2018)
17. Glasser, M.F., Sotiropoulos, S.N., Wilson, J.A., Coalson, T.S., Fischl, B., Andersson, J.L., Xu, J., Jbabdi, S., Webster, M., Polimeni, J.R., et al.: The minimal preprocessing pipelines for the human connectome project. Neuroimage **80**, 105–124 (2013)
18. Fedorov, A., Beichel, R., Kalpathy-Cramer, J., Finet, J., Fillion-Robin, J.C., Pujol, S., Bauer, C., Jennings, D., Fennessy, F., Sonka, M., et al.: 3d slicer as an image computing platform for the quantitative imaging network. Mag. Reson. Imaging **30**(9), 1323–1341 (2012)
19. Kingma, D.P., Ba, J.: Adam: A method for stochastic optimization (2014). arXiv:1412.6980
20. Dice, L.R.: Measures of the amount of ecologic association between species. Ecology **26**(3), 297–302 (1945)
21. Huttenlocher, D.P., Klanderman, G.A., Rucklidge, W.J.: Comparing images using the hausdorff distance. IEEE Trans. Pattern Anal. Mach. Intell. **15**(9), 850–863 (1993)
22. Zhou, Z., Siddiquee, M.M.R., Tajbakhsh, N., Liang, J.: Unet++: Redesigning skip connections to exploit multiscale features in image segmentation. IEEE Trans. Med. Imaging **39**(6), 1856–1867 (2019)
23. Asman, A.J., DeLisi, M.P., Mawn, L.A., Galloway, R.L., Landman, B.A.: Robust non-local multi-atlas segmentation of the optic nerve. In: Medical Imaging 2013: Image Processing, vol. 8669, p. 86691L. International Society for Optics and Photonics (2013)
24. Harrigan, R.L., Panda, S., Asman, A.J., Nelson, K.M., Chaganti, S., DeLisi, M.P., Yvernault, B.C., Smith, S.A., Galloway, R.L., Mawn, L.A., et al.: Robust optic nerve segmentation on clinically acquired computed tomography. J. Med. Imaging **1**(3), 034006 (2014)
25. Yang, X., Cerrolaza, J., Duan, C., Zhao, Q., Murnick, J., Safdar, N., Avery, R., Linguraru, M.G.: Weighted partitioned active shape model for optic pathway segmentation in mri. In: Workshop on Clinical Image-Based Procedures, pp. 109–117. Springer, Berlin (2014)

Exploring DTI Benchmark Databases Through Visual Analytics

William A. Romero R., Daniel Althviz Moré, Irvin Teh, Jürgen E. Schneider, Magalie Viallon, and Pierre Croisille

Abstract Diffusion MRI studies include tests on standardised phantoms and measurements on the output images to assess and benchmark the imaging system. These tests are an essential methodological step to guarantee the reproducibility of measurement outcomes. However, in longitudinal and multi-centre studies, analysis tasks become more complex with the increase in the sources and volume of data, as well as the parameters of interest. To manage this complexity, Visual Analytics (VA) allows researchers to explore large amounts of data easily and quickly by providing key information in a readily interpretable format and reducing the cognitive load of information. This paper presents CMRDiffMonitor, a VA tool for monitoring and benchmarking Diffusion Tensor Imaging (DTI) databases developed in the context of a multi-centre MRI project. Through an interactive dashboard, CMRDiffMonitor enables users to capture a snapshot of the study which includes: a temporal overview of the data sets (monitoring of the system stability across sites) and statistical mea-

W. A. Romero R. (✉) · M. Viallon · P. Croisille
Université de Lyon, UJM-Saint-Etienne, INSA, CNRS UMR 5520, INSERM U1206, CREATIS, F-42023, Saint Etienne, France
e-mail: William.Romero@creatis.insa-lyon.fr

M. Viallon
e-mail: Magalie.Viallon@creatis.insa-lyon.fr

P. Croisille
e-mail: Pierre.Croisille@creatis.insa-lyon.fr

W. A. Romero R.
Siemens Healthcare SAS, Saint-Denis, France

D. Althviz Moré
Systems and Computing Engineering Department, Universidad de los Andes, Bogotá, Colombia
e-mail: d.althviz10@uniandes.edu.co

I. Teh · J. E. Schneider
Leeds Institute of Cardiovascular and Metabolic Medicine, University of Leeds, Leeds, UK
e-mail: I.Teh@leeds.ac.uk

J. E. Schneider
e-mail: J.E.Schneider@leeds.ac.uk

N. Gyori et al. (eds.), *Computational Diffusion MRI*, Mathematics and Visualization,
https://doi.org/10.1007/978-3-030-73018-5_23

surements in Regions of Interest on Diffusion Anisotropy Indices (benchmarking of the different imaging sequences and systems).

1 Introduction

Clinical research involves multi-centre studies with the objective to evaluate MRI protocols on a large number of patients and imaging systems, and consequently demonstrate the utility of the developed tools. Therefore, ensuring the stability of the systems involved [14] enables the evaluation (intra- and inter-centre) of reliable and reproducible results. Moreover, Quality Assurance (QA) protocols include tests on a standardised phantom and measurements on the output images to identify the causes of failure and the potential corrective actions [7, 22].

The scenario of a multi-centre study relies on the capacity to manage and process data. As it is well-discussed by Burmeister et al., data collection and processing are the first stages of the workflow [6]. Research scientists need to extract meaningful information, find interesting insights and make correlations; in this direction, data analysis tools must not only extract parameters, but also ease the way to explore, report and share results. In this scenario, Visual Analytics (VA) techniques have the potential to enhance data exploration based on interactive visualisation. Consequently, research scientists may benefit from having a graphical representation to verify protocol compliance, identify anomalies, and validate imaging sequence outcomes across different systems and time-points in a concise and accessible way.

This paper describes CMRDiffMonitor, a VA tool for monitoring and benchmarking in vitro Diffusion MRI databases in order to enhance the ability to track, review and assess image acquisition sequences in longitudinal and multi-centre studies. CMRDiffMonitor is the result of an incremental development cycle in the context of a multi-centre study of cardiac diffusion MR sequences in an isotropic phantom. The tool has enabled participant researchers to streamline the analysis task as well as to report partial results of the research project as presented in [26, 27]

The next section gives an overview of the technological context and similar approaches. The subsequent sections describe in detail the problem domain and the system architecture in which the tool has been deployed. The paper concludes with a description of how the tool supports analysis tasks on in vitro DTI databases.

2 Related Work

The advent of data management solutions, workflow engines, and online interactive visualisation applications has led to the implementation of specialised databases and platforms such as the Cardiac Atlas Project [12], OpenNeuro [21], and Brainlife.io [5], to mention but a few. Bearing in mind the idea of reproducible computational research [11], research communities in Biomedical Sciences and Engineering have been contributing with exemplar infrastructures and technologies [4, 8, 10] in order to:

- collect, standardise and share data,
- execute complex computational methods (workflows) on different computing resources (grid, supercomputer, GPU-based cluster, etc.),
- manage workflows outputs or data formating (new utilisable data), and
- provide online data visualisation.

Following these ideas, the Human Heart Project [1] has been implemented to provide access to medical databases and computer-based tools in order to share data easily, test computational methods and enhance collaboration in the research community of cardiovascular imaging. The back-end infrastructure of this solution is based on Girder [18], a web-based data management platform, and the Virtual Imaging Platform (VIP) [3, 13], a scientific gateway for medical simulation and image data analysis.

DTI benchmark databases are used in the context of the harmonisation of imaging sequences [25–27]; a standard reference object (phantom) is used to assess and benchmark the performance of an MRI scanner. Understanding how images are produced by a set of parameters and conditions enables MRI and medical researchers to calibrate the imaging system as well as identify sources of variability and corrective actions. In the case of a multi-centre evaluation [25–27], the benchmark database allows the assessment of imaging sequences (product and custom-made) and scanner's stability (acquisitions over time, intra-scanner analysis), and highlight the factors that may influence reliable and predictable results across the different imaging systems (inter-scanner analysis).

Several approaches to automatic data processing and visualisation are presented by every single platform publication [4, 8, 10]. The work presented by Davids et al. [9] provides insights into automatic processing of phantom measurements, based on the quality parameters recommended by the American College of Radiology accreditation [22]. However, there is no discussion about interactive visualisation tools for the interpretation of the results or traceability of data. One such solution, outlined by Burmeister et al. [6], describes in detail the design guidelines in the development of a platform for data preprocessing, cohort exploration and result reporting. In spite of the cohort-driven nature of the solution, there are valuable and practical lessons about the implementation of VA tools for data stratification, hypothesis testing and data analysis.

In addition, a Quality Control (QC) system of neuroimaging data can be found in [17]. This solution has been implemented to calculate QC metrics for various modality images (sMRI, fMRI, DTI or CT), report image QC rating and visual assessment performed by users. A general overview of the user interaction follows: data selection (project, modality, data set, image, status, etc), QC Settings (i.e. metric threshold) and visual inspection. To this extent, the system provides an evaluation of data sets to users. Therefore, users know the quality of the data they are using for analysing or as input in image processing workflows. In contrast to this work,

CMRDiffMonitor is a domain and data specific tool focused on a multi-centre study of cardiac DTI sequences in an isotropic phantom. There is no data quality ranking or visual assessment by users.

3 Use Case

The main motivation to implement CMRDiffMonitor comes from a multi-centre study (11 geographically dispersed research sites) to investigate intra- and inter-site variation in DTI parameters in an isotropic phantom [2].

The experimental protocol is set up to scan a standardised phantom (7 tubes filled with 0–20% polyvinylpyrrolidone [20], Fig. 1) using product and custom DTI acquisition sequences with parameters matching typical cardiac DTI protocols: TR = 3000 ms, TE = 85 ms, in-plane resolution = 2.5 mm, slices = 3, thickness = gap = 8 mm, blow= 100 $\frac{s}{mm^2}$, bhigh= 450 $\frac{s}{mm^2}$, number of diffusion directions = 6, bandwidth $\sim$ 3000 Hz, parallel imaging = 2x, and triggered with simulated ECG = 60 bpm. The phantoms are chilled in ice water and imaged at 0 °C. Subsequent scans are performed under the same conditions in a time lapse between 1 and 30 days.

MRI researchers have become increasingly interested in the utilisation of interactive visualisation tools to accelerate data analysis and enhance collaborative work. This research project brought attention to Visual Analytics as a tool to track and review the variability of Diffusion Anisotropy Indices (DAI) from different image acquisition sequences across participant sites in time. The goal was to explore simple VA techniques to provide the full picture of the multi-centre study. Therefore, the project established the following requirements:

R1. Collect and organise DICOM data sets from each site.
R2. Process data sets to calculate DAI maps: Mean Diffusivity (MD) and Fractional anisotropy (FA).
R3. Segment DAI maps and calculate statistics (minimum, maximum, mean, standard deviation, etc.) in the Regions of Interest (ROI).
R4. Export DAI/ROI statistics in CSV format.
R5. Display a timeline of the data set uploads of each site.
R6. Display statistical results by filtering sites, DAI, ROI, acquisition sequence and parameters.

The baseline of the solution is the Human Heart Project infrastructure. Requirement R1 is managed by the Girder platform [18] which provides user authentication and data management functionalities. Requirements R2, R3 and R4 are managed by CMRDiffTools [23]. Figure 2 outlines the main components, workload process and data flow. Figure 3 presents the image processing workflow performed by CMRDiffTools.

Dynamic graphics (requirements R5 and R6) were requested in order to basically provide two points of view of the original data (series of 2D MR images): first, the temporal dimension of data sets; and second, the statistical measurements on the DAI

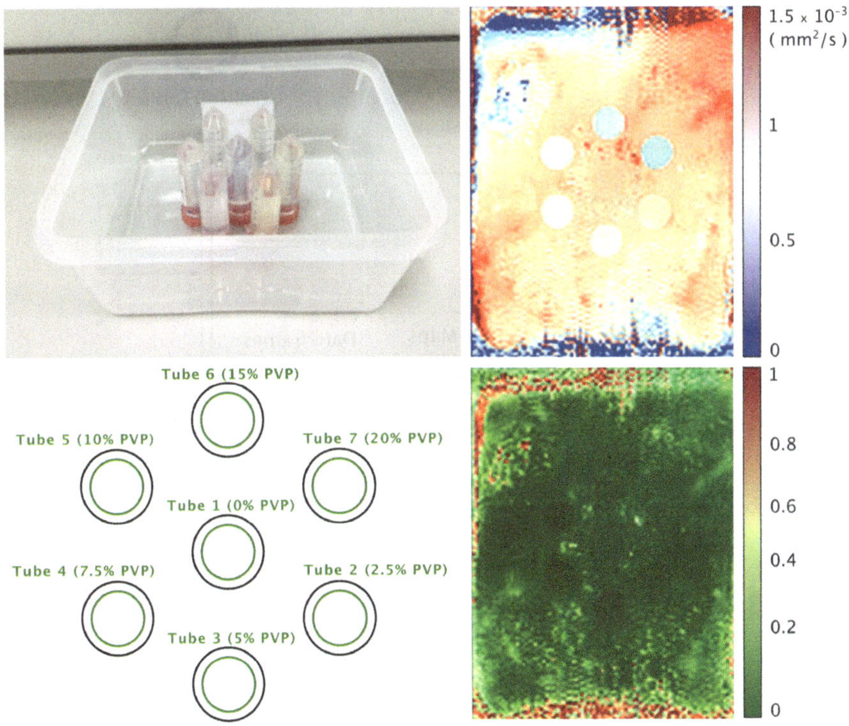

Fig. 1 Isotropic Phantom, ROI and DAI maps. Photograph of phantom before filling with ice and water (top left); layout of tubes and Regions of Interest (green circles, bottom left); MD and FA maps (top right and bottom right respectively). Background image texture in the MD and FA maps is caused by the ice

maps. The temporal dimension would give researchers the capacity to evaluate the imaging system stability (monitor), while statistical measurements would quantitatively assess the reproducibility across the different scanners (benchmarking). The next section presents the VA tool developed to address requirements R5 and R6.

4 Implementation

To fulfil requirements R5 and R6, a plug-in for the Girder platform [18] has been developed using Plotly [15], and open-source graphing library. The plug-in is fed with the statistical reports (CSV files, R4) in order to display a dashboard composed of 4 interactive areas (Fig. 4):

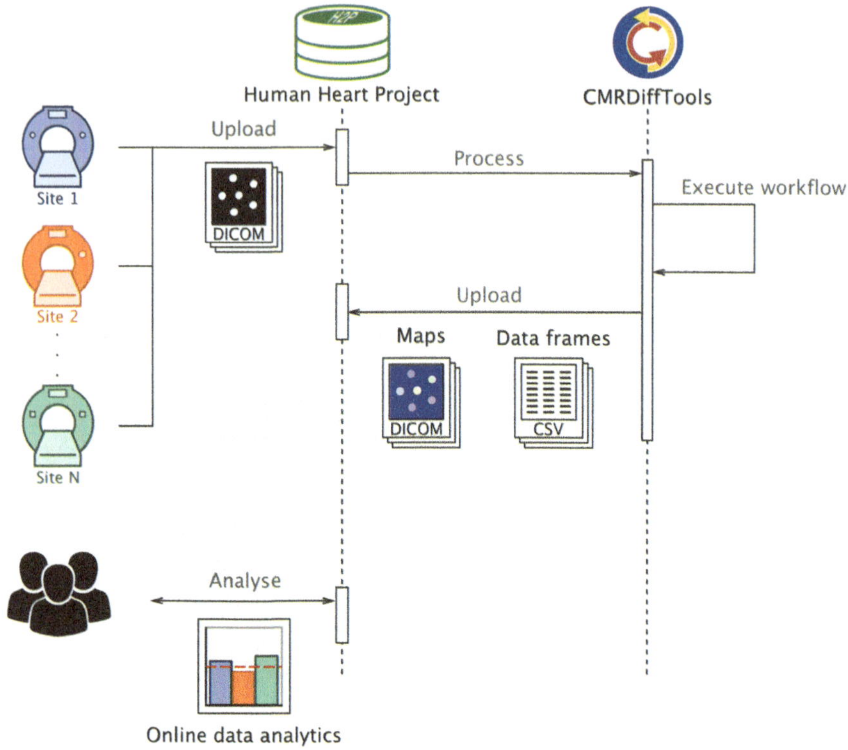

Fig. 2 Sequence diagram: main components and data flow

1. The **filters panel** is a dynamic interface where the user can select filter conditions: DAI measure, ROI, acquisition sequence, acquisition parameters and field strength.
2. **Timeline view** shows the performed acquisitions (per acquisition sequence) over the course of time.
3. **Sites panel** lists all the sites in the study. The user can select the sites of interest.
4. **Measure view** displays the DAI value across the different sites for a selected ROI.

The Measure view is interactively updated according to the user selection on the filters panel, timeline view and sites panel. The user interaction follows the "Visual Information Seeking Mantra: overview first, zoom and filter, then details-on-demand" [24] in order to provide an effortless pathway to extract information. As an example of details-on-demand features, the mouse over action displays details of the data according to the view. For instance, the mouse pointer over a bar on the measure view will display: centre, sequence name, sequence parameters, acquisition date, mean value, and standard deviation.

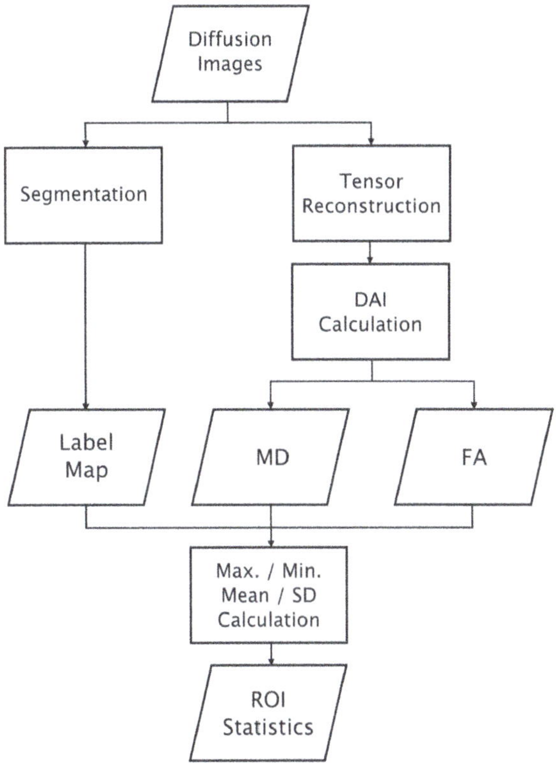

Fig. 3 Image processing workflow

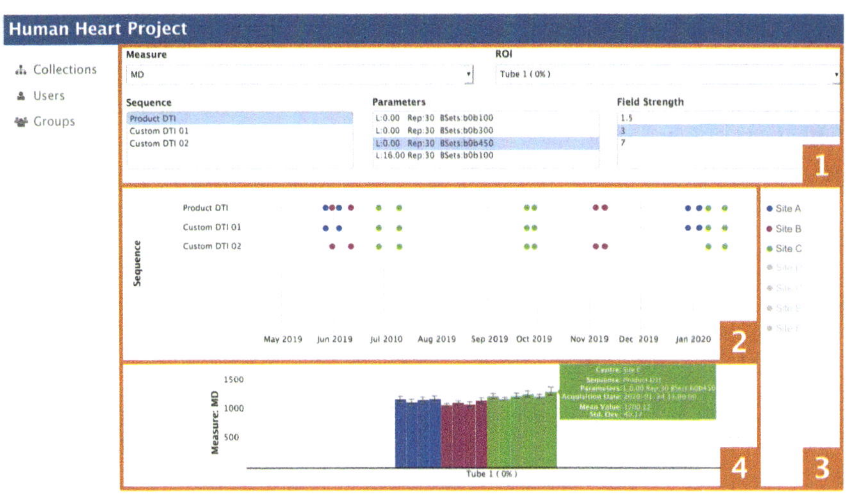

Fig. 4 Graphical User Interface (GUI) of the Girder plug-in. Components: 1 data filters, 2 timeline view, 3 list of sites, and 4 measure view

5 Discussion

According to Keim et al. [16], "Visual analytics combines automated analysis techniques with interactive visualisations for an effective understanding, reasoning and decision making on the basis of very large and complex data sets". The developed VA tool, CMRDiffMonitor, allows MRI researchers to see, in a concise and accessible way, the current status of the experiment. In a single eye glance, research scientists can:

- Verify protocol compliance by checking time gaps on the **timeline view**.
- Identify anomalies by looking for unusual values on the **measure view**, using the historical registry **timeline view**.
- Compare the outcomes with similar imaging systems from different sites by using the **filters panel** to select: the type of sequence, the b values used to calculate DAI maps (i.e. b0b450, b100b450, b300b450, etc.) and the field strength (1.5T, 3T or 7T).
- Verify and validate a hypothesis. The dashboard summarises the results for all sites.
- Export experiment-specific figures for reporting.

The plug-in integrates simple visualisation techniques to empower research scientists to explore data, analyse and share results. Moreover, the GUI reduces the cognitive load of information, all the image post-processing settings are managed by the back-end components (Girder, CMRDiffTools, etc.) as well as data sets details, driving the user attention to one task: analyse the outcomes from different imaging systems and cardiac DTI sequences.

As an example of reporting, representative MD across tubes between 2 different scans (time delay within 30 days) are shown in Fig. 5. At 0% PVP, the average MD (mean $\pm$ standard deviation across scanners) between scans were $(1.149 \pm 0.032) \times 10^{-3} \frac{\mathrm{mm}^2}{\mathrm{s}}$ and $(1.159 \pm 0.049) \times 10^{-3} \frac{\mathrm{mm}^2}{\mathrm{s}}$ respectively, while the coefficient of variation at Scan 1 was $1.9 \pm 1.4\%$. Ground truth diffusivity D_{ref} (H_2O), corresponding to 0% PVP, extrapolated from [19] was $1.113 \times 10^{-3} \frac{\mathrm{mm}^2}{\mathrm{s}}$.

Partial results of the analysis of product DTI sequences have been reported in [26, 27].

The current plug-in version is the result of a first development cycle. Indeed, improvements and new functionalities will be implemented according to the development of the project. Nonetheless, lessons learnt may help towards future projects supported by the Human Heart Project.

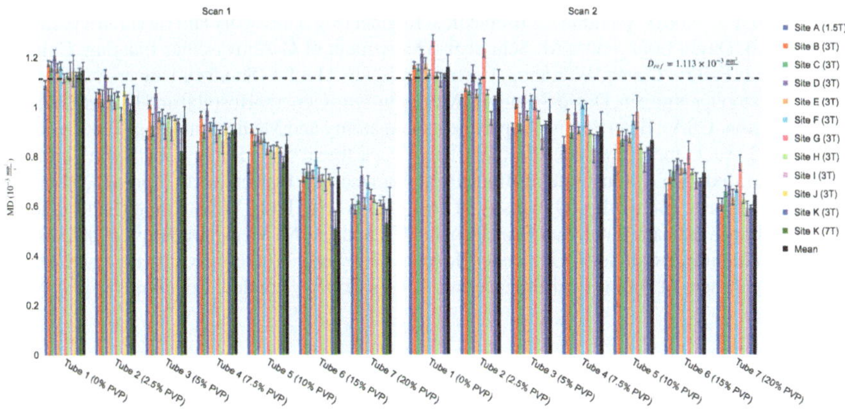

Fig. 5 MD across sites and tubes between two different scans (mean ± standard deviation across ROI). Figure reproduced from [27]

6 Conclusions and Future Work

DTI benchmark databases are used in the context of the imaging sequence harmonisation as well as QA protocols. A standard reference object (phantom) is used to assess and benchmark the performance of an MRI scanner. Understanding how images are produced by a set of parameters and conditions enables MRI and medical researchers to calibrate the imaging system as well as identify sources of variability. VA tools can streamline the analysis of extensive data sets of in vitro MR images through interactive data exploration in order to filter and display parameters of interest in an easily interpretable format.

In the context of a multi-centre study to investigate the inter- and intra-site variation of cardiac DTI sequences, CMRDiffMonitor, VA tool has been developed. This solution enables MRI researchers to track and review the variability of DAI from different image acquisition sequences. The CMRDiffMonitor's dashboard displays a temporal overview of the data sets (monitoring of the system stability) and the statistical measurements in ROIs on a DAI map (benchmarking of the different imaging sequences and systems). Future work will focus on the visualisation of the source image (DAI map) based on the user selection over a statistical measurement.

Acknowledgements This work was performed within the framework of the RHU MARVELOUS (ANR-16-RHUS-0009) of l'Université Claude Bernard Lyon 1 (UCBL), within the program "Investissements d'Avenir" operated by the French National Research Agency (ANR). This study was supported by the British Heart Foundation (PG/19/1/34076). The authors acknowledge the contributions of the members of the SCMR diffusion working group as they provided data for the multi-centre study: Erica Dall'Armellina (Leeds Institute of Cardiovascular and Metabolic Medicine, University of Leeds, Leeds, UK), Daniel Ennis (Department of Radiology, VA Palo Alto Health Care System, Palo Alto, USA), Pedro F. Ferreira, Andrew Scott, Sonia Nielles-Vallespin (Royal Brompton Hospital and Imperial College, London, UK), Prateek Kalra, Arunark Kolipaka (Department of Radiology, The Ohio State University Wexner Medical Center, Columbus, USA), Sebastian Kozerke, Christian T. Stoeck (Institute for Biomedical Engineering, University and ETH Zurich, Zurich, Switzerland), David Lohr, Laura M. Schreiber (Department of Cardiovascular Imaging, Comprehensive Heart Failure Center, Würzburg, Germany), Kevin Moulin (Stanford University, Stanford, USA), Christopher Nguyen, David Sosnovik (Massachusetts General Hospital and Harvard Medical School, Boston, USA), Beau Pontre (Department of Anatomy and Medical Imaging, The University of Auckland, Auckland, New Zealand), Cyril Tous (Department of Radiology, Radiation-Oncology and Nuclear Medicine and Institute of Biomedical Engineering, Université de Montréal, Montréal, Canada), Elizabeth Tunnicliffe (University of Oxford, Oxford, UK), Vicky Wang (San Francisco Veteran Affairs Medical Center, San Francisco, USA), Andreas M. Weng (Department of Diagnostic and Interventional Radiology, University Hospital Würzburg, Würzburg, Germany), and Alistair Young (Department of Biomedical Engineering, King's College London, London, UK).

References

1. The human heart project. http://humanheart-project.creatis.insa-lyon.fr/. Accessed 17 March 2020
2. Multi-centre evaluation of cardiac dti sequences. http://humanheart-project.creatis.insa-lyon.fr/MultiCentreEvaluationCDTI.html. Accessed 17 March 2020
3. The virtual imaging platform. http://vip.creatis.insa-lyon.fr/. Accessed 17 March 2020
4. Afgan, E., Baker, D., Batut, B., Van Den Beek, M., Bouvier, D., Čech, M., Chilton, J., Clements, D., Coraor, N., Grüning, B.A., et al.: The galaxy platform for accessible, reproducible and collaborative biomedical analyses: 2018 update. Nucleic Acids Res. **46**(W1), W537–W544 (2018)
5. Avesani, P., McPherson, B., Hayashi, S., Caiafa, C.F., Henschel, R., Garyfallidis, E., Kitchell, L., Bullock, D., Patterson, A., Olivetti, E., et al.: The open diffusion data derivatives, brain data upcycling via integrated publishing of derivatives and reproducible open cloud services. Sci. Data **6**(1), 1–13 (2019)
6. Burmeister, J., Bernard, J., May, T., Kohlhammer, J.: Self-service data preprocessing and cohort analysis for medical researchers. In: 2019 IEEE Workshop on Visual Analytics in Healthcare (VAHC), pp. 17–24. IEEE (2019)
7. Chen, C.C., Wan, Y.L., Wai, Y.Y., Liu, H.L.: Quality assurance of clinical mri scanners using acr mri phantom: preliminary results. J. Digital Imaging **17**(4), 279–284 (2004)
8. Das, S., Glatard, T., Rogers, C., Saigle, J., Paiva, S., MacIntyre, L., Safi-Harab, M., Rousseau, M.E., Stirling, J., Khalili-Mahani, N., et al.: Cyberinfrastructure for open science at the montreal neurological institute. Frontiers Neuroinf. **10**, 53 (2017)
9. Davids, M., Zöllner, F.G., Ruttorf, M., Nees, F., Flor, H., Schumann, G., Schad, L.R., Consortium, I., et al.: Fully-automated quality assurance in multi-center studies using mri phantom measurements. Mag. Reson. Imaging **32**(6), 771–780 (2014)
10. Dinov, I.D., Petrosyan, P., Liu, Z., Eggert, P., Hobel, S., Vespa, P., Woo Moon, S., Van Horn, J.D., Franco, J., Toga, A.W.: High-throughput neuroimaging-genetics computational infrastructure. Frontiers Neuroinf. **8**, 41 (2014)
11. Donoho, D.L.: An invitation to reproducible computational research. Biostatistics **11**(3), 385–388 (2010)
12. Fonseca, C.G., Backhaus, M., Bluemke, D.A., Britten, R.D., Chung, J.D., Cowan, B.R., Dinov, I.D., Finn, J.P., Hunter, P.J., Kadish, A.H., et al.: The cardiac atlas project–an imaging database for computational modeling and statistical atlases of the heart. Bioinformatics **27**(16), 2288–2295 (2011)
13. Glatard, T., Lartizien, C., Gibaud, B., Da Silva, R.F., Forestier, G., Cervenansky, F., Alessandrini, M., Benoit-Cattin, H., Bernard, O., Camarasu-Pop, S., et al.: A virtual imaging platform for multi-modality medical image simulation. IEEE Trans. Med. Imaging **32**(1), 110–118 (2012)

14. Ihalainen, T.M., Lönnroth, N.T., Peltonen, J.I., Uusi-Simola, J.K., Timonen, M.H., Kuusela, L.J., Savolainen, S.E., Sipilä, O.E.: Mri quality assurance using the acr phantom in a multi-unit imaging center. Acta Oncologica **50**(6), 966–972 (2011)
15. Inc., P.T.: Collaborative data science. https://plot.ly. Accessed 17 March 2020
16. Keim, D., Kohlhammer, J., Ellis, G., Mansmann, F.: Mastering the information age: solving problems with visual analytics (2010)
17. Kim, H., Irimia, A., Hobel, S.M., Esquivel Castelo-Blanco, R.I., Duffy, B., Zhao, L., Crawford, K.L., Liew, S.L., Clark, K., Law, M., et al.: Loni qc system: a semi-automated, web-based and freely-available environment for the comprehensive quality control of neuroimaging data. Frontiers Neuroinf. **13**, 60 (2019)
18. Kitware: Girder: a data management platform. https://girder.readthedocs.io/. Accessed 17 March 2020
19. Mills, R.: Self-diffusion in normal and heavy water in the range 1–45. deg. J. Phys. Chem. **77**(5), 685–688 (1973)
20. Pierpaoli, C., Sarlls, J., Nevo, U., Basser, P., Horkay, F.: Polyvinylpyrrolidone (pvp) water solutions as isotropic phantoms for diffusion mri studies. In: ISMRM, vol. 17, p. 1414 (2009)
21. Poldrack, R.A., Barch, D.M., Mitchell, J., Wager, T., Wagner, A.D., Devlin, J.T., Cumba, C., Koyejo, O., Milham, M.: Toward open sharing of task-based fmri data: the openfmri project. Frontiers Neuroinf. **7**, 12 (2013)
22. of Radiology, A.C., et al.: Phantom Test Guidance for the Acr Mri Accreditation Program. Reston, Va: ACR (1998)
23. Romero R., W.A., Viallon, M., Froeling, M., Stoeck, C., Kozerke, S., Tunnicliffe, E., Scott, A., Ferreira, P., Aliotta, E., Ennis, D., Moulin, K., Croisille, P.: CMRDiffTools: a processing and analysis tool for cardiac diffusion mr images. In: ISMRM (2018)
24. Shneiderman, B.: The eyes have it: A task by data type taxonomy for information visualizations. In: Proceedings 1996 IEEE Symposium on Visual Languages, pp. 336–343. IEEE (1996)
25. Tax, C.M., Grussu, F., Kaden, E., Ning, L., Rudrapatna, U., Evans, C.J., St-Jean, S., Leemans, A., Koppers, S., Merhof, D., et al.: Cross-scanner and cross-protocol diffusion mri data harmonisation: A benchmark database and evaluation of algorithms. NeuroImage **195**, 285–299 (2019)
26. Teh, I., Romero, W., et al.: Multi-centre evaluation of diffusion tensor imaging (dti) in an isotropic phantom: Towards validation of cardiac dti sequences. In: SCMR 23rd Annual Scientific Sessions (2020)
27. Teh, I., Romero, W., et al.: Reproducibility of diffusion tensor imaging (dti) on 12 clinical scanners: Towards validation of cardiac dti sequences. In: ISMRM (2020)

Index

N. Gyori et al. (eds.), *Computational Diffusion MRI*,
Mathematics and Visualization, https://doi.org/10.1007/978-3-030-73018-5

The manufacturer's authorised representative in the EU is Springer Nature Customer Service Centre GmbH, Europaplatz 3, 69115 Heidelberg, Germany. If you have any concerns regarding our products, please contact ProductSafety@springernature.com

Printed and bound by CPI Group (UK) Ltd, Croydon, CR0 4YY
28/09/2026
02224091-0014